ELECTRICAL ASPECTS
OF COMBUSTION

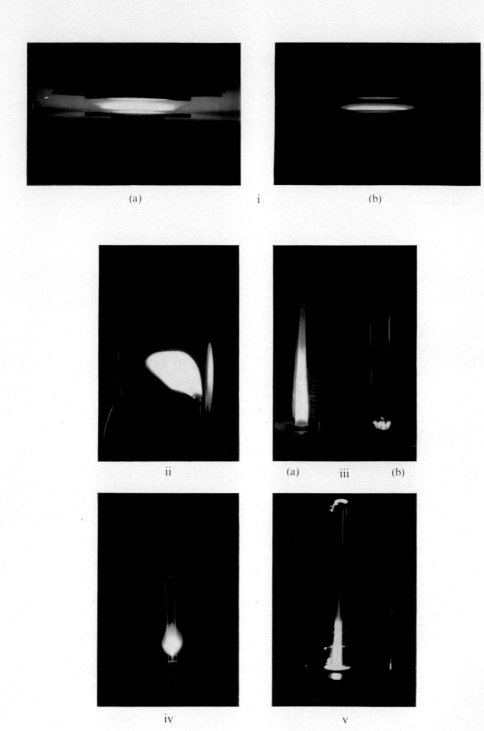

(a)   i   (b)

ii   (a)   iii   (b)

iv   v

(i) (a) and (b) Effect of a field on carbon luminosity in counterflow diffusion flame. (ii) Hydrocarbon diffusion flame deflected towards the more distant electrode, in the direction of movement of the larger charge-carrier. (iii) (a) and (b) The electrical aeration of a diffusion flame: (a) with no field; (b) 30 kV applied along tube. (iv) Using field-induced recirculation of hot products to increase combustion intensity. (v) Flame-arc combination within water-cooled jacket.

ELECTRICAL ASPECTS OF COMBUSTION

JAMES LAWTON,

Ph.D., D.I.C., A.M.I.Chem.E., A.Inst.P.

Lecturer in Combustion Physics

and

FELIX J. WEINBERG,

D.Sc., Ph.D., D.I.C., F.Inst.P., M.Inst.F.

Professor of Combustion Physics

Department of Chemical Engineering and Chemical Technology
Imperial College, University of London

CLARENDON PRESS · OXFORD
1969

Oxford University Press, Ely House, London W. 1

GLASGOW NEW YORK TORONTO MELBOURNE WELLINGTON
CAPE TOWN SALISBURY IBADAN NAIROBI LUSAKA ADDIS ABABA
BOMBAY CALCUTTA MADRAS KARACHI LAHORE DACCA
KAULA LUMPUR SINGAPORE HONG KONG TOKYO

SET ON MONOPHOTO FILMSETTER AND PRINTED BY
J. W. ARROWSMITH LTD., BRISTOL, ENGLAND

Preface

ONE of the few compensations of writing a book lies in the pleasure of being able to acknowledge publicly the debt owed to some of the individuals who have selflessly given of their time and enthusiasm to help the work along in various ways.

We are greatly indebted to Professor T. M. Sugden, F.R.S. and Dr. I. R. Hurle of Shell Research, Thornton, for helpful comments on Chapter 6. We should like to thank Professor A. R. Ubbelohde, C.B.E., F.R.S. and Professor A. G. Gaydon, F.R.S. for their interest and encouragement throughout the writing of this monograph. We wish to express our gratitude to many of our former and present research students, associates, and colleagues, particularly Dr. K. G. Payne, Dr. P. J. Mayo, Dr. G. Munday, Mr. M. J. R. Schwar, Mr. D. Boothman, and Mr. J. B. Cox for discussions, references, and other useful material. We owe a special debt of gratitude to our wives and to Miss Pat Taylor, without whose unfailing patient help and readiness to rescue us from the series of emergencies that tend to arise in the processing of a manuscript, the work would have been delayed even further.

Imperial College
December 1967

JAMES LAWTON
FELIX WEINBERG

Contents

List of Symbols

A Area (subscripts: a—apparent: r—real: e—for entrainment)
Langevin function
a Separation of electrode from ion source (Subscript: q—at quenching)
a_i Constant in series
B Magnetic flux (in Wb/m^2)
Pre-exponential term in thermionic emission
Constant
b   as suffix: burned state, at blow-off
b_1, b_2 Constants
C Specific heat (subscripts: p—at constant pressure: v—at constant volume: R—for reactants: P—for products)
Capacitance
C_D Discharge coefficient
c Velocity of:
light *in vacuo* (subscript: p—in plasma)
molecules (subscripts: x, y, z,—components along cartesian coords: i—of ion)
Concentration (subscripts: j—of jth species; s—of fuel at stoichiometric; o—of fuel at burner)
D Diffusion coefficient (subscripts: j—of jth species: a—ambipolar: +, −, e—respectively positive, negative ion and electron)
Dielectric constant
d Diameter: distance
E Field strength (subscripts: x, y, r—component along that coordinate; ⊥—perpendicular; B—at breakdown; o—at ion source; oc—on open circuit)
Activation energy
e Electronic charge
F Force (also /area and /volume) (subscripts: e—electric: m—magnetic: B—at breakdown)
f Fraction (of full charge; of energy converted; of volume; f'—of molecules which are ionized in a particle)
Frequency
G Flux:
of mass (into flame—mass burning rate)
of number (e.g. of particles)
g Velocity gradient (subscripts: f—at flash back: b—at blow-off)
H Magnetic field
Enthalpy (subscripts: u—unburned: b—burned) $\dot{H}$—flux of enthalpy
h Planck's constant
Specific enthalpy (suffix 1—at stagnation)
i Current (subscripts: $\parallel, \perp$—components respectively parallel and perpendicular thereto: p—to probe; +, −, e,—positive, negative, and electron; r—random)
as suffix: ion
J Mechanical equivalent of heat
Total flux of lines of electric field
J_+ $i_+/(i_+)_r$
j Current density (subscripts: +, −, i, e, p—carried by respectively, positive and negative charges, ions, electrons, and particles, x, y, r,—components along those coordinates: s—at saturation: r—random: p—to probe: ⊥—perpendicular)
j $\sqrt{-1}$
j', j'' Current per unit length of cylinder and total current from sphere, respectively

K	Mobility (subscripts: $+$, $-$, e, i, p, p_1—of, respectively, positive and negative charge carrier, electron, ion, particle, fully-charged particle: m—magnetic)
k	Boltzmann constant
	Reaction rate constant (subscripts: i—ionization: 3X—three-body)
	Polytropic expansion coefficient
L	Length, width
	Inductance
M, M'	Molecular mass, molecular weight (subscripts: R—of reactants: i—of ion: r—reduced, i.e. $M_r = M_1 M_2/(M_1 + M_2)$)
m	Mass (subscript: j—of jth species, for unit conversion)
N	Number of charges on particle (subscript: 1—at full charge)
n	Number concentration (subscripts: $+$, $-$, i, e, p—of, respectively, positive and negative charge, ions, electrons, and particles: o—unperturbed: s—at sheath edge)
P	Pressure (subscripts: u, b—unburned and burned: c—excess, B—at breakdown)
	as suffix: products
p	Order: of reaction exponent
	of interference fringes
	as suffix: at constant pressure: for particle: for probe: in plasma
Q	Collision cross-section
	Heat of combustion (subscripts: T, u—at temperature T, unburnt)
	Q of circuit
q	Charge
R	Gas constant
	Resistance
	Coefficient of reflection
	Radius: of curvature
	as suffix: reactants
r	Radial coordinate: radius (subscripts: i—of ion, p—of particle, probe: s—sheath: e—extra-sheath; o—initial, etc; c, n—rate of absolute and net ion generation per unit volume)
S	Velocity perpendicular to flame (subscripts: u—with respect to unburned reactants, i.e. burning velocity; T, L—turbulent and laminar burning velocities)
T	Temperature (subscripts: u and b—unburned and burned: $+$ and e—ion and electron, w—of wall)
t	Time (subscript: o—total burning time)
U	Energy; of molecule; internal (subscripts: u and b—unburned and burned)
	Gas velocity
u	Velocity of sound (subscript: b—in burned gases)
	as suffix: unburned state
V	Potential (subscripts: B—at breakdown: i—ionization: exc—excitation: p—probe: s—sheath; f—floating; e and $+$—electron and ion temperature in electron volts; V_a and V_d—electron affinity and dissociation energy)
	Volume
$\dot{V}$	Volumetric flow rate
v	Velocity of gas: of ions (subscripts: u and b—unburned and burned: x and y—components along these coordinates: r—relative: B—at breakdown: e—entrainment: $\perp$—perpendicular)
W	Power: per unit mass: per unit volume: per unit electrode area
w	Rate of reaction/unit volume
x, y, z	Cartesian coordinates
	as suffixes: components along x, y, z
X, Y, Z	Finite distances along x, y, z
x	Fraction of molecules ionized
y_p	Absolute value of ratio of probe potential in volts to electron temperature in eV
α	Recombination coefficient: ionizing collisions/cm
	Phase angle
β	Hall parameter
	Arbitrary constants (numbered)

β' Attenuation (dB/cm)

$\Gamma(x)$ Gamma function

γ C_p/C_v, ratio of specific heats
                Surface tension
                Electrons emitted from surface per collision
                Polarizability
                Electron attachment rate constant in gas phase

δ Fraction of energy lost or exchanged on collision
                Fraction of operating potential/open circuit potential

ε Ratio of electron temperature to positive ion temperature
                Arbitrary constants (numbered)

η Viscosity
                Electrical output/mechanical input

θ Angle

λ Mean free path (subscripts: $+$, $-$, e, eff—respectively positive ion, negative ion, electron, effective)
                Thermal conductivity
                Wavelength

λ_D Debye length

μ Refractive index
                $(K_+ \cdot i_+)/(K_e \cdot i_e)$
                Permeability
                Fermi energy

ν Collision frequency (subscripts: e—electron: i—ion)

ρ Density (subscripts: u and b—unburned and burned)

ρ_p Ratio of probe radius to Debye length

$\bar{\rho}_p$ $\rho_p(1 + T_e/T_+)^{\frac{1}{2}}$

Σ, σ Electrical conductivity (subscripts; x—real part; y—imaginary part)

σ Surface charge density

τ Mean free time (subscript: e—electron)
                Half-life of ionization

ϕ Work function

ϕ' Work function modified by Schottky effect

χ Mass susceptibility
                Attenuation coefficient

ψ Moles seed/moles inert

Ω Ratio of actual to resonant frequency of a.c. circuit (ω/ω_0)

ω Angular velocity
                Statistical weight
                Frequency (subscripts: p—plasma: b—cyclotron: o—resonant)

1. Introduction

ALTHOUGH the foundations of this subject were laid as early as 1600 when W. Gilbert, physician to Queen Elizabeth I, demonstrated that flame gases would discharge an electroscope, it attained the present prominence only within the last few years. The reason for the recent rapid growth in interest and research is undoubtedly due largely to its various practical consequences and potential applications. Examples of these include: 'direct' generation of electricity from rapidly flowing ionized flame gases; the control of some combustion processes by means of applied fields; modifying flame carbon formation and deposition; the interaction of radio waves with rocket exhausts; the use of ionization probes in timing detonations, in flame detection, and in gas chromatography; and many others.

At the time of writing, no book or review is in existence which covers comparable ground. We feel that this absolves us from any obligation to start off—as has become conventional in recent years—with an apologia for adding to an already crowded literature. One reason why classical reviews[1,2] are less than relevant today is that the subject has grown so quickly, not only in substance but particularly in the concept of what is interesting and worthy of study. It is a measure of this development that the entire scope of the monograph of H. A. Wilson,[1] despite its title, corresponds to only a small part of one of our present chapters. Several combustion texts carry brief summaries on flame ionization and the effects of fields; one of the most extensive (21 pages) and recent surveys is to be found in the second edition of Gaydon and Wolfhard's *Flames*.[3] In addition, many useful reviews of individual aspects of this subject have been published. They are referred to in the appropriate chapters.

One difficulty in reporting on a rapidly developing subject is the liability of omitting the substance of the very latest publications. We should like to think that no major aspect has been left out if it has appeared in print before the summer of 1966†. Another source of concern to us is that, although we have consciously tried to strike a balance between the various parts of the subject, it is nevertheless possible that workers in adjacent fields may consider the treatment biased towards our own research interest—perhaps as regards its stress on the effects of fields on flame processes, or its detailed discussion of augmented flames. Part of our effort to avoid one-sidedness

† Particularly relevant developments since then have been added during the production of the book up to late 1968.

lies in the generality of the introductory chapters, which are intended to provide an adequate introduction to any branch of the subject, irrespective of the reader's background. The first three chapters that follow introduce the subject from three independent directions; they summarize relevant aspects of mechanisms of ionization, of combustion, and of the effects of fields on ions in gases. Where appropriate, the treatment is in the 'textbook' manner with references given only where they aid calculation or provide additional information. Parts dealing with less well-established concepts are written in the manner of a review, particularly where no previous review is available in the literature. For example, much of the chapter on combustion is intended as a text, whilst the part that deals with augmenting combustion temperatures and hence ionization, is essentially a survey of the literature. These summaries are incomplete to the extent that from each is omitted all that does not have a direct bearing on subsequent chapters: to use the same example, augmenting by electrical discharges becomes irrelevant beyond the point where the contribution of combustion to the total enthalpy becomes inappreciable. There is therefore no discussion of discharges as such.

Subsequent chapters include comprehensive surveys of experimental methods used in the study of electrical properties of flames and of the present state of knowledge as regards ionization in combustion processes. This is followed by a summary of the practical consequences and of potential applications. Next, the theory of the distributions of space-charge, current, and potential are established for various relevant combustion systems under the application of fields. This leads to the calculation of maximum realizable current densities and to the prediction of absolute maxima to practical effects of applying fields to flame processes.

An appreciable part of the monograph is based on original work that has not been reviewed before. The treatment is a development of a post-graduate course of lectures that we have given periodically in the Department of Chemical Engineering and Chemical Technology of Imperial College.

REFERENCES

1. WILSON, H. A. *The electrical properties of flames and incandescent solids*. London University Press (1912).
2. WILSON, H. A. *Rev. Mod. Phys.* **3**, 156 (1931).
3. GAYDON, A. G. and WOLFHARD, H. G. *Flames*, pp. 302–23. Chapman and Hall, London (1960).

2. Processes of Generation, Attachment, and Recombination of Charged Species

THE purpose of this chapter is to provide fundamental information regarding ionization and allied processes in the homogeneous gas phase and in gas–solid systems, both of which are important in combustion applications. It is convenient to divide the chapter into two parts. Part I, which deals with the homogeneous gas phase, contains the following: a brief account of the relevant kinetic theory, a discussion of processes of ionization, a section on processes involving change of charge carrier (electron attachment, charge exchange, and clustering), a discussion of charge recombination, and an account of equilibrium ionization. Part II is divided into three sections, which deal with electron emission from surfaces, thermal ionization of particles, and particle-charging in the presence of applied electric fields.

PART I

HOMOGENEOUS GAS-PHASE

Kinetic theory

Collisions between molecules are conveniently described in terms of a collision cross-section. In the case of hard spheres of radii r_1 and r_2, this is a well-defined quantity $\pi(r_1 + r_2)^2$. Although molecules in general are neither spherical nor 'hard', the concept of collision cross-section remains valuable so long as account is taken of its dependence upon relative velocity: in some cases this may be done by using a suitably averaged value.

If the collision cross-section is independent of velocity, under conditions of equilibrium, the collision frequency v_1 of a molecule of type 1 with those of type 2 can be expressed as[1]

$$v_1 = n_2 Q_{12} \sqrt{\left(1 + \frac{M_1}{M_2}\right)} \bar{c}_1 \tag{2.1}$$

where n_2 is the concentration of molecules of type 2, Q_{12} the collision cross-section, $\bar{c}$ the mean velocity, and M the mass. In the special case of electrons, $M_1 \ll M_2$ and to a good approximation

$$v_e = n_2 Q_{e2} \bar{c}_e$$

By definition, the mean free path λ_{12} is equal to $\bar{c}_1/v_1$ (the distance travelled in unit time divided by the number of collisions). Hence

$$\lambda_{12} = 1/\{n_2 Q_{12}\sqrt{(1 + M_1/M_2)}\} \tag{2.3}$$

Treatments of this kind, in which average values of collision cross-section are used, are valuable for obtaining estimates of transport properties, such as diffusion coefficients and ionic mobility, because all the molecules contribute to the transport process. However, in processes involving the surmounting of energy barriers, for example processes of chemical reaction involving an energy of activation, only those molecules with sufficient energy can take part. Under these circumstances the distribution of molecular velocities is crucial. The distribution function $f(c_x, c_y, c_z)$, where c_x, etc. are the components of velocity, is defined so that the product $f(c_x, c_y, c_z)\,dc_x\,dc_y\,dc_z$ is equal to the fraction of molecules with velocities in the range c_x to $c_x + dc_x$, c_y to $c_y + dc_y$ and c_z to $c_z + dc_z$. If the integral is taken over all possible velocities, all the molecules must be included, i.e.

$$\iiint_{-\infty}^{\infty} f(c_x, c_y, c_z)\,dc_x\,dc_y\,dc_z = 1. \tag{2.4}$$

Gas molecules in equilibrium have a Maxwellian distribution,

$$f(c_x, c_y, c_z) = \left(\frac{M}{2\pi kT}\right)^{\frac{3}{2}} \exp\left\{-\frac{M}{2kT}(c_x^2 + c_y^2 + c_z^2)\right\}. \tag{2.5}$$

However, expression (2.5) should not be used indiscriminately. Electrons, for example, in the presence of electric fields can have distributions widely different from this; see Chapter 4.

Consider collisions between molecules of type 1 whose velocities lie in the range c_x to $c_x + dc_x$, etc., and molecules of type 2 whose velocities lie in the range c'_x to $c'_x + dc'_x$, etc. The relative velocity v_r is given by

$$v_r^2 = (c_x - c'_x)^2 + (c_y - c'_y)^2 + (c_z - c'_z)^2. \tag{2.6}$$

In unit time, a molecule of type 1 sweeps out a volume $v_r Q_{12}$ relative to molecules of type 2, i.e. it collides with $n_2 v_r Q_{12} f(c'_x, c'_y, c'_z)\,dc'_x\,dc'_y\,dc'_z$ molecules of type 2. If $p(v_r)$ is the probability of reaction, the number of collisions leading to reaction per unit volume per unit time amongst particles of the chosen velocities is

$$v_r n_1 n_2 p(v_r) Q_{12}(v_r) f(c_x, c_y, c_z) f(c'_x, c'_y, c'_z)\,dc_x\,dc_y\,dc_z\,dc'_x\,dc'_y\,dc'_z.$$

In the general case, the total reaction rate is found by integration between plus and minus infinity for each of the six components of velocity. In the case of reactions involving electrons and molecules, for example ionization or excitation, the integral can be simplified. Electrons, because of their small mass, move much faster than molecules. Therefore, to a good approximation, $v_r = c'$, where c' is the electron velocity and c'_x, c'_y, c'_z are its components. Thus, the integration over c_x, c_y, and c_z is seen to involve only $f(c_x, c_y, c_z)$, and, from eqn (2.4), is equal to unity. Moreover, because the electron velocities have no preferred direction, one may integrate with respect to velocity alone, in which case $dc'_x\, dc'_y\, dc'_z$ is replaced by $4\pi c'^2\, dc'$ and the integration is carried out between zero and infinity. Thus, w, the net rate of reaction per unit volume, is given by

$$w = n_1 n_2 \int_0^\infty 4\pi p(c')Q(c')c'^3 f(c')\, dc'. \tag{2.7}$$

Another useful case is that of reactions between molecules with Maxwellian velocity distributions and at the same temperature. For these[2]

$$w = \frac{4n_1 n_2}{\sqrt{(2\pi M_1)}} \left(\frac{M_r}{M_1 kT}\right)^{\frac{1}{2}} \int_0^\infty p(U) . Q(U) . U . \exp\left(\frac{-M_r U}{M_1 kT}\right) dU, \tag{2.8}$$

where $M_r = M_1 M_2/(M_1 + M_2)$ and $U = \frac{1}{2}M_1 v_r^2$. Often descriptions of rates of reaction are made using the concept of an activation energy (see, for instance *Physical Chemistry*[3]). Thus, it is assumed that reaction occurs if the energy of collision is in excess of some critical value. For a Maxwellian distribution the probability of a pair of molecules, one of type 1 and the other of type 2, having an energy of relative motion in excess of E is $e^{-E/kT}$. The reaction rate is found by multiplying the total collision frequency by this factor, assuming a constant collision cross-section. Thus from eqn (2.1)

$$\text{rate of reaction} = n_1 n_2 Q_{12} \sqrt{\left|\left(\frac{M_2 + M_1}{M_2}\right)\right|} \bar{c}_1\, e^{-E/kT}$$

$$= n_1 n_2 \sqrt{\left|\left(\frac{8kT}{\pi M_r}\right)\right|} Q_{12}\, e^{-E/kT}. \tag{2.9}$$

It must be emphasized that eqn (2.9), although much used, represents only a rough description of the true state of affairs. Apart from the assumption of a constant collision cross-section and the use of the concept of an activation energy above which the reaction proceeds with a probability of unity, the factor $e^{-E/kT}$ does not allow for the faster moving pairs having a greater than average collision frequency.

The rate constant, k_2, for bimolecular reactions is defined by the relation

$$w = k_2 n_1 n_2 \tag{2.10a}$$

and can be evaluated from eqns (2.7), (2.8), and (2.9). Clearly, it attains its maximum value when reaction occurs at each collision, i.e.

$$(k_2)_{max} = Q_{12}\left(\frac{M_2 + M_1}{M_2}\right)^{\frac{1}{2}} \bar{c}_1. \qquad (2.10b)$$

At room temperature $(k_2)_{max}$ is of the order of 10^{-11} cm^3/s for molecules.

Ionization processes

Ionization usually requires considerable energy, the ionization potentials for most atoms and molecules falling within the range 4–20 eV. Table 2.1 contains the ionization potentials for a variety of species to be found in combustion systems. The data was taken from the compilation of Vedeneyev et al.[4]

TABLE 2.1

Ionization potentials

	eV		eV	
H	13·595	O_2	12·2	$\pm 0·2$
N	14·53	CO_2	13·84	$\pm 0·11$
O	13·614	NO	9·25	$\pm 0·02$
Cl	13·01	CH	11·13	$\pm 0·22$
Br	11·84	CH_2	11·82 and 10·396	
Li	5·390	CH_3	9·905 $\pm 0·075$	
Na	5·138	CH_4	13·06	$\pm 0·06$
K	4·339	CH_3O	9·2	
Rb	4·176	C_2	12·0	$\pm 0·6$
Cs	3·893	C_2H	11·3	
Ca	6·111	C_2H_2	11·41	$\pm 0·02$
Sr	5·692	C_2H_4	10·5	$\pm 0·1$
Ba	5·21	C_2H_6	11·65	
Pb	7·415	C_3H_8	11·14	$\pm 0·07$
H_2	15·427	CHO	9·88	$\pm 0·05$
OH	13·18 $\pm 0·1$	—	—	
H_2O	12·6 $\pm 0·01$	—	—	
CO	14·05 $\pm 0·05$	—	—	

There are many processes that result in ionization. The most important, in the present context, are discussed below:

(i) Ionization by collision

$$A + B \rightarrow A^+ + B + e^-,$$

$$A + e^- \rightarrow A^+ + e^- + e^-.$$

(ii) Electron transfer

$$A + B \rightarrow A^+ + B^-.$$

(iii) Ionization by transfer of excitation energy

$$A + B^* \rightarrow A^+ + B + e^-.$$

(iv) Chemi-ionization

$$A + B \rightarrow C^+ + D + e^-,$$

$$A + B \rightarrow C^+ + D^-.$$

(i) *Ionization by collision.* If a particle of mass M_1 collides head-on with a particle of mass M_2, the initial relative velocity being v_r, the maximum amount of kinetic energy convertible into internal energy can be calculated.

From a momentum balance in the frame of reference in which M_2 is initially stationary

$$M_1 v_r = M_1 v_1 + M_2 v_2, \tag{2.11}$$

i.e.

$$v_2 = (M_1 v_r - M_1 v_1)/M_2, \tag{2.12}$$

where v_1 and v_2 are the velocities of M_1 and M_2 respectively, after collision. The amount of kinetic energy converted into internal energy is

$$U = \tfrac{1}{2}(M_1 v_r^2 - M_1 v_1^2 - M_2 v_2^2). \tag{2.13}$$

Substituting for v_2 from (2.12) into (2.13) and differentiating U with respect to v_1 at constant v_r, the condition for the maximum U is found to be

$$v_1 = v_2 = M_1 v_r/(M_1 + M_2), \tag{2.14}$$

i.e.

$$U_{max} = \tfrac{1}{2}\frac{M_1 M_2}{(M_1 + M_2)} v_r^2 \tag{2.15}$$

—a fraction $M_2/(M_1 + M_2)$ of the original kinetic energy of M_1. If $M_1 \ll M_2$, as, for example, when an electron strikes a molecule, it is possible for all but a minute fraction of the energy to be converted from kinetic to internal. On the other hand, when atoms or molecules of approximately equal masses collide, only about half the relative kinetic energy can be converted. It is to be expected, therefore, that ionization by electron collision sets in once the electron energy exceeds the ionization potential, V_i, higher energies being necessary for ionization by collisions between molecules. Figure 2.1 shows ionization efficiency of electrons as a function of electron energy for a variety of molecular and monatomic gases. The ionization efficiency is defined as the number of ionizing collisions per electron in a 1-cm length of path at

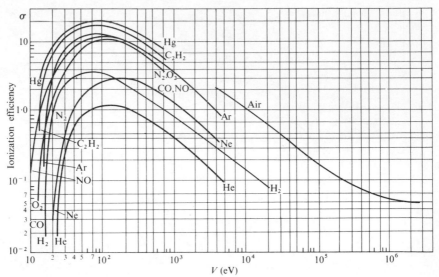

FIG. 2.1. Ionization efficiency, σ, versus electron energy, V. From Kondratiev.[5]

0°C and 1 mmHg. Conversion to a collision cross-section in cm^2 is effected by multiplying by 0.283×10^{-16}. Typically, ionization sets in at the ionization potential and the cross-section rises approximately linearly to a maximum of between 0.3×10^{-16} and 3×10^{-16} cm^2 at around 2 to $5V_i$, after which it falls slowly.

Data for ionization by collision with atoms and ions are shown in Figs. 2.2 and 2.3. Where data are available, they confirm the expectation that the critical energy at which ionization sets in is higher than in the case of collisions with electrons. It has been found that for collisions between helium atoms the critical energy is $2V_i$ as predicted by eqn (2.15).[7] The maximum ionization cross-sections are of the same order as for electrons. However, the maxima occur at energies well in excess of 10^3 eV.

It is clear that in the energy range of interest in combustion systems, i.e. temperatures from 0.1 to 1.0 eV, electron collisions are very much more efficient than molecular collisions in inducing ionization. Further data are given by Kondratiev,[5] von Engel,[6] Brown,[8] McDaniel,[9] and Massey and Burhop.[10]

(ii) *Electron transfer.* In this case an electron is transferred from one atom or molecule to another. This kind of reaction is likely to be endothermic, since ionization potentials are usually much larger than electron attachment energies. The energy deficit must be made up by conversion of kinetic into internal energy; the limitations of this process have just been discussed. An example of such a reaction is

$$K + Cl \rightarrow K^+ + Cl^-,$$

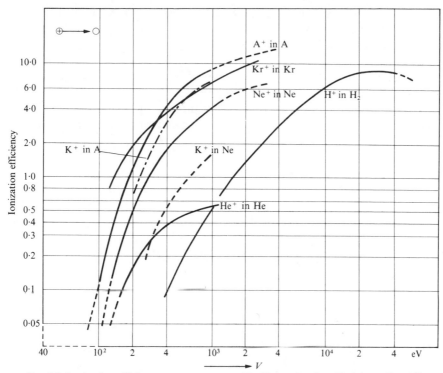

FIG. 2.2. Ionization efficiency, σ, versus energy, V, of impacting ion. From von Engel.[6]

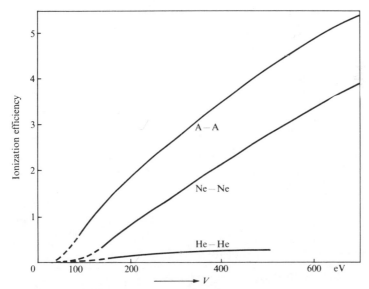

FIG. 2.3. Efficiency of ionization, σ, of atoms colliding in their own gas as a function of their energy, V. From von Engel.[6]

which has been postulated to account for the ionization of potassium in flames containing chlorine.[11] This reaction is endothermic only to the extent of 0·6 eV and is therefore energetically quite feasible at flame temperatures.

(iii) *Ionization by transfer of excitation energy.* In processes of this kind the excitation energy transferred from one species to another is sufficient to cause ionization of the recipient. A number of instances of this type of reaction are known, e.g.

$$Ne^* + Ar = Ne + Ar^+ + e^- \qquad \Delta H = -0.79 \text{ eV}.$$

The excess energy is carried away by the electron. In cases where there is an energy deficit, the extra energy must come from the conversion of kinetic into internal energy. When the reaction is close to thermoneutral, the probability of ionization can be high—of the order of 0·1.[10,12,13]

(iv) *Chemi-ionization.* In this case there is a chemical rearrangement which releases energy leading to ionization of the product species; for example,

$$Hg(6^3P_1) + Hg(6^3P_0) \rightarrow Hg_2^+ + e^- \qquad \Delta H \simeq 0 \quad [14]$$

$$CH + O \rightarrow CHO^+ + e^- \qquad \Delta H \simeq 1 \text{ eV}. \quad [15]$$

In the first example, most of the energy comes from excitation. In the second, however, which is believed to be important in the reaction zones of hydrocarbon flames, the chemical rearrangement is thought to be sufficiently exothermic to cause ionization of CHO with only small contributions from either excitation of one of the species or the energy of motion. Bascombe, Green, and Sugden[16] have estimated a rate constant of 3×10^{-12} cm³/s for this reaction, which indicates that reactions of this kind can be fast.

Electron attachment, charge exchange, and clustering

These three processes are similar inasmuch as they all involve a change in the nature of the charge carrier.

Electron attachment

Many species are capable of forming stable negative ions. Some values of electron affinity are given in Table 2.2.[4]

In view of the exothermic nature of attachment reactions, some means is required by which the energy can be dissipated, and this sets certain limitations on the type of mechanism.

There are three principal gas-phase mechanisms.

(i) Radiative attachment

$$e^- + A \rightarrow A^- + h\nu.$$

TABLE 2.2

Electron affinities

	eV		eV
O	1·47	OH	1·73†–2·65
F	3·5 ±0·04	H₂O	∼0·9
Cl	3·75±0·05	O₂	0·87±0·13
Br	3·55±0·05	HO₂	3·04
Cl₂	≤1·7	O₃	2·89
Br₂	2·6	CO₂	∼3·8 ‡
I₂	2·4	CN	3·7 ±0·2 and 3·2±0·2
		NO₂	1·62

† The lower value appears more reliable.
‡ There is still considerable uncertainty over this.

(ii) Three-body attachment

$$e^- + A + X \rightarrow A^- + X$$

or

$$e^- + A \rightarrow (A^*)^-$$

followed by

$$(A^*)^- + X \rightarrow A^- + X.$$

(iii) Dissociative attachment

$$AB + e^- \rightarrow A + B^-.$$

(i) *Radiative attachment.* Although radiative attachment is known to occur, it is a relatively slow process. For example, in the range 0·1–2·0 eV, the second-order rate coefficients for radiative attachment are of order 10^{-15} cm³/s and 2×10^{-16} cm³/s for O and H respectively; that for O₂ varies between 10^{-18} and 3×10^{-16}.[17] Processes (ii) and (iii) are more likely in combustion systems.

(ii) *Three-body attachment.* Reactions of this type are described by a reaction rate constant k_{3X} defined by the equation

$$w = \frac{-\mathrm{d}n_e}{\mathrm{d}t} = k_{3X}n_e n_A n_X. \tag{2.10c}$$

The suffix X indicates the dependence of the constant upon the nature of the third body. If the third body removes the excess energy by absorbing it internally, the cross-sections can be quite high. The great dependence of reaction rates upon the type of third body is shown by studies on O_2, in the range 77–300°K. The rate constants are very insensitive to temperature, and have the approximate values 2×10^{-30} cm^6/s, 10^{-31} cm^6/s, and 10^{-32} cm^6/s, respectively, for O_2, N_2, and He acting as the third body.[18,19] At n.t.p. there are approximately 3×10^{19} molecules/cm^3; thus if the third bodies were present in such concentration, the effective two-body coefficients would be 6×10^{-11}, 3×10^{-12}, and 3×10^{-13} cm^3/s. Clearly then, three-body attachment processes are likely to be important at high pressure.

(iii) *Dissociative attachment.* In this case the excess energy is taken up by dissociation of the colliding molecules, for example

$$O_2 + e^- \rightarrow O + O^-.$$

In order for the reaction to occur, the electron requires a minimum energy of $(V_d - V_a)$, where V_d is the dissociation energy of the neutral molecule and V_a the electron affinity of the attaching species. Figure 2.4 shows values of attachment cross-sections for H_2 and O_2, and indicates the dissociative processes believed to be operative.

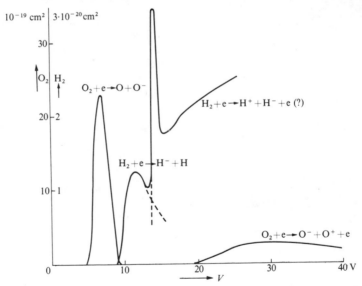

FIG. 2.4. Measured electron dissociative attachment cross-sections in O_2 and H_2. From von Engel.[6]

The occurrence of one or more maxima in the cross-section as the energy is increased is frequently observed. Table 2.3 summarizes some of the data on two-body attachment cross-sections by giving the values of the maxima in the cross-section curves and the energy at which they occur. All the cases listed are believed to involve dissociation, except that of SF_6.

TABLE 2.3

Two-body electron attachment cross-sections

Species	1st Peak		2nd Peak	
	Cross-section (Å^2)	Energy (eV)	Cross-section (Å^2)	Energy (eV)
O_2	0·13	6·2[20]	—	—
CO	0·027	10·1[21]	—	—
CO_2	0·0051	7·8[22]	—	—
SF_6	5·7	0·00[20]	—	—
CCl_4	1·3	0·02	1·0	0·6[20]
CF_3I	0·78	0·05	0·32	0·9[20]
CCl_2F_2	0·54	0·15[20]	—	—
BCl_3	0·28	0·4[20]	—	—
HBr	0·58	0·5[20]	—	—
HCl	0·039	0·6[20]	—	—
H_2O	0·048	6·4[20]	0·013	8·6

Charge exchange processes

The simplest type of charge exchange process is one in which an electron is transferred from a neutral atom or molecule to a positive ion. In more complex processes, chemical rearrangement may accompany charge transfer.

Electron transfer. This can be represented schematically as

$$A + B^+ \rightarrow A^+ + B \qquad \Delta H = V_i(A) - V_i(B).$$

In processes of this kind, very little kinetic energy is exchanged (the effect this has on ionic mobility is discussed in Chapter 4). When collisions between an atomic ion and its parent atom occur, $\Delta H = 0$, and symmetrical resonant electron transfer occurs with high probability. Figure 2.5 shows typical curves of resonant electron transfer cross-sections. The values are referred to 1 mmHg and 0°C, and can be converted to absolute cross-sections in cm^2 by multiplying by $0·283 \times 10^{-16}$. The cross-section falls with increasing energy. In the energy range below 1·0 eV, which is of particular interest here, cross-sections are very large.

Cross-sections for charge exchange between diatomic ions and their parent molecules are shown in Fig. 2.6 and are of the same order of magnitude as for the atomic case.

When collisions between dissimilar ions and atoms or molecules occur, ΔH is usually not zero. In this case, cross-sections for charge exchange are small, below the kilovolt range, except in some cases in which ΔH is small (i.e. about 0·1 eV).

Further data are given by von Engel,[6] Brown,[8] and Hasted.[19]

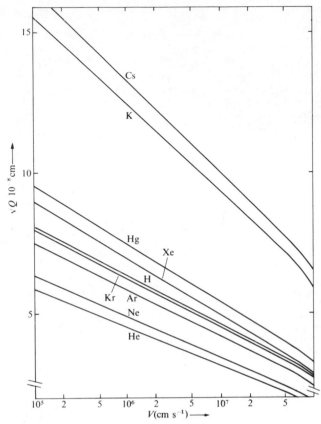

FIG. 2.5. Calculated symmetrical resonant charge transfer cross-sections. From Hasted.[19]

Charge transfer with chemical rearrangement. Many ion/molecule reactions involve a chemical rearrangement, e.g.

$$H_2^+ + O_2 \rightarrow HO_2^+ + H.$$

Exothermic reactions involving atom interchange usually have small activation energy and therefore exhibit rate constants insensitive to temperature changes.[23] Most of these are very fast and are found to lie in the range 10^{-8}–10^{-11} cm^3/s, those involving exchange of H and H$^+$ being particularly rapid. A number of rate constants for low energy encounters, most below 1·0 eV, are to be found in Table 2.4.

Very little work has been done on negative ions, but it seems that cross-sections lying between 10^{-15} and 10^{-14} cm^2 are to be found at energies less than 1·0 eV, indicating that reactions with negative ions are also fairly fast.[32,33] Further information can be found in Massey and Burhop,[10] Hasted,[19,23] Franklyn and Munson,[24] Biondi,[29] and Pahl.[34]

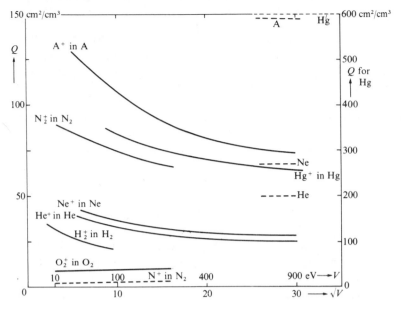

FIG. 2.6. Charge transfer cross-section Q referred to 1 mm Hg and 0°C (dotted lines on right are classical cross-sections). From von Engel.[6]

TABLE 2.4

Rate constants for charge exchange reactions involving chemical rearrangement at energies below 1·0 eV

Reaction	Rate constant (cm^3/s)
$CH_4^+ + CH_4 \rightarrow CH_5^+ + CH_3$	
$CH_3^+ + CH_4 \rightarrow C_2H_5^+ + H_2$	$\sim 10^{-9(24)}$
$CH_2^+ + CH_4 \rightarrow C_2H_3^+ + H_2 + H$	
or $\quad C_2H_4^+ + H_2$	
$CH_4^+ + O_2 \rightarrow CH_3O^+ + OH$	$2\cdot6 \times 10^{-11(24)}$
$H_2^+ + H_2 \rightarrow H_3^+ + H$	$\sim 10^{-9(25)}$
$H_2^+ + O_2 \rightarrow HO_2^+ + H$	$7\cdot6 \times 10^{-9(26)}$
$CO^+ + D_2 \rightarrow COD^+ + D$	$1\cdot63 \times 10^{-9(26)}$
$H_2O^+ + H_2O \rightarrow H_3O^+ + OH$	$8\cdot5 \times 10^{-10(27)}$
$O^+ + O_2 \rightarrow O_2^+ + O$	$2\cdot5 \times 10^{-11(28)}$
$N_2^+ + O_2 \rightarrow N_2 + O_2^+$	$2 \times 10^{-10(29)}$
$N^+ + O_2 \rightarrow NO^+ + O$	$5 \times 10^{-10(30)}$
$O^+ + N_2 \rightarrow NO^+ + N$	$5 \times 10^{-10(31)}$

Clustering

In the presence of gases of high dielectric constant, ions tend to form clusters that are held together by the interaction of the divergent electric field around the ion with the dipole, either permanent or induced, in the dielectric molecules (see Chapter 4 for theory). For example, when only 0·008 per cent water vapour is present in argon, very nearly all lithium ions passing through the gas are found to have a lower mobility than in the absence of water vapour because of cluster formation.[35] It is interesting to note that only two mobilities were detected in this work, i.e. there were no clusters of intermediate size, the ion being either unclustered or clustered to the fullest extent. This can be explained in the following way. Clustering is an exothermic process, therefore the attachment of the first molecule requires the presence of a third body and is consequently relatively slow. On the other hand, once a molecule is attached, subsequent attachment can proceed by rapid bimolecular collision because of the existence of many degrees of freedom to absorb the liberated energy. This implies that clusters of intermediate size have only short lifetimes and are unlikely to be detected.

The large decrease in entropy on cluster formation and their relatively weak binding energies make clusters unstable at high temperatures. Thus direct sampling of flame gases into a mass spectrometer[15] has shown that there is negligible hydration of ions in the high temperature regions, in spite of the presence of large concentrations of water vapour, whereas there is a high degree of hydration of ions that pass through the cool boundary layer that surrounds the sampling orifice of the mass spectrometer. The disruption caused by increase in energy is also shown by applied electric fields. As the field is increased, the fraction of clustered ions falls, indicating that the increased energy of the ions causes break up of the cluster.[35,36]

A theoretical approach to clustering has been proposed by Bloom and Margenau.[37] Their calculations of the mobility of ions in inert gases are in general agreement with measurements. An extensive survey of experimental results has been given by Loeb.[38]

Recombination

Positive ions can recombine with either electrons or negative ions. The processes of interest in the present context are:

  (i) Three-body recombination

$$A^+ + e^- + M \rightarrow A + M$$

$$A^+ + B^- + M \rightarrow A + B + M;$$

  (ii) Dissociative recombination,

$$AB^+ + e^- \rightarrow A + B; \text{ and}$$

(iii) Mutual neutralization,

$$A^+ + B^- \to A^* + B^*.$$

Dielectronic and radiative recombination also exist but are unimportant in combustion systems and are just mentioned for completeness. They are discussed by McDaniel[9] and Massey and Burhop.[10]

(i) *Three-body recombination.* In this process the recombination energy is dissipated by collisions with neutral species. Thomson's theory of this process[39] is as follows. The average relative kinetic energy of an ion pair separated by an infinite distance is $\frac{3}{2}kT$. As the ions approach one another, their relative kinetic energy increases at the expense of electrical potential energy. At some distance r_1 the gain in kinetic energy is equal to $\frac{3}{2}kT$, i.e.

$$e^2/r_1 = \tfrac{3}{2}kT. \tag{2.16}$$

If one of the ions suffers a collision at r_1 which restores the kinetic energy to its thermal value, the ions will now have only just enough kinetic energy to escape one another's coulomb field. However, if the collision occurs when they are closer than r_1, the residual thermal energy, $\frac{3}{2}kT$, will be insufficient to permit separation. The ions will then go into closed orbits around one another and recombination will result.

By consideration of the flux of ions across a sphere of radius r_1, and the probability of collision while within the sphere, Thomson deduced the recombination coefficient, α,

$$\alpha = \frac{2 \cdot 23 e^4}{(kT)^{\frac{3}{2}}} \sqrt{\left(\frac{M_+ + M_-}{M_+ \cdot M_-}\right)} (w_+ + w_- + w_+ \cdot w_-) \, \text{cm}^3/\text{s} \tag{2.17}$$

where

$$w_+ = 1 + 2\left(\frac{\exp(-g_+^2)}{g_+^2} + \frac{\exp(-g_+)}{g_+} - \frac{1}{g_+^2}\right) \tag{2.18}$$

and

$$g_+ = \frac{4}{3} \frac{e^2}{kT(\lambda_+)_{\text{eff}}} \tag{2.19}$$

and

$$(\lambda_+)_{\text{eff}} = \frac{M_+}{M} \lambda_+ \quad \text{or} \quad \frac{M}{M_+} \lambda_+,$$

whichever is the greater, e being in e.s.u. The use of a modified mean free path was introduced by Massey and Burhop[10] to allow for the inefficiency of energy exchange between particles of different masses, M being the mass of the neutral species.

At low pressures, $r_1 \ll \lambda_{\text{eff}}$ and eqn (2.17) reduces to

$$\alpha = \frac{2 \cdot 97 \, e^6}{(kT)^{\frac{5}{2}}} \sqrt{\left(\frac{M_+ + M_-}{M_+ \cdot M_-}\right)} \left\{\frac{1}{(\lambda_+)_{\text{eff}}} + \frac{1}{(\lambda_-)_{\text{eff}}}\right\} \text{cm}^3/\text{s}. \qquad (2.20)$$

$$\propto P. \qquad (2.21)$$

The temperature dependence of α is rather uncertain. For solid sphere collisions it varies as $T^{-\frac{7}{2}}$ at constant pressure. If the long-range dipole effects dominate, as would be the case in molecular gases and in most of the inert gases, it can be described approximately by an inverse third power law. There is not very much information on the influence of temperature, but Gardner[40] found that the temperature dependence in oxygen in the range 150–400°K could be represented by an exponent of the magnitude indicated by the Thomson theory. Equation (2.20) predicts a linear dependence on pressure. This is supported by experiment at lower pressures.

At high pressures, i.e. $r_i \gg \lambda_{\text{eff}}$, eqn (2.17) becomes

$$\alpha = \frac{4 \cdot 45 \, e^4}{(kT)^{\frac{3}{2}}} \sqrt{\left(\frac{M_+ + M_-}{M_+ \cdot M_-}\right)} \text{cm}^3/\text{s}, \qquad (2.22)$$

i.e. α is independent of pressure. This predicted independence at high pressure is not observed. It is found that α goes through a maximum value as the pressure is increased. The reason for the failure of the theory at high pressure, i.e. when $r_1 \gg \lambda_{\text{eff}}$, is that it does not take account of further collisions that send the ions out of their closed orbits. This process becomes increasingly important as the pressure rises.

Before treating the high pressure case, it is important to consider the form of the expression for α when electrons rather than negative ions are involved. If δ is equal to the fraction of the excess energy of the electron exchanged on collision, $\delta = 2M_e/M \simeq 10^{-5}$ for elastic collisions. Values of this order of magnitude are found in the case of low-energy impact of electrons with atoms. For inelastic collisions, which are the usual type of collision between electrons and molecules, measured values of δ are of the order of 10^{-2}; see Chapter 4. In order to apply the Thomson theory, therefore, an effective mean free path for energy transfer must be used,

$$\lambda_{\text{eff}} = \lambda_e/\delta. \qquad (2.23)$$

Thus, for electrons in the low-pressure regime,

$$\alpha \simeq \frac{2 \cdot 97 \, e^6}{(kT)^{\frac{5}{2}}} \frac{\delta}{\sqrt{(M_e)}} \frac{1}{\lambda_e} \propto P. \qquad (2.24)$$

In atomic gases, α is much smaller than the corresponding value for recombination of ions, whilst in molecular gases the calculated α for electrons is of the same order of magnitude as for ions. Table 2.5 shows some values of α at s.t.p. for electrons in atomic and molecular gases, with the estimated pressure limits of the Thomson theory.

TABLE 2.5

Three-body recombination coefficients for electrons in different gases. From Massey and Burhop[10]

Gas	α at n.t.p. (cm^3/s)	Estimated saturation pressure (mm Hg)
Helium	$6\cdot8 \times 10^{-9}$	$2\cdot8 \times 10^4$
Argon	$6\cdot8 \times 10^{-11}$	$2\cdot8 \times 10^5$
Air	$1\cdot7 \times 10^{-7}$	10^4
Hydrogen	$1\cdot6 \times 10^{-7}$	10^4

A satisfactory approach to the high-pressure regime has been given by Langevin.[41] Positive and negative ions attract one another by virtue of their coulomb fields; thus at separation r, the field strength at ion A due to ion B or vice versa is e/r^2. If K_+ and K_- are their respective mobilities (see Chapter 4), they will drift together at a speed $(K_+ + K_-)e/r^2$. The number of negative ions crossing a sphere of radius r drawn around a positive ion is therefore $4\pi(K_+ + K_-)n_- e$ per s. Thus

$$\frac{dn_+}{dt} = 4\pi(K_+ + K_-)n_+ n_- e, \qquad (2.25)$$

i.e.

$$\alpha = 4\pi(K_+ + K_-)e \propto 1/P, \qquad (2.26)$$

since

$$K \propto 1/P. \qquad (2.27)$$

This expression would be expected to apply when $\lambda \ll r_1$. Figure 2.7 shows how the two theories compare with experiment for air and carbon dioxide at 0°C. The agreement is good in the regions where the assumptions are correct, i.e. $r_1 \ll \lambda$ for Thomson's theory and $r_1 \gg \lambda$ for Langevin's theory. A more general analysis that accounts for the pressure dependence in the intermediate range, where $r_1 \approx \lambda$, has been given by Natanson.[42]

(ii) *Dissociative recombination.* In this process the energy liberated by the recombination of an electron with a polyatomic ion causes the neutralized molecule to break up. The recombination coefficient in this case is independent of pressure and values of the order of 10^{-7} cm^3/s have been estimated from theory.[43,44] Measurements of α for recombination of electrons with diatomic ions of the rare gases have been made by Oskam and Mittelstadt in

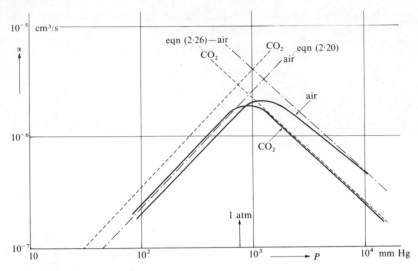

FIG. 2.7. Recombination coefficient for ions in different gases as a function of pressure. From von Engel.[6]

the pressure range 5–45 mm Hg.[45] Values were found to be independent of pressure suggesting a dissociative process

$$Ne_2^+(2\cdot2\pm0\cdot2\times10^{-7}), \qquad Ar_2^+(6\cdot7\pm0\cdot5\times10^{-7}),$$
$$Kr_2^+(1\cdot2\pm0\cdot1\times10^{-6}), \text{ and } Xe_2^+(1\cdot4\pm0\cdot1\times10^{-7}).$$

Very little is known about the temperature dependence of this kind of recombination coefficient.

(iii) *Mutual neutralization*. In this process the products carry away the excess energy as excitation. Under suitable conditions the recombination coefficient, which is independent of pressure, can be quite large. Thus a value of 10^{-7} cm^3/s was found for ion–ion recombination in iodine :[46] however, three-body processes became dominant above a pressure of a few millimetres of mercury.

Thermal ionization in the gas phase

When the various processes of charge generation, transfer, and recombination have taken the system to equilibrium, the concentration of the different species, charged and uncharged, can be deduced from thermochemical data, without reference to the rate processes. The equation describing the equilibrium state of ionizing molecules can be deduced by considering the ionization process as a simple chemical reaction whose heat of reaction is the ionization energy and whose products are ions and electrons, for example

$$\underset{(n_A)}{A} \rightleftharpoons \underset{(n_+)}{A^+} + \underset{(n_e)}{e^-} \qquad \Delta H = +eV_i$$

hence,

$$\frac{n_+ n_e}{n_A} = \frac{\omega_+ \omega_e}{\omega} \left(\frac{M_+ M_e}{M} \frac{2\pi kT}{h^2} \right)^{\frac{3}{2}} \exp\left(\frac{-eV_i}{kT} \right), \qquad (2.28)$$

where ω is the statistical weight.

Equation (2.28) is usually known as the Saha or the Saha–Lindemann equation. For practical purposes, M_+ and M are considered equal, also $\omega_e = 2$; moreover, in the absence of electron attachment and charge transfer $n_+ = n_e$. Thus taking a pure gas, if x is the fraction of the original molecules that are ionized, i.e. $x = n_e/(n_A + n_+)$,

$$\frac{x^2}{1-x^2} P = \frac{2\omega_+}{\omega} \left(\frac{2\pi M_e}{h^2} \right)^{\frac{3}{2}} (kT)^{\frac{5}{2}} \exp\left(\frac{-eV_i}{kT} \right) = \frac{\omega_+}{\omega} F(T). \qquad (2.29)$$

If an inert gas is seeded with an easily ionized material, for example argon with caesium,

$$\frac{\psi x^2 P}{(1-x)\{\psi(1+x)+1\}} = \frac{\omega_+}{\omega} F(T), \qquad (2.30)$$

where $\psi = $ (moles seed)/(moles inert) and x is the fraction of seed ionized. It is easy to show that x increases with increasing temperature and decreasing ionization potential, and with decreasing seed fraction. Usually $\psi < 0.05$, whether this is a result of deliberate seeding or of the incidental production of materials of low ionization potential by chemical reaction; the occurrence of NO in air at elevated temperatures is an example of the latter. One can therefore write eqn (2.30), to a first approximation, as

$$x^2 P/(1-x) = \frac{\omega_+}{\omega\psi} F(T). \qquad (2.31)$$

At constant temperature and low values of ψ, $x \simeq 1$, i.e.

$$n_e \propto \psi x = \psi. \qquad (2.32)$$

In this case there is a linear increase of electron concentration with seed fraction. At higher values of ψ the right-hand side becomes small, i.e. $x \ll 1$. Hence

$$\psi x^2 \propto F(T), \qquad (2.33)$$

i.e.

$$n_e \propto \psi x \propto \psi^{0.5} \qquad (2.34)$$

In this case the electron concentration increases only as the square root of the seed fraction. It is also interesting to consider the variation of electron concentration with increasing temperature. Initially x increases rapidly with temperature, until it approaches unity. Further increases in temperature then lead to quite negligible increases in x. However the gas density is reduced. The electron concentration, therefore, eventually begins to fall according to the ideal gas law, i.e. as $1/T$.

Table 2.6 shows the fraction of alkali metals ionized in an inert atmosphere as a function of temperature and partial pressure.

TABLE 2.6

Fraction, x, *of metal atoms ionized for various partial pressures,* P, *of added metal.* From Gaydon and Wolfhard[47]

Metal	K			Na			Ca	
V_i (eV)	4·33			5·14			6·11	
P (atm)	10^{-2}	10^{-4}	10^{-6}	10^{-2}	10^{-4}	10^{-6}	10^{-4}	10^{-6}
$T(°K)$								
1000	3×10^{-11}	3×10^{-10}	3×10^{-9}	4×10^{-12}	4×10^{-11}	4×10^{-10}	3×10^{-13}	3×10^{-12}
1500	27×10^{-7}	27×10^{-6}	27×10^{-5}	12×10^{-8}	12×10^{-7}	12×10^{-6}	55×10^{-9}	55×10^{-8}
2000	25×10^{-5}	25×10^{-4}	0·025	27×10^{-6}	27×10^{-5}	27×10^{-4}	3×10^{-5}	3×10^{-4}
2500	47×10^{-4}	47×10^{-3}	0·42	7×10^{-4}	7×10^{-3}	0·07	0·0014	0·014
3000	0·028	0·27	0·94	0·0062	0·062	0·53	0·019	0·19
3500	0·11	0·75	0·99	0·031	0·30	0·95	0·12	0·77

apparently simple cases of caesium chloride injected into hot inert gas, there will be at least three simultaneous reactions:

$$CsCl \rightleftharpoons Cs + Cl,$$

$$Cs \rightleftharpoons Cs^+ + e,$$

$$Cl^- \rightleftharpoons Cl + e.$$

The first removes some of the caesium, the last some of the electrons. All must be accounted for in the equilibrium equations.

In combustion products the situation is very much more complex because of the large number of reactions to be considered. Techniques for solving the systems of simultaneous equations arising from multiple equilibria are discussed in Chapter 3.

The case of electron attachment may be treated by Saha's equation, because ionization and electron detachment are formally identical, as can be seen from the equations above. Ionization potentials and electronegativities of a number of materials are given in Tables 2.1 and 2.2 respectively.

Care must be taken as regards units when employing the Saha equation. Thus ionization potentials and electronegativities are usually expressed in volts, in which case the electronic charge should be in coulombs ($1·602 \times 10^{-19}$) and Boltzmann's constant in joules/°K ($1·38 \times 10^{-23}$). If it is desired to use CGS units, the potential must be divided by 300 to convert to electrostatic volts (see Table 4.1), e expressed in e.s.u. (4.8×10^{-10}), and k in ergs/°K (1.38×10^{-16}).

Excited states of the ionizing material are likely to play an important role in very high temperature systems. Their principal effect is to reduce the ion concentration from that calculated ignoring their existence. The reason for this is that, although eqn (2.30) applies whether there is excitation or not, when excitation does occur, there are fewer ground state species and therefore ψ has effectively been reduced.

PART II

GAS–SOLID SYSTEMS

Electron emission from surfaces

There are four processes by which the supply of energy to a surface can cause electrons to be emitted. These are referred to as thermionic, impact, photo, and field emission. Thermionic emission is the one of greatest importance in the present context: the others, however, can become important in certain circumstances.

Thermionic emission

(i) *In the absence of applied fields.* When electrons are emitted from a surface they leave behind a positive charge that tends to draw them back. Consequently, only electrons with sufficient energy are capable of escaping entirely from the condensed phase. At high temperatures, the number of electrons of sufficient energy may be large enough to give rise to emission current densities in excess of tens of amperes per square centimetre. The minimum energy required to remove an electron from a surface is $e\phi$, where ϕ is the work function, which is usually expressed in volts. The remarks concerning units for the Saha equation apply here also. In subsequent equations ϕ is in e.s. volts, unless stated otherwise. The rate of emission can be expressed in terms of this parameter[48] as

$$ j = BT^2 \exp\left(\frac{-e\phi}{kT}\right) \text{ A/cm}^2. \tag{2.35} $$

The pre-exponential term B, like ϕ, is characteristic of the material. Some values are given in Table 2.7.[6] Comparison of the work functions of various materials with the corresponding ionization potentials (Table 2.1) shows that ϕ is considerably smaller than V_i.

Table 2.7

Work functions and emission constants

Material	$B\,(A/cm^2)$	ϕ volts†
C	48	4·4
W	70	4·5
Na	—	2·28
K	—	2·25
Cs	—	1·94
CaO	—	1·93
SrO	—	1·43
BaO	40	1·7 (1·0)
CuO	—	5·3

† Divide by 300 to obtain e.s.V.

The expression for the emission current requires modification[48] when the emitter is a positively charged particle of small finite radius, r_p, such as is encountered in suspensions of powdered fuel, for example. Two additional terms then arise in the work function. The surface charge Ne e.s.u. attracts the emitted electrons, increasing the work function by an amount Ne/r_p e.s.u. V: in addition, the interaction between the departing electron and the dipole it induces in its parent particle further increases the work function by $e/2r_p$ e.s.u. V. Thus for particles charging as a result of electron emission

$$j = BT^2 \exp\left[\frac{-e}{kT}\left\{\phi + (N + \tfrac{1}{2})\frac{e}{r_p}\right\}\right] \text{A/cm}^2. \qquad (2.36)$$

The refractories BaO and SrO are particularly effective emitters because of their low work-functions and high boiling-points. This facility for emission may be associated with, for example, the observed effect of barium on carbon formation in flames (see Chapter 7). Very low work functions are to be found in the case of oxide coated cathodes.[50,51]

(ii) *In the presence of applied fields.* If an electric field is applied to a thermionic emitter in a direction that assists removal of electrons, the effective work-function of the material is decreased. This is usually known as the Schottky effect.[52]

When an electron is at a distance x from the surface of a plane conductor there is a force of attraction of $e^2/4x^2$ dyn, for values of x large compared to atomic dimensions. If an electric field of strength E is applied to aid emission, there will be a point at a distance x_0 from the surface beyond which the net force on the electron is directed away from the surface, i.e. the external field dominates over the attractive field if $x > x_0$. At $x = x_0$

$$E = e/4x_0^2, \qquad (2.37)$$

i.e.

$$x_0 = \frac{1}{2}\sqrt{\left(\frac{e}{E}\right)}. \qquad (2.38)$$

Electrons reaching this point will have escaped.

The minimum energy for escape is thus reduced in two ways. On the one hand, the attractive field is opposed by the applied field E. On the other, the electrons need to do work against the attractive forces only up to $x = x_0$, and not $x = \infty$ as before. Thus the modified work function, ϕ', is given by

$$\phi' = \phi - \int_{x_0}^{\infty} \frac{e}{4x^2}\,dx - \int_0^{x_0} E\,dx, \qquad (2.39)$$

i.e.

$$\phi' = \phi - \sqrt{(eE)}. \qquad (2.40)$$

The emission equation becomes,

$$j = BT^2 \exp\left[-\frac{e}{kT}\{\phi - \sqrt{(eE)}\} \right] \text{A/cm}^2. \qquad (2.41)$$

The Schottky effect becomes significant when $(eE)^{\frac{1}{2}}/kT$ is greater than about 0·25, at which value the electric field causes the emission to increase by 27 per cent. At 2000°K this corresponds to a field of 23 kV/cm.

In the aforegoing case the energy of the escaping electrons derives in part from their thermal motion and in part from the applied field. However, when the applied field is sufficiently high, large electron currents can be drawn out of surfaces that are far too cold to emit appreciable currents thermionically. In this process, known as cold emission, all but a negligible fraction of the energy of the escaping electron comes from the applied field. It has been shown by Fowler and Nordheim[53] that the current density under conditions of cold emission is given by

$$j = \frac{6(\mu/\phi)^{\frac{1}{2}}}{\mu+\phi} E^2 \exp\left(\frac{-6\cdot 8 \times 10^7\ \phi^{\frac{3}{2}}}{E} \right) \mu\text{A/cm}^2, \qquad (2.42)$$

where μ and ϕ are the Fermi energy and the work function respectively (measured in V) and E is in V/cm. The equation indicates that cold emission currents should not become measurable until fields in excess of 10^6 V/cm are attained. In practice, however, owing to surface irregularities and impurities, appreciable currents can be obtained at very much lower mean field strengths than this, especially if electropositive materials are adsorbed.[54]

It is also interesting to consider what happens when the applied field opposes emission. The energy required to reach a point at distance x is then $e(\phi + Ex)$. Thus the emission current passing beyond a distance x is less by the Boltzmann exponential factor, $\exp(-Eex/kT)$, than that crossing the surface. If the field extends indefinitely from the surface, no electron can escape because there will always be a value of x such that $e(\phi + Ex)$ is greater than its initial energy. In practice, however, the field E will be applied between surfaces a finite distance apart and the product Ex cannot exceed the applied potential. A concept of the order of magnitude of practical effects can be obtained by considering the special case of an adverse potential of 2 V at 2000°K. Here the current escaping is reduced to about 10^{-5} of its original value.

The case of emission from particles in the presence of applied fields is discussed at the end of this chapter in conjunction with other mechanisms of particle charging.

Impact and photo-emission

Emission of electrons from a surface occurs also as a result of impact by electrons, ions, and neutral species in metastable electronically excited states (unexcited neutral species are only effective at very high energy and are therefore not considered here). The energy requirements for emission by an ion or an excited species are, respectively,

$$\text{Kinetic energy} + V_i \geq 2\phi,$$

$$\text{Kinetic energy} + V_{exc} \geq \phi,$$

In the case of the ion, energy is required not only to cause emission of an electron but also to remove a second electron from the surface in order to neutralize the ion's own charge. The effectiveness of collisions in producing secondary electrons is measured in terms of a parameter, γ, which is the average number of electrons emitted per collision. For positive ions the maximum value of γ lies between about 1 and 10, occurring at energies around 10^5 eV. However, quite high values are found in the range $0-10^3$ eV. Hagstrum[55] found the following values at or near zero kinetic energy:

$$Ne^+ - 0.29, \ He^+ - 0.21, \ Ar^+ - 0.09, \ Kr^+ - 0.05, \text{ and } Xe^+ - 0.02.$$

These values vary by less than 25 per cent in the range $0-10^3$ eV. It will be noticed that γ increases with increasing ionization potential, as might be expected, except in the case of He^+ and Ne^+. The reason for the anomalous behaviour of these ions is unknown.

Metastable atoms are particularly effective. Dorrestein[56] estimated values of γ of 0.2 and 0.48, respectively, for He (2^3S) and He (2^1S) impinging on clean platinum; a value of 0·4 was found for metastable argon atoms striking a caesium surface.

When electrons strike the surface of a conductor or semiconductor, γ has a maximum value of about 1·0, occurring in the energy range 200–800 eV. For insulators, the maximum value of γ lies between 2 and 7 and occurs at between 300 and 2000 eV.[57,58] There is very little information at energies of the order of 1 eV. Some additional data for Ag, Ba, BaO, Cu, W, Pd, and Pt surfaces are given by Bruining[59] and Farnsworth.[60]

Photoemission is characterized by a clearly defined threshold energy, which is governed by quantum requirements. Thus in order for emission of an electron to occur

$$hv \geq e\phi.$$

Visible radiation is of the order of 2 to 3 eV and at such energies most materials show very little emission; $\gamma \simeq 10^{-6} - 10^{-3}$ electrons/photon. It is not until the far ultraviolet that γ reaches its maximum value, which lies usually between 0·1 and 1·0. Emission can be greatly increased by special surface coatings, as for example when metal oxides are coated with thin layers of alkali metal.[61,62] Data for various surfaces are available.[6,61]

Thermal ionization of particles

Various workers have put forward statistical-mechanical treatments of the ionization of particles, in which allowance is made for the effect of the residual charge upon the effective work-function. The treatment followed here is that due to Einbinder[63] as modified by Smith.[49] These workers treated the emission of electrons from solid particles by direct analogy with the ionization of gaseous atoms and molecules, the ionization potential being replaced by the effective work function, $\{\phi + (N + \frac{1}{2})(e/r_p)\}$, which is the work necessary to remove an electron from a particle carrying charge Ne. Hence, using Saha's equation for the single step ionization process

$$\underset{(n_p)}{A\,(\text{solid})} \rightleftharpoons \underset{(n_+)}{A^+\,(\text{solid})} + \underset{(n_e)}{e^-\,(\text{gas})}$$

it follows that

$$\frac{n_e^2}{n_p} = 2\left(\frac{2\pi M_e k}{h^2}\right)^{\frac{3}{2}} T^{\frac{3}{2}} \exp\left\{-\frac{e}{kT}\left(\phi + \frac{e}{2r_p}\right)\right\}, \qquad (2.43)$$

setting $M_p = M_+$ and $\omega_p = \omega_+$, where subscript p stands for particle.

The positive charge left on the particle after the emission of the first electron makes further emission more difficult. Einbinder[63] has considered the case of multiple-charged particles as represented by the reaction scheme

$$A \rightleftharpoons A^+ + e^-$$

$$\vdots$$

$$A^{(N-1)+} \rightleftharpoons A^{N+} + e^- \qquad N^{\text{th}} \text{ reaction.}$$

For the Nth reaction,

$$\frac{n_{N} + n_e}{n_{(N-1)+}} = 2\left(\frac{2\pi M_e k}{h^2}\right)^{\frac{3}{2}} T^{\frac{3}{2}} \exp\left\{\frac{-e}{kT}\left(\phi + \frac{N - \frac{1}{2}}{r_p}e\right)\right\} \qquad (2.44)$$

Einbinder's treatment is inexact, in that he omitted the term $e^2/2r_p$ from eqn (2.44) and neglected to take account of the formation of negatively charged particles by electron attachment. These are not serious omissions and they were corrected by Smith[49] who, using Einbinder's method of solution of the system of simultaneous equations, was able to obtain relatively simple expressions for the overall gas-phase electron concentration. For the usual case when the term $e^2/2r_p kT$ is small, i.e. less than about 3, Smith showed that

$$\frac{n_e}{n'_e} = n_p \frac{r_p kT}{e^2 n'_e} \ln\left(\frac{n'_e}{n_e}\right), \qquad (2.45)$$

where

$$n'_e = 2\left(\frac{2\pi M_e kT}{h^2}\right)^{\frac{3}{2}} \exp(-e\phi/kT), \qquad (2.46)$$

i.e. the concentration of electrons in equilibrium with the bulk solid. The general expression relating n_e to work function, temperature, and particle concentration is given by Smith.[49]

Other treatments of the problem have been given by Arshinov and Musin,[64] Sodha, Palumbo, and Daley,[65] Soo,[66] and Soo and Dimick.[67] The work of Soo is of particular interest because it explicitly takes into account thermal ionization of the gas and because it does not rely upon the analogy with thermal ionization of gas molecules. The objection to the analogy lies in the assumption that the electrons are either free in the gas or bound in the solid. In fact, each particle will be surrounded by a cloud of loosely bound charges. Soo's treatment, however, does not make allowance for the distribution of different degrees of charging amongst the particles.

Particle charging in regions of space charge

When a potential difference is applied between an ion source, for example a flame, and an electrode, either positive or negative charge will flow from the source to the electrode, the sign depending upon the direction of the electric field. Unless convective and diffusive ionic velocities are comparable to the field-induced velocities, the conduction region will be one in which there exists a unipolar positive or negative space charge and an electric field. This section is concerned with the charging of particles that are present in such a region. Unlike in the case of electron emission due to impact, here the energy of the arriving charge on impact is immaterial and charging occurs because ions and electrons reach the surface of the particle from the gas phase and remain there. A special case arises when thermionic emission is important either as a result of high gas-temperature or of low work-function of the particle, or both. These processes are discussed below.

(a) *Charging without thermionic emission*

Two theories have been developed[68,69,70] one of which is appropriate for particles greater than about $0.5\ \mu m$ radius (bombardment charging), the other for those of radius less than about $0.1\ \mu m$ (diffusion charging).

(i) *Bombardment charging* $(>0.5\ \mu m$ radius). In a uniform field, lines of force crowd into any conducting particle inducing a distribution of charge across its surface. The effect for a sphere can be shown to be equivalent to that which would be due to two 'image charges' within the sphere, acting as a dipole of strength $E_0 r_p^3$ (where E_0 = field in absence of particle) in e.s.u. Ions travelling along lines of force onto the surface will gradually neutralize this effect, until an equilibrium charge is collected that makes the lines of force by-pass the particle. The equilibrium charge is thus a function of the unperturbed field intensity.

Figure 2.8 shows the configuration of lines of force under three conditions: (a) uncharged, (b) partially charged, and (c) fully charged.

At the surface of the particle

$$E = \{3E_0 \cos \theta - (Ne/r_p^2)\}. \tag{2.47}$$

At $\theta = \theta_0$, $E = 0$, i.e.

$$\cos \theta_0 = Ne/3E_0 r_p^2. \tag{2.48}$$

The total flux J (the product of field and area perpendicular to it) entering the particle between $\theta = 0$ and $\theta = \theta_0$ is given by the surface integral

$$J = \int_0^{\theta_0} \{3E_0 \cos \theta - (Ne/r_p^2)\} 2\pi r_p^2 \sin \theta \, d\theta, \tag{2.49a}$$

i.e.

$$J = 3\pi r_p^2 E_0 \{1 - (Ne/3E_0 r_p^2)\}^2. \tag{2.49b}$$

The flux originates from an area A in the undisturbed region in which there is a concentration of ions, n_i, of mobility K. Ions that enter this area will

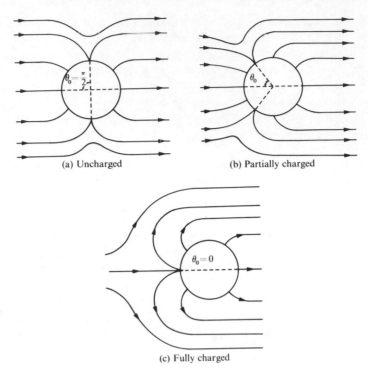

(a) Uncharged   (b) Partially charged

(c) Fully charged

FIG. 2.8. Field distribution round a conducting sphere. From Gugan, Lawton, and Weinberg.[69]

terminate on the particle. This presupposes that the relevant dimensions are large in comparison with the mean free path so that, irrespective of inertia, ion trajectories follow lines of force. This assumption is consistent with the size of particles considered in this regime. Since ion mobility is always much greater than that of the particles, we can neglect the particle velocity by comparison. Thus, the current flowing to one particle at any given time t is given by

$$KAE_0 n_i e = KJn_i e = d(Ne)/dt, \qquad (2.50)$$

i.e.

$$d(Ne)/dt = 3\pi r_p^2 (Kn_i E_0 e)\{1-(Ne/3E_0 r_p^2)\}^2 \qquad (2.51)$$
$$= 3\pi r_p^2 j_i \{1-(Ne/3E_0 r_p^2)\}^2,$$

where j_i is the current density due to the ions in the unperturbed gas. In the equilibrium condition $\theta_0 = 0$, i.e. no more current flows to the particle. Thus, setting $\cos \theta_0 = 1$ in eqn (2.48) gives the equilibrium charge as

$$(Ne) = (Ne)_1 = 3E_0 r_p^2. \qquad (2.52)$$

Setting $3E_0r_p^2 = (Ne)_1$ in eqn (2.51)

$$d(Ne)/dt = 3\pi r_p^2 j_i[1 - \{Ne/(Ne)_1\}]^2$$

$$= 3\pi r_p^2 j_i(1 - f)^2, \tag{2.53}$$

where $f = (Ne)/(Ne)_1$, the fraction of the equilibrium charge attained at time t.

For a nonconducting particle of dielectric constant D the above results (both the equilibrium charge and the rate of charging) must be multiplied by $D/(D+2)$. Most of the materials likely to be present as particulate matter either are conductors or may be considered conducting by comparison with the surrounding gas, at the ion densities and field strengths involved. Although the discussion will therefore continue to be based on $D \to \infty$, the correction factor (if required) is a simple constant throughout.

In a field of 5 kV/cm, which is a moderate value for a space charge region, conducting particles of radius 0·1, 1, 10 and 100 μm reach saturation at 5, 10^3, 10^5, and 10^7 electronic charges, respectively.

(ii) *Diffusion charging* ($<0·1$ μm radius). In this case the particle is charged by ions conveyed not along lines of force but by random diffusion to its surface. The random flux of ions across any unit area is given by kinetic theory as $n_i \bar{c}/4$. Initially, this is the flux density reaching the particle. However, once the particle has attained a charge, it begins to repel the oncoming ions. The potential of the charged particle V with respect to an ion at infinity is equal to $(N - \frac{1}{2})e/r_p$ e.s. V. Only ions with energy in excess of this value will reach the particle; therefore only a fraction $e^{-eV/kT}$ of the ions approaching will reach the surface. The charging rate is, accordingly, given by

$$\frac{dN}{dt} = \pi a^2 \bar{c} n_i \exp\{-e^2(N - \frac{1}{2})/r_p kT\} \tag{2.54a}$$

Integrating, and setting $N = 0$ at $t = 0$,

$$N = \frac{r_p kT}{e^2} \ln\left(1 + \frac{\pi r_p \bar{c} n_i e^2 t}{kT}\right) + \frac{1}{2}. \tag{2.54b}$$

In this case there is no saturation charge inherent in the process itself; the particle continues to charge indefinitely, given infinite time. In practice, the charge on the particle would be limited by breakdown of the gas and emission of the surface charges under the action of the intense local field.

Taking an ion concentration of 10^{10} cm^{-3}, which is a high space charge density, and a temperature of 1000°K, particles of radius 0·1, 1, 10, and 100 μm are calculated to collect $1·4 \times 10$, $1·9 \times 10^2$, $2·3 \times 10^3$, and $2·8 \times 10^4$ electronic charges respectively in 0·1 s. The charge increases by about 25 per cent when the time is lengthened by a factor of 10. A similar increase is found if the ion concentration is increased instead by the same factor. Clearly the numerical values are very insensitive to time and to ion concentration.

If the particle emits electrons whilst in a region of negative space charge, it may become positive and attract negative charges from the gas phase. If, in this case, $\lambda > r_p$ and $E_0\lambda \ll kT/e$, the current from the gas to the particle is given by[71]

$$\frac{dN}{dt} = \pi r_p^2 \bar{c} n_i [1 + e^2(N + \tfrac{1}{2})/r_p kT] \tag{2.55}$$

To be precise, both bombardment- and diffusion-charging occur simultaneously; Murphy, Adler, and Penney[70] have discussed the difficulties involved in the complete solution. However, as can be seen from the charges calculated for the two cases, in practice, diffusion-charging dominates for particle radii below about 0·1 μm and bombardment-charging dominates above about 0·5 μm—the charges for intermediate radii being estimated by taking the larger of the two calculated values. White[68] has compared the calculated values with those obtained experimentally by himself and others and found good agreement for the case of bombardment charging. Murphy et al.[70] found that the diffusion theory predicts values that are about one-half of those determined experimentally.

(b) Charging in presence of thermionic emission

Charging by thermionic emission and by acquisition of ions and electrons can occur simultaneously. For clarity, however, the case in which thermionic processes dominate is treated first. The field configuration around the particle, before the loss of any electrons, is shown in Fig. 2.8(a). Half the lines of force reaching the surface assist emission, the rest oppose it. As the particle becomes progressively charged, the changes in field-configuration are identical with those described for bombardment charging (Fig. 2.8(b) and (c)). Progressively fewer lines of force reaching the particle are directed so as to aid emission, until a charge of $3E_0 r_p^2$ e.s.u. is acquired when all the lines of force oppose electron removal.

Emission from plane surfaces has been discussed already. The case of spheres in linear fields is a much more complex problem to which no general solution exists at present. Nevertheless, estimates of rates of charging can be made for most cases of practical interest.

The net rate of charging is the sum of the emission currents from the regions of surface experiencing adverse and assisting fields, i_1 and i_2, respectively,

$$\frac{dNe}{dt} = 3E_0 r_p^2 \frac{df}{dt} = i_1 + i_2. \tag{2.56}$$

It is convenient to consider the two contributions separately.

(i) *Assisting fields.* If the field strength at the surface satisfies the inequality

$$e(eE)^{\frac{1}{2}}/kT < 0.25 \tag{2.57}$$

the Schottky effect is small, causing an increase in emission of less than 27 per cent. At $10^{3°}$K this corresponds to a field of about 10 e.s.u. V/cm (3×10^3 V/cm) and at $2 \times 10^{3°}$K to a field of about 40 e.s.u. V/cm (12×10^3 V/cm). The rate of emission is then approximately equal to the usual thermionic emission. Writing

$$j_0 = BT^2 \left(\exp \frac{-e\phi}{kT} \right) \tag{2.58}$$

it follows that

$$i_1 = j_0 2\pi r_p^2 \int_0^{\theta_0} \sin \theta \, d\theta \tag{2.59}$$

or, since $f = \cos \theta_0$,

$$i_1 = 2\pi r_p^2 j_0 (1 - f). \tag{2.60}$$

On the other hand, when the inequality is not satisfied, the Schottky effect must be taken into account. Then

$$i_1 = 2\pi r_p^2 j_0 \int_0^{\theta_0} \exp\left[\frac{e}{kT} \sqrt{\{3E_0(\cos\theta - \cos\theta_0)e\}} \right] \sin\theta \, d\theta \tag{2.61}$$

$$= \frac{4\pi r_p^2}{3E_0} j_0 \left(\frac{kT}{e} \right)^2 \left[1 - \exp\left[\frac{e}{kT} \sqrt{\{3E_0(1-f)e\}} \right] \left[1 - \frac{e}{kT} \sqrt{\{3E_0(1-f)e\}} \right] \right]. \tag{2.62}$$

The calculation of the Schottky effect was based upon a one-dimensional model. This is valid so long as x_0 is much less than the particle radius, i.e. from eqn (2.38)

$$r_p \gg \tfrac{1}{2}\sqrt{(e/E)}. \tag{2.63}$$

Since the Schottky effect is small if $e(eE)^{\frac{1}{2}}/kT$ is less than about 0.25, the inequality may be written as a function of temperature

$$r_p \gg 1.65 \times 10^{-3}/T \text{ cm.} \tag{2.64}$$

This means that eqn (2.62) applies if $r_p \gg 10^{-6}$ cm, since there is very little emission below 1000°K. Clearly this is valid for the bulk of practical situations.

If the particle is within a negative space charge it may acquire a net negative charge that will also aid emission. In the case when $E_0 r_p \ll kT/e$, i.e. when the Schottky effect due to the applied field is small, that due to the accumulated charge leads to a net current given by[71]

$$i_1 = 4\pi r_p^2 A T^2 \exp\left[\left[-\frac{e}{kT}\left[\phi - \left\{(1-n)x + (n-\tfrac{1}{2}) + \frac{x^4}{2(1-x^2)}\right\}\frac{e}{r_p}\right]\right]\right] \qquad (2.62b)$$

where

$$x = 1 - \frac{1}{2\sqrt{n}} \qquad (2.62c)$$

for $n \geq 4$ to within 2 per cent, and for $n = 1, 2$ and 3, x assumes the values 0, 0·61, and 0·7.

(ii) *Adverse fields.* If the product eV/kT, where V is the adverse potential, is greater than 4·6, the emission current density is reduced to less than $\frac{1}{100}$ of its initial value. This is equivalent to

$$V > 3{\cdot}96 \times 10^{-4} T \text{ volts.} \qquad (2.65)$$

On the other hand, if eV/kT is less than 0·25, more than 70 per cent of the emission current can escape. This condition is given by

$$V < 2{\cdot}15 \times 10^{-5} T \text{ volts.} \qquad (2.66)$$

Thus, for practical purposes, a potential satisfying eqn (2.65) effectively suppresses emission, whilst one satisfying eqn (2.66) has negligible effect. From these considerations it is possible to deduce approximate solutions to the problem of rates of charging against adverse fields which are valid in particular circumstances, as discussed below.

The influence of the particle on the surrounding field is important up to about one radius from its surface, the lines of force converging on the particle in this region. Therefore, to an order of magnitude, the retarding potential is $E_0 r_p$. Accordingly, if the product $E_0 r_p$ satisfies (2.65), the adverse field is effective in suppressing emission, while if it satisfies (2.66), it has a negligible effect. Thus two cases may be distinguished for practical purposes: if

$$E_0 r_p > 3{\cdot}96 \times 10^{-4} T \text{ volts}$$

$$i_2 \simeq 0; \qquad (2.67)$$

if

$$E_0 r_p < 2{\cdot}15 \times 10^{-5} T \text{ volts and } Ne \leq (Ne)_1$$

$$i_2 \simeq 2\pi r_p^2 j_0 \int_\pi^{\theta_0} \sin\theta \, d\theta = 2\pi a^2 j_0 (1 + f). \qquad (2.68)$$

In the latter case emission is eventually limited by the potential due to charge accumulation on the particle as a whole; thus when $Ne \geq (Ne)_1$, following the same basic argument that was used to obtain eqn (2.54), the rate of charging is given approximately by

$$i_2 = j_0 4\pi r_{\mathrm{p}}^2 \exp\left\{\frac{Ne^2 - (Ne^2)_1}{r_{\mathrm{p}}kT}\right\}. \tag{2.69}$$

It is very difficult to estimate the emission current for intermediate values of $E_0 r_{\mathrm{p}}$. However, it is fortunate that the ranges of $E_0 r_{\mathrm{p}}$ where the present treatment is appropriate cover the most important practical cases. The values inserted in the inequalities, although realistic, are, of course, arbitrary. They are intended merely to illustrate a method by which approximate solutions can be obtained. The appropriate expressions for i_1 and i_2, for any particular case, can be inserted into eqn (2.56) and the variation of charge with time can be found by integration, thus

$$3E_0 r_{\mathrm{p}}^2 \int_0^f \frac{\mathrm{d}f}{i_1(f) + i_2(f)} = t. \tag{2.70}$$

When there is appreciable accretion of charge from the gas phase, by bombardment or diffusion, the net rate of charging is found by adding this to the emission current.

REFERENCES

1. JEANS, J. *An introduction to the kinetic theory of gases*. Cambridge University Press (1962).
2. ROSE, D. J. and CLARK, M. *Plasmas and controlled fusion*, Chap. 4. Wiley, New York (1961).
3. MOORE, W. J. *Physical chemistry*, 2nd edn. Longmans, Green, London (1956).
4. VEDENEYEV, V. I., GURVICH, L. V., KONDRATIEV, V. N., MEDVEDEV, V. A., and FRANKEVICH, YE. L. *Bond energies, ionization potentials and electron affinities*. Arnold Press, London (1966).
5. KONDRATIEV, V. N. *Chemical kinetics of gas reactions*. Pergamon Press, London and New York (1964).
6. VON ENGEL, A. *Ionized gases*. Clarendon Press, Oxford (1965).
7. HORTON, F. and MILLEST, D. *Proc. R. Soc.* **A185**, 381 (1946).
8. BROWN, S. C. *Basic data of plasma physics*. Chapman and Hall, London (1961).
9. MCDANIEL, E. W. *Collision phenomena in ionized gases*. Wiley, London and New York (1964).
10. MASSEY, H. S. W. and BURHOP, E. H. S. *Electronic and ionic impact phenomena*. Clarendon Press, Oxford (1952).
11. HAYHURST, A. N. Ph.D. Thesis, Cambridge University (1964).
12. KRUITHOF, A. A. and DRUYVESTEYN, M. J. *Physica* **4**, 450 (1937).
13. KRUITHOF, A. A. and PENNING, F. M. *Physica* **4**, 430 (1937).
14. STEUBING, W. *Phys. Z.* **10**, 787 (1909).

15. GREEN, J. A. and SUGDEN, T. M. *9th Int. Symp. Combust.*, p. 607. Academic Press, London and New York (1963).
16. BASCOMBE, K. N., GREEN, J. A., and SUGDEN, T. M. *Joint symposium on mass spectrometry—ASTM and Institute of Petroleum.* Pergamon Press, London (1962).
17. BRANSCOMBE, L. M. *Atomic and molecular processes* (edited by D. R. Bates), Chap. 4, p. 134. Academic Press, London and New York (1962).
18. CHANIN, L. M., PHELPS, A. V., and BIONDI, M. A. *Phys. Rev.* **2**, 344 (1959).
19. HASTED, J. B. *Physics of atomic collisions.* Butterworths (1965).
20. BUCHEL'NIKOVA, N. S. *Zh. éksp. teor. Fiz.* **35**, 1119 (1958).
21. CRAGGS, J. D. and TOZER, B. A. *Proc. R. Soc.* A**247**, 337 (1958).
22. CRAGGS, J. D. and TOZER, B. A. *Proc. R. Soc.* A**254**, 229 (1960).
23. HASTED, J. B. *Adv. Electronics Electron Phys.* **13**, 1 (1960).
24. FRANKLYN, J. L. and MUNSON, M. S. B. *10th Int. Symp. Combust.*, p. 561. Combustion Institute, Pittsburgh (1965).
25. GIOMOUSIS, G. and STEVENSON, D. P. *J. chem. Phys.* **29**, 294 (1958).
26. STEVENSON, D. P. and SCHLIESSER, D. O. *J. chem. Phys.* **24**, 926 (1956); **23**, 1353 (1955); **29**, 282, 294 (1958).
27. TALROSE, V. L. and FRANKEVICH, E. L. *Zh. fiz. Khim.* **34**, 1275 (1960).
28. DICKENSON, P. G. H. and SAYERS, J. *Proc. phys. Soc.* **76**, 137 (1960).
29. BIONDI, M. A. *Adv. Electronics Electron Phys.* **18**, 67 (1963).
30. FITE, W. L., RUTHERFORD, J. A., SNOW, W. R., and VAN LINT, V. *Discuss. Faraday Soc.* **33**, 264 (1962).
31. LANGSTROTH, G. F. O. and HASTED, J. B. *Discuss. Faraday Soc.* **33**, 298 (1962).
32. HENGLEIN, A. and MUCCINI, G. A. *J. chem. Phys.* **31**, 1426 (1959).
33. CURRAN, K. *Phys. Rev.* **125**, 910 (1960).
34. PAHL, M. *Ergebn. exact. Naturw.* **34**, 182 (1962).
35. MUNSON, R. J. and TYNDALL, A. M. *Proc. R. Soc.* A**172**, 28 (1939).
36. MUNSON, R. J. and HOSELITZ, K. *Proc. R. Soc.* A**172**, 43 (1939).
37. BLOOM, S. and MARGENAU, H. *Phys. Rev.* **85**, 670 (1952).
38. LOEB, L. B. *Basic processes of gaseous electronics.* University of California Press (1961).
39. THOMSON, J. J. and THOMSON, G. P. *Conduction of electricity through gases*, Vol. I. Cambridge University Press (1928).
40. GARDNER, M. E. *Phys. Rev.* **53**, 75 (1938).
41. LANGEVIN, P. *Annls Chim. Phys.* **28**, 287, 433 (1903).
42. NATANSON, G. L. *Soviet Phys. tech. Phys.* **4**, 1263 (1959).
43. BATES, D. R. *Phys. Rev.* **78**, 492 (1955).
44. BATES, D. R. and DALGARNO, A. *Atomic and molecular processes* (edited by D. R. Bates), p. 245. Academic Press, New York (1962).
45. OSKAM, H. J. and MITTELSTADT, V. R. *Phys. Rev.* **132**, 1445 (1963).
46. YEUNG, T. H. and SAYERS, J. *Proc. phys. Soc.* **71**, 341 (1958).
47. GAYDON, A. G. and WOLFHARD, H. G. *Flames*, 2nd edn, p. 304. Chapman and Hall (1960).
48. DUSHMAN, S. *Phys. Rev.* **21**, 623 (1923).
49. SMITH, F. T. *J. chem. Phys.* **28**, 746 (1958).
50. FRIEDENSTEIN, N. *Rep. Prog. Phys.* **9**, 298 (1948).
51. EISENSTEIN, A. S. *Adv. Electronics* **1**, 1 (1948).

52. SCHOTTKY, W. *Annln Phys.* **44**, 1011 (1914).
53. FOWLER, R. H. and NORDHEIM, L. *Proc. R. Soc.* A**119**, 173 (1928); A**124**, 694 (1929).
54. COBINE, J. D. *Gaseous conductors.* Dover, New York (1958).
55. HAGSTRUM, H. D. *Phys. Rev.* **96**, 325 (1954).
56. DORRESTEIN, R. *Physica* **9**, 433, 447 (1942).
57. MCKAY, K. G. *Adv. Electronics* **1**, 65 (1948).
58. BRUINING, H. *Secondary electron emission.* Pergamon Press, London (1954).
59. BRUINING, H. *Physica* **5**, 913 (1938).
60. FARNSWORTH, H. E. *Phys. Rev.* **25**, 41 (1925).
61. SOMMER, A. *Photoelectric cells.* 2nd edn. Methuen, London (1951).
62. GORLICH, P. *Adv. Electronics* **11**, 1 (1959).
63. EINBINDER, H. *J. chem. Phys.* **26**, 948 (1957).
64. ARSHINOV, A. A. and MUSIN, A. K. *Dokl. Akad. Nauk. SSSR* **122**, 848 (1958).
65. SODHA, M. S., PALUMBO, C. J., and DALEY, J. T. *Br. J. appl. Phys.* **14**, 916 (1963).
66. SOO, S. L. *J. appl. Phys.* **34**, 1689 (1963).
67. SOO, S. L. and DIMICK, R. C. *10th Int. Symp. Combust.*, p. 699. The Combustion Institute, Pittsburgh (1965).
68. WHITE, H. J. *Trans. Am. Inst. elect. Engrs,* Pt. 2, **70**, 1186 (1951).
69. GUGAN, K., LAWTON, J. and WEINBERG, F. J. *10th Int. Symp. Combust.*, p. 709. The Combustion Institute, Pittsburgh (1965).
70. MURPHY, A. T., ADLER, F. T., and PENNEY, G. W. *Trans. Am. Inst. elect. Engrs,* Pt. I, **78**, 318 (1959).
71. LAWTON, J. *Combust Flame* **12**, 534 (1968).

3. Combustion Processes and Burner Systems

Definition and presentation of subject

THE subject of combustion and flames is covered in detail in several excellent books and monographs. It is characteristic of its many-sidedness that the treatment and bias in emphasis given to its several branches can vary as much as it does from one treatise to another. Those that provide some part of the background relevant here are Gaydon and Wolfhard's *Flames*,[1] Lewis and von Elbe's *Combustion, flames and explosions of gases*,[2] Fristrom and Westenberg's *Flame structure*,[3] and, to some extent, Minkoff and Tipper's *Chemistry of combustion reactions*.[4] A book by one of us—Weinberg: *Optics of flames*[5]—contains a large section (about seventy-five pages) devoted to flame processes, the treatment of which is particularly appropriate to the present monograph because it is similarly designed to provide the basis for physical aspects of combustion.

This chapter consists of the tersest summary of the subject which is adequate to introduce the reader to those concepts, terminology, and techniques of combustion which are used in subsequent parts of the book. In addition to its cursory nature, as a summary of combustion it is both unbalanced and unorthodox in its sequence. It is unbalanced because topics relevant to electrical aspects have been given disproportionate weight, as regards both the space devoted to them and the coverage of original literature. Whilst established combustion concepts have been treated in a 'textbook manner', with reference given only where they provide additional data, topics such as, for example, 'augmented flames' have been dealt with much more exhaustively because they are relevant here and because they have not been reviewed elsewhere. The sequence is unorthodox in that the various flame mechanisms have been separated from burner types. The latter form a large and separate section, intended to acquaint the reader with the tools available for this kind of research. This chapter is therefore not to be regarded as a general review of the field of combustion, outside the context of the present monograph.

We shall treat all highly exothermic reactions in gases as flame phenomena. These can proceed in a large variety of ways depending on whether the reactants are initially pre-mixed or separate, whether they are stationary

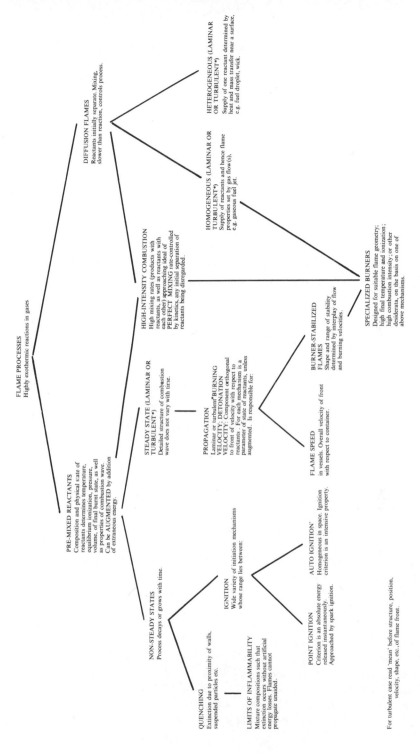

FLAME PROCESSES
Highly exothermic reactions in gases

DIFFUSION FLAMES
Reactants initially separate. Mixing,
slower than reaction, controls process.

HETEROGENEOUS (LAMINAR
OR TURBULENT*)
Supply of one reactant determined by
heat and mass transfer near a surface,
e.g. fuel droplet, wick.

HOMOGENEOUS (LAMINAR OR
TURBULENT*)
Supply of reactants and hence flame
properties set by gas flow(s),
e.g. gaseous fuel jet.

PRE-MIXED REACTANTS
Composition and physical state of
reactants determines temperature,
equilibrium ionization, pressure,
volume, of final burnt state, as well
as properties of combustion wave.
Can be AUGMENTED by addition
of extraneous energy.

HIGH-INTENSITY COMBUSTION
High mixing rates (products with
reactants, as well as reactants with
each other) approaching ideal of
PERFECT MIXING rate-controlled
by kinetics, any initial separation of
reactants being disregarded.

STEADY STATE (LAMINAR OR
TURBULENT*)
Detailed structure of combustion
wave does not vary with time.

PROPAGATION
Laminar or turbulent.'BURNING
VELOCITY: DETONATION
VELOCITY: Component orthogonal
to front of velocity with respect to
reactants. For each mechanism is a
parameter of state of reactants, unless
augmented. Is responsible for:

SPECIALIZED BURNERS
Designed for suitable flame geometry;
high final temperature and ionization;
high combustion intensity; or other
desiderata, on the basis on one of
above mechanisms

BURNER-STABILIZED
FLAMES
Shape and range of stability
determined by interplay of flow
and burning velocities.

FLAME SPEED
in vessels. Overall velocity of front
with respect to container.

NON-STEADY STATES
Process decays or grows with time.

IGNITION
Wide variety of initiation mechanisms
whose range lies between:

AUTO IGNITION'
Homogeneous in space. Ignition
criterion is an intensive property.

POINT IGNITION
Criterion is an absolute energy
released instantaneously.
Approached by spark ignition.

QUENCHING
Extinction due to proximity of walls,
suspended particles etc.

LIMITS OF INFLAMMABILITY
Mixture compositions such that
extinction occurs without artificial
energy losses. Flames cannot
propagate unaided.

For turbulent case read 'mean' before structure, position,
velocity, shape, etc., of flame front.

FIG. 3.1. Subdivision and interrelation of flame processes.

or in laminar or turbulent flow, and whether the reaction front is accompanied by a shock wave or an electric discharge. Figure 3.1 provides a general framework, showing the interrelation of the various subdivisions.

THE FINAL STATE

When all the heat of reaction is available to raise the temperature and volume (or pressure) of the products, i.e. when the process occurs rapidly and in the absence of appreciable radiation or other energy losses, the calculation of the state of the products is a purely thermo-chemical one. Rates of reaction or burning velocities are then not required, because all the quantities involved are perfect differentials and the final state is independent of the actual temperature-composition path by which it was attained. The calculation consists essentially of equating the initial heat of reaction to the rise in enthalpy (at constant pressure) or internal energy (at constant volume) of the same mass of products. Thus,

$$Q_u = \int_{T_u}^{T_b} C_p \, dT = (H_b - H_u)_P \quad \text{at constant pressure} \qquad (3.1)$$

and

$$Q_u = \int_{T_u}^{T_b} C_v \, dT = (U_b - U_u)_P \quad \text{at constant volume.} \qquad (3.2)$$

When the initial temperature T_u is not the standard value at which data are tabulated, Kirchhoff's equation

$$Q_{T_2} - Q_{T_1} = \int_{T_1}^{T_2} (C_R - C_P) \, dT \qquad (3.3)$$

may be used between any two temperatures T_1 and T_2 to deduce Q_{T_u} from that at the standard conditions (suffixes R and P denote reactants and products).

The main difficulty in carrying out such calculations lies in the arithmetical complexities that arise when several of the molecular species dissociate. The product composition is then no longer immediately obvious and can only be inferred by solving several (often a large number) of simultaneous equilibrium equations at an assumed final temperature. Dissociation affects eqns (3.1) and (3.2) in two ways; in that the enthalpy (or internal energy) and its variation with temperature is changed from that of the undissociated products, and in that the heat of reaction is diminished by the energy required to break the bonds of the dissociating products.

When dissociation occurs readily, the energy tied up in this manner causes product enthalpy to rise steeply for very small changes in final temperature.

It is for this reason that the highest final temperatures occur in flames whose products do not dissociate readily (for example cyanogen and oxygen). Such compositions can be used to increase equilibrium ion concentrations, particularly by arranging also for one product to have a low ionization energy. A most striking example of this principle is provided by the work of Friedman and Macek[6] who set out to produce a high electron-concentration in the products of a solid propellant-type composition. This was compounded of tetracyanoethylene, hexanitroethane, and caesium azide in proportions stoichiometric to CO, N_2, and Cs. In addition to a minimum of dissociation and an ingredient of low ionization-potential, the authors aimed for avoidance of species of high electron affinity, as well as for certain desirable physical and combustion properties of the mixture. The low electron-affinity requirement precluded the use of either conventional perchlorate oxidizers or hydrocarbon fuels (which form Cl^- and OH^- respectively). In fuel-rich mixtures the CN^- radical tends itself to attach electrons but for the optimum mixture, burning to the above-mentioned principal products, this does not occur, and a calculated final temperature of 3660°K was attained together with an electron density of 10^{15} (measured) or 10^{16} (calculated) electrons/cm^3. Under these conditions, the calculated product composition is (in mole fractions) $CO-0.5882$, $N_2-0.3636$, $Cs-0.0286$, $Cs^+-0.0057$ $(=[e])$, $O-0.0052$, $CO_2-0.0018$, $NO-0.0010$, $N-0.0003$, $O_2-0.0001$, and less than 0.0000 for CN, CN^-, $CsCN$, C(gas), C_2, and C_3. It will be seen that as many as *one in every six* Cs atoms is ionized, at atmospheric pressure.

Returning to normal hydrocarbon flames, the dissociation equilibria that must be taken into account in the presence of nitrogen are

$$CO_2 \rightleftharpoons CO + \tfrac{1}{2}O_2, \qquad H_2O \rightleftharpoons H_2 + \tfrac{1}{2}O_2, \qquad H_2O \rightleftharpoons \tfrac{1}{2}H_2 + OH,$$

$$\tfrac{1}{2}H_2 \rightleftharpoons H, \qquad \tfrac{1}{2}O_2 \rightleftharpoons O, \quad \text{and} \quad \tfrac{1}{2}N_2 + \tfrac{1}{2}O_2 \rightleftharpoons NO.$$

This applies only when no solid carbon is formed and even then the number of product species is ten, i.e. OH, H, O, NO, H_2, and CO, in addition to the expected main products of CO_2, H_2O, N_2, and O_2. This implies that ten simultaneous equations—six corresponding to the above equilibria and four arising from the conservation of the atoms of each element—must be solved to calculate composition for a given temperature. The dissociation equilibrium constants at various temperatures, as well as the other thermodynamic data required, have been summarized in a number of compilations.[1,2,7-10] The method of calculation then is cumbersome rather than difficult. In principle, the procedure always amounts to calculating product composition at an estimated first-approximation temperature and using this to calculate the corresponding relationship between enthalpy (or internal energy) and temperature for the products, as well as the net heat of reaction. This leads to a second value of temperature which would be correct only if it coincided with that used originally. The discrepancy is employed as the basis for the next approximation, and so on. Because of the large number of simultaneous equations, the composition at an estimated temperature is itself usually calculated by a method of successive approximations. These are the principles of the calculation and much work has been done to facilitate the process. Schemes based on partial or complete solutions of particular cases,[10,11] on graphical methods for successive approximations,[12,13] on explicit algebraic solutions for errors in first approximations,[14,15] and on programming digital computers[16] are available in the literature.

The final volume (at constant pressure, as on a burner) or the final pressure (at constant volume, as in a confined deflagration) is calculated from the final temperature and composition using an equation of state.

As regards equilibrium product ionization, the Saha equation discussed in the preceding chapter is one of the dissociation equations and, strictly speaking, it (or they) ought to be solved simultaneously with the set. However, under normal conditions of flame temperature and ionization potentials of the species involved, the extent of thermal ionization is not high enough for an appreciable part of the energy to be tied up in that form. For normal flames, ionization levels can therefore be generally calculated by the method given in Chapter 2, using a final temperature that has been deduced without regard to ionization. This is not necessarily the case for flames whose enthalpy has been augmented in some way, as discussed below, or for detonations. In recent years, several compilations of thermodynamic data at the higher temperatures involved have become available.[7,17−19]

Increasing the energy content and final temperature of the product gases beyond the magnitudes attainable by the burning of stoichiometric mixtures is an obvious method of increasing equilibrium ionization. The additional energy can be supplied either from an external source, or by recirculating some of the heat of combustion using the principle of a heat exchanger. The latter method does not, of course, increase the ultimate product temperature because the terminal heat flux must be equal to the influx of chemical energy into the system. In between the initial and final states, however, enthalpies, temperatures, and ion concentrations (and also reaction rates) are increased on 'borrowed' heat and the desired state can be established over a limited distance. This is illustrated in Fig. 3.2.

The 'heat exchanger' is purely schematic; it could consist, for example, of a generator of electrical power at the downstream end, supplying a discharge that releases the generated energy into the reactants, or of a shock wave maintained by the expansion of products. The question of how the energy can be recirculated is, in fact, crucial to the whole concept. Even leaving aside considerations of pre-ignition in the upstream part of the heat exchanger, the walls of any *material* exchanger would melt before the

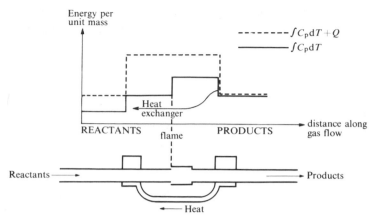

FIG. 3.2. Transient increase of enthalpy of flame gases on heat 'borrowed' from products.

energy recirculated became appreciable, on the scale of the heat release available from stoichiometric fuel-oxidant mixtures. Incidentally, among the few favourable systems, from this point of view, are flames into which the two reactants counterflow from opposite directions, one of them being a solid fuel producing a refractory ash. It has been shown that this type of mechanism is responsible for the attainment of higher than overall equilibrium temperatures in the smouldering zones of cigarettes.[20] Here tobacco, preheated by the hot smoke, reacts with air preheated by the hot ash, the 'burner system' operating as a counterflow heat exchanger.

In general, however, as regards the transition from a flame to an appreciably ionized plasma, there is an enormous gap between the maximum temperature at which the gas can be contained (and heat be transferred) without melting the container and the minimum temperature at which adequate control by means of electro-magnetic fields can be exercised on the gas as a whole. In attempting to narrow this gap from the flame end, we must conclude that the additional energy supplied or recirculated cannot be conducted through solid walls as heat in sufficient quantity—in fact it cannot be supplied as heat at all. The possible methods for augmenting the enthalpy of flame gases appreciably are adiabatic compression, electric discharges, and radiation. Of these radiation is not, at the time of writing, a serious contender, except in the case of absorbing reactants (for example, dust flames), although the advent of powerful continuous lasers holds out some promise for the future. The other two, however, have been used.

Detonation is the process of initiating and maintaining combustion by a compression wave. Its mechanism is discussed briefly in the next section. From the present point of view, the additional enthalpy given to the reactants by the shock wave is derived from an external source (e.g. the spark, etc., initiating the detonation) only in the very early stages. In the steady state, enthalpy may again be thought of as being recirculated, the mechanism being the degradation of the kinetic energy. The expansion of the shock- and combustion-heated products provides the thrust that drives the leading shock wave. By way of a rough analogy, we may think of the steady-state phenomenon as a rocket, propelled by the expansion of the reaction-heated products, ignition occurring behind the shock front formed ahead of its nose. Attempts at stabilizing detonations, in which energy is supplied to the reactants (or, at least, to one reactant) also by the initial compressor, will be discussed in the section dealing with burners.

The calculation of the 'final' state conditions—this is taken here as the equilibrium state beyond the detonation front, i.e. the state of maximum enthalpy—differs from the aforegoing only in that the changes in kinetic energy and pressure are now appreciable and must be taken into account. The conservation of energy here demands

$$H_u + \frac{v_u^2}{2} + Q = H_b + \frac{v_b^2}{2}, \tag{3.4}$$

using the previous notation and referring to unit mass of the mixture. The velocity, v, of the gas must be taken relative to the moving front around which the conservation equations are set up. If the gas is at rest initially, v_u is numerically equal to the detonation velocity, in the opposite direction. Thus the dynamics of the system is now involved and the equations of conservation of mass and momentum,

$$\rho_u v_u = \rho_b v_b \tag{3.5}$$

and

$$P_u + \rho_u v_u^2 = P_b + \rho_b v_b^2 \tag{3.6}$$

respectively, must be solved simultaneously with eqn (3.4). The method of solution is discussed in some detail, for example, in Gaydon and Hurle's *The shock tube in high temperature chemical physics*,[18] which includes a worked example.

A general result, which is interesting from the present point of view, is that the detonation velocity is equal to the sum of the speed of sound in the burnt gas, u_b, and the local particle velocity, v_b. Thus $v_u > v_b$, and it is the difference between the two which determines the magnitude of the 'enthalpy recirculated'—see eqn (3.4). This is, in fact, equal to $\{(v_b + u_b)^2/2\} - (v_b^2/2)$. Moreover, from calculations for an ideal gas it transpires that the particle velocity is equal to u_b so that the additional enthalpy is, approximately $\frac{3}{2}u_b^2$. (This concept is not to be confused with the 'von Neumann spike'[18] in pressure and density which occurs *between* the shock and the reaction fronts.)

The various methods for augmenting flames by electrical discharges will be discussed under the heading of burners. As regards calculation of their final state, at the time of writing the velocities involved do not necessitate the inclusion of kinetic energy terms.† The calculation is thus formally the same as for flames, except for the addition of W_e to Q in eqn (3.1), where W_e is the electrical energy dissipated per unit mass of gas and the mechanical equivalent of heat must be taken into account. In general it is not correct to assume that W_e can simply be calculated from {(potential × current)/mass flow rate}. This will tend to give too high a value for two reasons. The first, which applies to practically all cases, is that a considerable amount of heat is generally lost to the electrodes. The second applies to a.c. and to direct currents fluctuating intentionally or unintentionally. The power factor, i.e. the cosine of the phase angle between the current and voltage vectors, tends to be important in these systems because the generators required to power the arcs frequently introduce large inductances into the circuit. In such cases it is desirable to use a watt-meter across the arc, as well as to measure heat losses to the electrodes calorimetrically (they are frequently water-cooled anyway).

In some cases it may be necessary or convenient to measure the final state parameters, rather than calculate them. Methods for measuring temperature and pressure in flame gases have been summarized from various points of view.[1–3,5,18,22] Methods for the measurement of ionization and ion species, which are the main concern here, are detailed in Chapter 5.

REACTION IN FLAMES, THEIR MECHANISMS OF PROPAGATION AND BURNING

The final state and the parameters that depend on it, such as detonation velocity and equilibrium product ionization, are the only combustion properties that are independent of reaction rates. Quantities such as burning velocity, ignition temperature, chemi-ionization, quenching distance, etc. are all related theoretically to reaction rates, but the subject of reaction kinetics in flames is very complex and, at present, imperfectly understood. Only relatively simple and, in practice, unimportant reactions are reasonably well understood quantitatively, whilst in hydrocarbon flames, the kinetics of individual steps are just beginning to emerge as a result of extensive and painstaking studies (see for example, Fristrom and Westenberg[3]).

The overall conversions proceed through steps involving a variety of radicals which can multiply very rapidly by branching reactions, such as, for example,

$$H + O_2 \rightarrow HO + O.$$

† It may be only a question of time, however, before detonations and discharges are combined[21] experimentally and this no longer applies.

This is the reason why flame reactions can occur at the exceedingly rapid rates that characterize them. This part of the subject is intimately connected with that of chemi-ionization in flames (see Chapter 6), because it appears that in many cases ions are the occasional by-product of certain radical reactions. These processes give rise to very much larger ion concentrations, in unseeded hydrocarbon flames, than are attained by the equilibrium ionization in hot products discussed in Chapter 2.

The most effective flame inhibitors and extinguishers are substances that break these radical chains. Notable amongst them are halogenated hydrocarbons and the halogens themselves, which attack species such as hydrogen atoms and fuel radicals and thus render them useless as chain-carriers. It is of interest to note that the effectiveness of inhibitors seems to be closely paralleled by their electron affinities.[23,128]

It is possible to set up a physical picture of the role of reaction rate in flame properties and, to some extent, make certain quantitative predictions from experiments not involving detailed analyses, by considering[24-28] overall conversion rates of reactants to products. The distribution of reaction rate (as exemplified by the rate of heat release, for instance) has certain common features for all exothermic reactions in gases. Figure 3.3 shows the general form of this relationship, which, except for detailed variations, is similar for all combustion reactions. The horizontal scale is the temperature of the reacting gas mixture, or a more general index of reaction progress such as the fraction of reaction completed; the latter being defined in terms of temperature rise as $(T - T_u)/(T_b - T_u)$. The reaction rate is subject to two opposing influences; the rise in temperature that tends to increase the rate, and the depletion of reactants (which at constant pressure is further aided by expansion of the gas) that tends to decrease it. On certain simplifying assumptions,

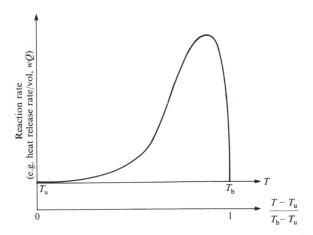

FIG. 3.3. Reaction rate as a function of reaction progress ('reactedness').

the general form of the relationship in terms of temperature becomes,

$$w Q \propto \left(\frac{T_b - T}{T^{\frac{1}{2}}} \right)^p \exp(-E/RT),$$

where p is an effective reaction order. Since the increase caused by the rising temperature is very largely exponential, it tends to dominate the shape of the curve. Thus the initial rise in reaction rate is almost entirely due to the Arrhenius factor. However, when the reactants are nearly completely consumed and the temperature has ceased to rise, the reaction rate plunges steeply, producing the near-vertical down-stroke on the graph. The maximum reaction rate thus occurs much closer to the final than to the initial state.

In the case of a flame-front in pre-mixed reactants, which is characterized by a particular temperature distribution, the temperature axis of Fig. 3.3 can be transformed to one of distance (moving along with the flame). In the case of a homogeneous reaction in a vessel, the reaction rate–time dependence follows from the temperature history together with the above relationship. Figure 3.3 is thus essentially a convenient representation of the overall kinetics in a particular mixture of reactants and its general properties will be applied to individual cases below.

Propagation of pre-mixed flame

The term 'propagation' here implies a velocity with respect to the reactants, not necessarily with respect to the laboratory. Once the non-steady (ignition) states are over, the flame remains similar to itself and propagates at a particular velocity into the cold gas. When it is stabilized on a burner at an appropriate flow velocity, the mechanism of propagation is no different, even though the flame appears stationary to an observer. The various types of burner and methods of stabilization will be discussed in a later section.

In steady-state propagation, the detailed structure of the front, for example the spatial distribution of temperature and composition, does not vary with time, provided the distance coordinate travels with the flame. The thicknesses of the zones over which variations occur are usually so small (except at very low pressures) that the distributions may be considered to be uni-dimensional. Events along such a single coordinate, y, perpendicular to the front are sketched in Fig. 3.4.[5] The flame propagates at a *burning velocity*, S_u, that is a characteristic property of the initial composition and state of the reactants. It is defined as the component, perpendicular to the flame front, of its velocity with respect to the cold reactants. It is a consequence of the rate at which the burning gas initiates reaction, by energy transfer, in the adjacent layers of cool reactants. As will be seen from Fig. 3.4, reaction rate does not become important until the temperature has been raised appreciably by heat transfer from hotter reacting layers of gas.

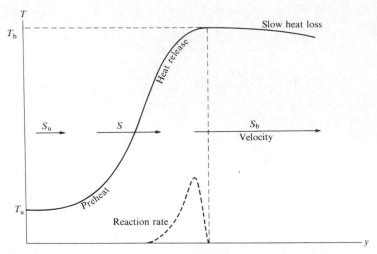

FIG. 3.4. Structure of premixed flame. From Weinberg.[5]

The steady-state conservation equations can be written in the following form.

Mass

$$\rho S = \text{constant} = \rho_u S_u = G, \tag{3.7}$$

where ρ = density of the gas, S its velocity perpendicular to the flame, and G = mass burning rate per unit area.

Heat

$$-\frac{\partial}{\partial y}\left(\lambda \frac{\partial T}{\partial y}\right) + G\frac{\partial(\bar{C}_p T)}{\partial y} - wQ = 0, \tag{3.8}$$

where λ = thermal conductivity, $\bar{C}_p$ = mean specific heat up to y, Q = heat of combustion, and w = reaction rate such that (wQ) = rate of generation of heat per unit volume. Radiation may be neglected, in gas flames.

Species

$$-\frac{\partial}{\partial y}\left(D_j \frac{\partial c_j}{\partial y}\right) + G\frac{\partial}{\partial y}\left(\frac{c_j}{\rho}\right) - w_j m_j = 0, \tag{3.9}$$

where D_j = diffusion coefficient of the jth species, c_j = its concentration, w_j = the rate of reaction, and m_j = mass of j produced per unit conversion so that $w_j m_j$ is the volumetric rate of generation of the jth species (i.e. sign becomes positive for a reactant). Momentum and kinetic energy terms are not important in ordinary deflagration because the energy associated with the pressure drop across such flames is insignificant by comparison with the change in enthalpy. The equations in this form lack some sophistication (thus $\bar{C}_p T$ is the local enthalpy; also the diffusion coefficient D_j must deal with concentration gradients that are partly due to density changes consequent upon the temperature distribution) but they have been set down in this manner because they convey a picture of the system and of the similarity between the heat and species equations more readily in this than in the detailed form. On an overall reaction model, i.e. when the disappearance of reactants releases heat and other products in proportions dictated by simple stoichiometry, the w of eqn (3.8) and the w_j of all eqns (3.9) coincide.

The complete solution of these equations involves substitution for w_j in terms of the temperature and composition dependences of the reaction rate(s). The system of simultaneous differential equations includes one for each of the species on which the reaction rate depends, plus that for heat conservation. Together with an equation of state and the appropriate boundary conditions they give a set of which the burning rate, G, is an eigenvalue: the distributions of all the variables follow. The main obstacle to their solution, at the time of writing, is still the incomplete knowledge of detailed kinetics under the extreme conditions that prevail in flame zones. Solutions have been produced for several artificial models. In particular, the assumption of simple overall kinetics, or indeed any reaction law that can be written in a temperature-explicit form, together with certain simplifications regarding the physical properties of the gas (such as constancy of specific heat, interrelations between $k/\rho C_p$ and D) greatly simplify the mathematics (see, for example, Spalding[29]). Whether the actual rate-laws can be reduced to such a form, however, still remains an open question. Complete solution has been attempted only in a few simple, and in practice unimportant, reactions. At present, the measurement of burning velocity is not only a more expedient but, in fact, a necessary alternative to calculation, and the flame equations have found more immediate application in the analysis of flame structure.

The subject of flame structure has been extensively discussed in a recent monograph[3] and, insofar as optical methods are used, also by one of us.[5] The subject will be briefly outlined here, because it is applicable equally to profiles of ion concentration. It uses the continuity equations, not to attempt complete solution of the system, but to allow the deduction of some quantities from the measurement of others. It follows from the above discussion that the deduction of reaction rate is the usual aim of the exercise. Inspection of eqns (3.8) and (3.9) indicates that the profile of reaction rate can be calculated from a knowledge of the distribution of the flux of the relevant quantity. In the case of laminar flames, the flux of a quantity whose distribution is known, is determined by the molecular diffusivity and the flow velocity in the convective term. Taking eqn (3.8) as an example, if the burning velocity is measured, together with the temperature distribution perpendicular to the flame front, and if the local thermal conductivity and specific heat can be calculated from a knowledge of temperature, the rate of heat release per unit volume can be determined at every point in the flame. Many such analyses have been carried out (summarized, for example, by Fristrom and Westen-berg[3]). Similarly, an equation of the form of (3.9) can be applied to ion concentration in the flame zones. The reaction rate $w_j m_j$, will then represent the net rate of charge generation per unit volume, i.e. the rate of creation minus the rate of recombination. For reasons discussed in subsequent chapters, non-equilibrium ionization is the main source of ionization in most flames and this causes ion concentration to peak in the reaction zone. Thus towards the end of this zone the only significant contribution to the net generation term comes from recombination, and from ambipolar dif-fusion at low pressures. Such data can be used to determine recombination and ambipolar diffusion coefficients. Wortberg[30] has compared the spatial distributions of the net rate of ion generation and that of heat release. The latter was determined from interferometric measurements of the temperature profile of the same flat flame. The results are illustrated in Fig. 6.1. The analogy with flame structure analyses generally having been drawn, the results will be discussed again in the context of flame ionization.

The analysis of flame structure—including that of ionization profiles—and many other concepts concerning pre-mixed flames involve the measure-ment of burning velocity. Methods of measuring this quantity have been extensively reviewed in the combustion texts mentioned earlier. For present purposes, a brief statement of the principles will suffice. S_u being the com-ponent perpendicular to the flame of the velocity at which the cold gas enters the front, it can be determined directly from knowledge of the flame's geometry (for example, from a photograph) and of the local flow velocity vector. In the case of a flame propagating into stationary reactants, any relative velocity is caused by flame movement, which is usually recorded using some device such as a rotating drum camera. In the case of burner-stabilized flames, the burning velocity may be deduced, as an average, from the total volumetric flow-rate V, or preferably measured locally. An accurate method of measuring local flow velocities involves the inclusion in the stream of particles of a few microns diameter, which follow the stream lines. Their trajectories can be displayed by photography under Tyndall-beam illumina-tion. By interrupting the convergent–divergent light beam (or the light path to the camera) at a known frequency, flow velocity can be determined, both in direction and magnitude, at every point across a flame diameter, from a single photograph. Burning velocity is then given by

$$S_u = v_u \sin \theta \tag{3.10}$$

where v_u is the initial flow velocity locally and θ the angle between it and the local direction of the flame-front. It is important that measurements of this kind should not be carried out in the presence of an applied electric field. As shown elsewhere in this monograph, even particles that do not emit electrons tend to become charged under these conditions and to acquire appreciable mobilities, which can make their trajectories deviate greatly from the streamlines.

When only the total volumetric flow rate is known, burning velocity may be deduced from the relationship

$$S_u = \frac{\dot{V}}{A} \tag{3.11}$$

which follows directly from the definition of burning velocity, A being the flame area. This method gives a mean burning velocity integrated over the entire flame front. Burning velocity is a constant for one reactant mixture only in the case of a free and one-dimensional flame, and the average tends to differ somewhat from this ideal because it usually includes zones of heat loss to the burner periphery, as well as a zone of steep curvature at the apex of the flame. When they can be used, the most accurate methods are those using burner-stabilized flames, in which flame geometry is determined from schlieren or dissected shadow records[5] and the local velocity from particle tracks.

Turbulent flames occur in pre-mixed reactant streams whenever the critical Reynolds number is exceeded. They differ from the equivalent laminar flames in appearance, the thin clear cone becoming a diffuse 'flame brush', which is often somewhat noisy. In the case of burner-stabilized flames it can be readily seen that the cone also becomes abruptly shorter when the critical Reynolds number is exceeded so that, on the basis of apparent area, the burning velocity of the turbulent flame is much increased: see eqn (3.11). The effect of turbulence on premixed flames may roughly be subdivided into three regimes. Very large intensities tend to disrupt the front completely and cause burning in isolated pockets. This regime is more relevant to the discussion of high-intensity combustion as below. At very small scales of turbulence, scales smaller than the flame thickness, the flame front 'knows' about the existence of turbulence only in that it is subjected to greater heat and diffusion losses by transport processes. Since it is reasonable to suppose that the chemistry is unaffected, the altered burning velocity can be derived from the change in transport coefficients brought about by the contribution of eddy diffusivity: S_u is proportional to the square root of the total effective diffusivity. However, scales small by comparison with flame thickness tend to be a somewhat academic problem and the most interesting regime, in the present context, is that of scales which are large by comparison with the flame thickness. The fluctuating velocity distributions across the reactant stream then corrugate the instantaneous flame front shape. Figure 3.5, a special form of a schlieren photograph from Fox and Weinberg,[31] shows what is happening on a time scale irresolvable by eye. The flame brush of apparently decreased frontal area is, in fact, a time-mean of a rapidly fluctuating, thin, but convoluted, flame front, the averaging being caused

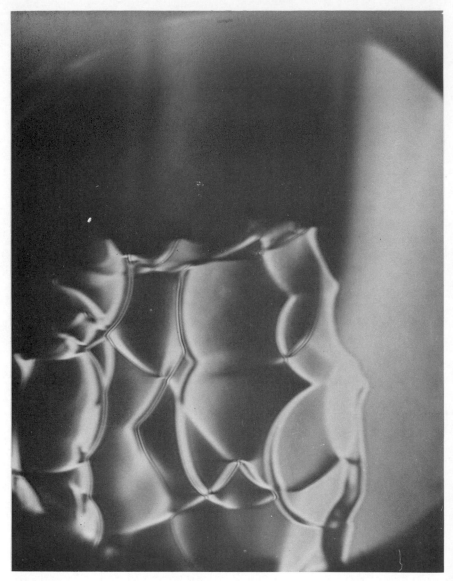

FIG. 3.5. Crest of premixed turbulent flame as outlined by a schlieren system of limited sensitivity range, using a spark source. From Weinberg.[5]

either by retinal image retention or by the finite exposure time of a photograph. The simple theory for this regime is that the ratio of turbulent to laminar burning velocity

$$\frac{S_T}{S_u} = \frac{A_r}{A_a} \qquad (3.12)$$

where A_r = real area taken instantaneously round all the corrugations and A_a = apparent mean area, for example, as seen by the eye. The implication is that the turbulent burning velocity is an erroneous value due to the inevitable error in defining the flame area; had the burning velocity been measured by dividing the volume flow by the true area A_r the laminar S_u would have been obtained. It will be shown later that electrical methods have an important and hitherto barely exploited part to play in research in this field, both because saturation currents (Chapter 5) are proportional to instantaneous flame area and because ionization probes, with suitable circuitry (Chapter 7), can simplify the statistical analysis of turbulent flame fluctuations very greatly. See refs. (36) and (37) of Chapter 7.

Detonation

Detonation differs entirely from deflagration—the process of flame propagation—in that initiation of a fast reaction in the cold reactants is by a compression wave and diffusive transport processes are unimportant by comparison. The conservation equations have already been given (eqns (3.4)–(3.6)) because they were required in the final state calculations. These equations contrast with those describing deflagration not only in that the one set neglects diffusive transport processes and the other mechanical forms of energy, but also in that the detonation equations are not written as differential continuity equations. The reason for this derives from the nature of shock waves, as follows.

The velocity of a pressure disturbance in a gas increases with amplitude, so that the peak of any composite disturbance will eventually catch up with and sweep up any 'foothills' that may originally have existed in the direction of propagation. The resulting shock-front is as near a discontinuity as a gas will sustain and the conservation equations are therefore written in the form of finite changes in mass, momentum, and energy, for infinitesimal increments in y. As burning velocities increase, so do the pressure changes, which in deflagration are insignificant; whilst for velocities well below the speed of sound pressure waves outstrip the flame front and leave it behind, at around 10^5 cm/s it is possible for a reaction front to coalesce with a shock wave. The shock front then acts as the flame's initiator, while the energy release from the flame prevents attenuation of the shock. This is the essence of the mechanism of propagation of detonations. As regards their 'classical' structure, first, it can be shown that the pressure and density behind a shock front driven at the detonation velocity is greater than those behind the reaction zone. This phenomenon, confined to the narrow region between the shock and reaction front is the 'von Neumann spike'. After the passage of the shock, the reactants take a short but finite time (similar to relaxation–dissociation times, of the order 10^{-6} s) in which conditions are somewhat indeterminate, in the present state of knowledge. Finally, the equilibrium state after the reaction zone is followed by a rarefaction region which, in a detonation tube, extends to the point of initiation.

Equations (3.4)–(3.6), together with the appropriate equation of state, contain too many variables to be solved explicitly. They permit of a range of final states and corresponding velocities for a given initial state. This is generally represented by a Hugoniot curve, a graph of pressure against specific volume ($1/\rho$) for a given set of initial conditions. One point on this locus—the Chapman–Jouguet state—is taken to represent normal detonation; it represents the thermodynamically most probable state and corresponds to the velocity of the subsequent expansion wave, thus giving a true steady state because it is never caught up by the adiabatic expansion. The Chapman–Jouguet state is calculable or obtainable from a simple geometrical construction on the Hugoniot curve (see for example p. 267 of ref. 18). The resulting detonation velocity is the speed of sound in the burnt gas multiplied by the density ratio (hot/cold). Taking into account that reaction rate is finite, and therefore distributed in distance, leads to a series of Hugoniot curves corresponding to various degrees of conversion. It appears that for fast (near-stoichiometric) mixtures, the degree of conversion at which the Chapman–Jouguet condition occurs is near enough the final state to allow this assumption to be made in calculating detonation velocities. For such mixtures the agreement between measured and calculated values is indeed much in advance of what can be done in problems involving reaction rates, e.g. in computing burning velocities. At increasing dilution, discrepancies increase as more of the reaction remains uncompleted by the onset of expansion, until eventually a 'limit of detonability' is reached.

It is is fortunate that this relatively simple 'classical' model predicts experimental velocities to such a good approximation; in recent years it has become obvious that the detailed mechanism is much more complex. The shock front is not a simple plane surface but contains individual interacting, 'spin heads' having velocity components in directions other than axial and leading to patterns of local temperature, velocity, etc. considerably in excess of those for the plane wave. This becomes readily apparent only for relatively weak detonations in which the phenomenon of 'spin' has now been accounted for theoretically in terms of individual detonation heads propagating spirally, at an angle to the overall propagation direction. It has also been deduced that for strong detonations these are less apparent, not because they do not exist, but rather because they are so numerous as to become indiscernible by simple methods.

Ion concentrations in detonation attain high levels. This is why measuring the propagation of detonations has provided one of the earliest and most suitable applications of ionization gaps, and also why detonations have been considered in various schemes for direct generation of electricity. This consideration provides additional impetus towards stabilizing detonations by methods discussed in the section on burners. Several workers have measured ion concentrations in detonations, for example Basu and Fay,[32] Jaarsma and Fuhs,[33] Edwards and Lawrence,[34] and Tung, Kelly and Toong.[35] Some calculations for equilibrium products are also available.[36,37] In H_2/O_2 mixtures, already in 1958 Basu and Fay[32] have shown that ion concentrations in the post-detonation zone can be accounted for on the basis of thermodynamic equilibrium, taking NO as the only sufficiently ionizable species (formed from a 'standard impurity' level of N_2). In the case of acetylene–oxygen detonations, ionization considerably in excess of calculated equilibrium values was observed and was tentatively attributed to solid carbon formation. The contribution of chemi-ionization has not been fully elucidated at the time of writing (and nor has the effect of multi-headed spin). Peaks in ion concentration are found when the reaction zone passes over the probe in the case of detonations in hydrocarbons, but they may not, in fact, reveal the full extent of chemi-ionization because of the exceedingly small thickness of the zone. Thus Kistiakowsky and collaborators[38,39] found that in shock-induced acetylene–oxygen reactions, chemi-ionization gives rise to ion concentrations several orders of magnitude higher than those expected on considerations of thermodynamic equilibrium. These authors have indeed proposed a detailed mechanism (see Chapter 6) based on their conductivity and mass spectrometric observations. In these highly diluted mixtures (e.g. 90 per cent Ar, 4 per cent C_2H_4, 6 per cent O_2) the shock plus reaction waves are not detonations and their reaction zones are thicker. There is, however, no reason to suppose that the role of chemi-ionization in true detonation is less important. The recent advances in detonation theory mentioned above suggest a further reason why ion concentrations much above equilibrium values, but localized to minute regions, should occur. The main difficulty in measuring them may be experimental in that probe sizes are finite and that the correct height of a vanishingly narrow peak, for example in conductivity, as displayed on a C.R.-oscilloscope cannot be reliably ascertained.

Augmented flames and high-intensity combustion

Detonation is not the only conceivable mode of propagation of a reaction-front through a gas which involves an igniting agency other than the normal diffusion of heat and radicals, which characterizes flame. Consider the following hypothetical experiment. An inflammable mixture is contained in a conducting tube, provided with an axial electrode. The tube and axial-electrode are connected to a power supply such that, when a flame passes, a discharge occurs through the highly ionized reaction zone. A solenoid may be thought of as wound around the tube so as to rotate the discharge at a rate sufficient to heat all the gas uniformly in the plane of the flame. This would result in a 'propagating augmented flame' and its rate of propagation would be greater than that of a normal flame, not only because of its higher T_b but also because of the additional diffusion of electrons, other labile species and heat, into the reactants. Flames of this kind are, in fact, more conveniently stabilized on burners and they will be discussed under that heading. Their 'mechanism of propagation' is mentioned here only because, like that of detonations, it differs from that of pure deflagrations in such a way that they are stabilizable at higher flow rates, give rise to higher combustion intensities, and make possible greater throughput rates for a given burner area, than do normal flames.

Augmented flames have attracted attention only relatively recently. The aim to increase the mass of reactants that can be burned in a given volume per unit time, however, has been a major preoccupation, at least since the advent of jet propulsion devices. Moreover, unlike in the case of augmented flames, the method of maximizing the input rate of reactants into a combustion chamber without blowing out the flame must here be achieved without adding energy to the system.

In considering how this may be achieved, it will be helpful to consider once more the shape of the graph in Fig. 3.3. The rate of heat release there, when integrated over the volume of the combustion chamber with its various states of reactedness, must equal the net heat carried away by the products over that brought in by the reactants, if steady-state burning is to be maintained, i.e.

$$\iiint wQ \, dV = \left[\iint GH \, dA \right]_u^b. \tag{3.13}$$

The obstacle to stabilization lies in the very small values of heat release-rate which, because of the exponential form of the law, occur at the lower values of reactedness. If the subsequent stages of the combustion process are dependent on these slow and unproductive early states, any appreciable flow velocity will push the entire reaction system out of the chamber volume. What is required, therefore, is some method of increasing reactedness early on, without having to rely on the initial reaction stages for this function. This is achieved when hot products are mixed in with the cold reactants. Indeed, on the assumption of a single overall reaction and constant specific heat, the results of such mixing are indistinguishable from those produced by reaction, since dilution of reactants and the corresponding rise in temperature obviously are interrelated in the same way, whether brought about by reaction or by inter-mixing. Under the conditions laid down by these assumptions, 'reactedness' becomes a unique and sufficient index of the state of the gas along the single path it must follow from reactants to products.

Irrespective of any such idealization, however, improvements in flame stability and volumetric efficiency, up to the theoretical maximum, are brought about by increasing mixing of products, including heat and intermediates, with the reactants. In all the mechanisms of propagation, initiation of the process is dependent on some such recirculation because reaction rate does not become appreciable without it, in the absence of extraneous supplies of energy. In the case of the laminar deflagration, such 'mixing' is due to molecular transport processes alone, as discussed above. This leads to a very high volumetric conversion rate only within the extremely thin reaction zone characteristic of such a flame, the mean free path of the mixing process being very short. The volume enveloped by such a flame is, for stable conditions, generally vastly greater than the reaction volume. When the flow becomes turbulent, the mean free path of the mixing process increases from that of a molecule to the scale of turbulence. As a result, the volume utilized for burning improves considerably, since more flame area occurs per unit volume as the thin laminar flame changes to a turbulent flame brush. However, even this mixing length is, under normal conditions of turbulence, very small when compared with the dimensions of a practical combustion chamber. The general solution to the problem therefore is the mixing of sufficient products with reactants with a mixing length comparable to the dimensions of the apparatus. This is usually achieved by inducing suitable vortices in the wake of obstacles, changes in the contours of combustion chambers, etc.

The optimum extent of mixing and practical methods of achieving this will be discussed in the section on burners. This is, of course, essentially a flow problem and the mixing is attained in practice at the cost of some of the stream momentum. The purpose of the above introduction was to show that, as far as the mechanism of burning is concerned, volumetric efficiency, flame stability, and hence maximum throughput can be increased by increasing mixing length and intensity, even at constant total enthalpy. Essentially the reactants must be brought to a high temperature rapidly and this can be achieved either by the supply of external energy, as in some augmented flames, or by adequate recirculation of the normal product enthalpy.

Diffusion flames

Diffusion flames occur in the mixing zone of initially separate reactants. Examples are the burning of a jet of fuel gas emerging into the atmosphere, or of fuel vapour from a wick, a droplet, or other open surface. Such flames, of course, do not propagate, because no appreciable volume of mixture is available, and their structure is very dependent on the rate of the mixing process. Their common feature is that reaction is conditional on and must follow the process of mixing, so that the slower of these two consecutive steps determines the rate of reactant consumption. Diffusion flames then may be *defined* as flames whose properties are determined by the rate of diffusion of the reactants into each other, because the rate of the chemical reaction is potentially so much faster. It is possible to increase rates of mixing by high intensity turbulence, or other means, to such an extent that this no longer applies, but we shall then not call such burning 'diffusion flames'.

Theoretical treatments of diffusion flames may be sub-divided into two categories; theories designed to calculate flame shape and those concerned with blow-out. The former are the simpler, because the idealization of a reaction rate infinite by comparison with mixing, and occurring within an infinitesimal thickness, is adequate for calculating flame shape.

The general concepts underlying this theory are as follows. If reaction is potentially much faster than mixing, the reaction zone will occur where conditions are most favourable for a rapid rate. If a fuel jet is allowed to discharge into the atmosphere, a steady-state mixing pattern develops. This can be characterized by a set of contours, each representing one constant fuel-air ratio. If an extended ignition source is now applied from above, a pre-mixed flame will flash back to the burner port along the locus of composition that gives the highest flame velocity. This, to a sufficient approximation, is the stoichiometric contour. Once established at that locus, the flame surface becomes the sink for all reactants and the source for products, including heat. So long as diffusive transport rates of the two reactants are not altered relative to one another by increased temperature, fuel and oxidant will continue to be conveyed to this surface in stoichiometric proportions. On the basis of such simplifications, the calculation of flame shape and, in particular, flame height, becomes a relatively simple matter and, in general, little allowance has been made for the variation in gas properties due to temperature changes. Thus, the height of the flame is the distance along the axis within which the fuel concentration has been depleted, by inter-diffusion, to the stoichiometric value. This gives

$$Z \simeq \frac{\dot{V}}{4\pi D c_s}\left(1 - \frac{c_s}{2c_o}\right) \simeq \frac{\dot{V}}{4\pi D c_s}, \tag{3.14}$$

where $\dot{V}$ = volumetric flow rate of fuel (even in the presence of other gases in the burner effluent), D = its diffusion coefficient and c = its concentration, the suffixes s and o denoting conditions in a stoichiometric mixture and in the burner mouth. The second approximation in eqn (3.14) has been made because (c_s/c_o) is usually a very small quantity (because the problem is most frequently concerned with hydrocarbons burning in air). It is not an acceptable approximation where the volumetric content of fuel in the stoichiometric mixture is not very small, for example in the case of H_2 burning in an atmosphere of O_2. Various attempts at refining this theory have been made from time to time, particularly as regards making some allowance for the variation of gas properties with height, or for the presence of small amounts of oxidant in the burner gas. The above expression, however, is adequate for present purposes and gives a reasonable estimate under conditions well within the laminar regime.

In the case of turbulent flow, the above result can be adapted simply by substituting the eddy diffusivity on the axis for the diffusion coefficient. The equivalent equation then becomes

$$Z \simeq 25d\left(\frac{c_0}{c_s}\right)\left(1 - \frac{c_s}{2c_0}\right) \simeq 25d\left(\frac{c_0}{c_s}\right). \tag{3.15}$$

This result has often been modified by the inclusion of an empirically adjustable factor for the deviation of conditions on the axis of the flame from those of fully developed turbulent flow in a tube. An important conclusion of the argument leading to eqn (3.15) is that the flame height, Z, is independent of velocity and proportional, for a given set of gases, only to the burner diameter. This arises because the increase in stream momentum is compensated for by the increase in turbulent diffusivity. For a more detailed discussion of these problems, using the same approach and nomenclature, the reader is referred to pp. 93–103 of ref. (5).

The other property of diffusion flames which has become of particular importance recently is their blow-out velocity. If a flame is stabilized, for instance, in the counter-flow regime of opposed jets of the two reactants, the velocity at which it is blown out must depend on the reaction rate within it. This has been used as a method for determining relative magnitudes of reaction rates and the velocity at which blow-out occurs has been termed[40–42] 'flame strength'. Flame structure analyses similar to those described in the contest of pre-mixed flames have been carried out on specially designed flat counter-flow diffusion flames.[43,44] These burner systems are relevant to the study of flame ions and will be discussed in a later section. As regards the mechanisms of burning, it will be apparent that the infinite reaction rate concept assumed for calculations of flame shape is incompatible with this use of diffusion flames. The infinite reaction rate within an infinitesimal thickness must here be replaced by a finite rate within a finite thickness (giving the same rate of conversion per unit flame area, which must be equal to the rate at which reactants are conveyed in, if no reactants escape and the flame is to remain stable). Theories that consider such detailed structure of diffusion flames have been developed by several workers[45,46] but are outside the scope of this monograph.

The term 'reaction' has hitherto been used to denote the main, heat-releasing conversion of reactants into products. This is not a single-step conversion, as indicated earlier, and individual reaction steps in flames have been investigated in some detail in recent years (see for example refs. 3 and 4). One aspect of this is particularly relevant in the context of diffusion flames and of this monograph: the formation of carbon, or soot. Although this is not confined to diffusion flames but occurs also in over-rich, or halogen-inhibited pre-mixed flames, as well as in certain detonation processes, conditions in the pyrolysis zones of diffusion flames are particularly conducive to it. This is so because, under these conditions, fuel approaches the flame and attains a high temperature largely in the absence of oxygen. The carbon particles so formed greatly increase the emissivity of the flame gases and diffusion flames are therefore generally more luminous than pre-mixed ones. The usefulness of a candle, for instance, depends on this feature, which is also responsible for radiation being the chief mode of heat transfer from certain furnace flames.

The process of soot formation has exercised the attention of chemists for many years. This has led to a large variety of proposed mechanisms. Examples are the theory that carbon is formed more or less directly from acetylene,[47,48] that it is formed by polymerization of C_2,[49] the hypothesis of its formation from atomic carbon,[1] its growth by decomposition of hydrocarbons on the surface of the growing particle,[50] etc. A summary of the main possibilities that have been proposed is illustrated schematically by Fig. 3.6.[51] The above are typical examples of proposed mechanisms rather than an exhaustive survey. For more detailed reviews the reader is referred to refs. (1), (4),

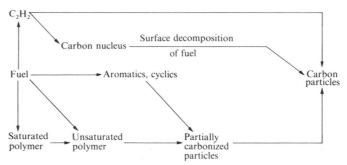

FIG. 3.6. Mechanisms proposed for carbon formation in flames. From Cullis and Palmer.[51]

(48), (52), and (53). (The last reference in fact consists of parts of the proceedings of a one-day symposium on carbon, held by the British Section of the Combustion Institute which, for simplicity, was published under the authorship of one of the introductory speakers who made an extensive contribution.) A recent, virtually encyclopaedic review,[51] which includes close to 200 references, is particularly to be recommended.

The reason for devoting some space to this topic here is that it is relevant to several aspects of the electrical properties of flames. Quite early on in the development of the subject of ionization in flames it was suggested that the excessive (over equilibrium) ionization may be due, at least in part, to the low work-function of carbonaceous agglomerates.[2,54] More recently it has been shown that thermionic emission does, in fact, play an important part in the charge acquisition of carbon particles themselves.[55,56] Several of the electrical properties of flames and, in particular, effects induced by applied electric fields, have been shown to be associated with the onset of carbon formation, for example ref. (57). In many cases this appears to be due chiefly to the large difference between the mobilities of charged carbon particles and those of molecular and smaller charge carriers. This tends to produce observable differential effects in otherwise symmetrical systems, see Chapter 7.

It has recently been found[55] that *all* carbon particles in flames are charged, at least in the presence of an electric field, and that they can therefore be manipulated by means of applied fields, both during their growth in the pyrolysis zone and during deposition. This has made possible the control of particle size, and of the total amount of carbon formed in a flame, as well as of its site and form of deposition. Subsequent phases of this work,[56] moreover, indicate that charge flows from the flame zone can affect nucleation of carbon formation in the pyrolysis zone. This work is detailed in Chapter 7, whilst the role attributed to carbon formation in ionization in flames is reviewed in Chapter 6. Some of these effects are conspicuous in the colour plate (Plate 1).

BURNERS

In this section, burners and the manner in which flames are stabilized on them are broadly reviewed. This is necessary for several reasons. Some burners are ideally suited to particular methods of measurement (for example ionization in flames), others provide the means of augmenting flame enthalpies and equilibrium ionization. Often conventional burners have been used to study effects of fields on flame properties such as shape, stability, etc., and an understanding of the normal behaviour of burners is a prerequisite to a discussion of such effects. The first part of this section is devoted to pre-mixed reactants and the second to diffusion flames.

Burners for pre-mixed reactants; simple tubes

Most flames are stabilized on burner tubes of circular cross-section. Their behaviour may be deduced from their property of forming a continuous sheet that propagates everywhere perpendicular to itself at the burning velocity, S_u, with respect to the cold reactants. This determines their orientation with respect to the approach flow velocity vector, v_u, as well as their effect on the gas flow. Considering an element of this front, the above inter-relation implies, as shown earlier, that

$$v_u \sin \theta_u = S_u, \tag{3.10}$$

where θ_u is the angle at which the flame orientates itself to the approach velocity vector v_u. Thus

$$\theta_u = \sin^{-1}(S_u/v_u). \tag{3.10a}$$

Mass conservation perpendicular to the flame front requires that

$$\rho_u v_u \sin \theta_u = \rho_b v_b \sin \theta_b, \tag{3.16}$$

whilst parallel to the element, there is no change in velocity, i.e.

$$v_u \cos \theta_u = v_b \cos \theta_b. \tag{3.17}$$

Dividing (3.16) by (3.17) gives

$$\frac{\tan \theta_b}{\tan \theta_u} = \frac{\rho_u}{\rho_b}. \tag{3.18}$$

This is sometimes referred to as the 'law of flow line refraction in a flame'. Flame shape, as well as the modification in flow pattern due to the presence of flame, is thus calculable if the reactant composition and the distribution of flow velocity are known. θ_u follows from the burning velocity of the particular mixture of reactants together with the local value of v_u, after which θ_b can be calculated using eqn (3.18) together with the value of ρ_u/ρ_b, the expansion ratio, obtained from a calculation of the final state.

The implications of these equations are as follows. The local ratio of (v_u/S_u) fixes the magnitude of the approach angle, but not its orientation: for a given ratio, the flame may be inclined to the flow lines either way. Thus, within any particular flow regime, V-shaped, M-shaped, or W-shaped flames, in addition to the normal flame form, can be obtained by varying the method of stabilization. Flames are straight-sided only if the velocity ratio is constant across the burner mouth. The types of burner required to satisfy this condition are discussed in subsequent sections. In general, however, the flow velocity varies across the burner mouth and the calculation of flame shape involves evaluation of θ_u at different points along the burner radius. In general also, as v_u is increased, θ_u falls and the flame tends to become increasingly parallel to the flow-lines until stabilization fails and blow-off occurs. Conversely, as v_u tends to S_u, the flame tends to become perpendicular to the flow lines. For $v_u < S_u$, no stable solution exists, because the flame must flash back. It will be shown below that, except on special burners, flames flash back already before this condition is reached over any significant proportion of the burner radius.

So far it has been tacitly assumed that the burning velocity remains constant and this does lead to correct predictions of flame shape, except at one or two points, which will now be considered in more detail. As a practical model we may use the velocity distribution characteristics of fully developed viscous (Poiseuille) flow

$$v_u = 2\bar{v}\left(1 - \frac{r^2}{r_1^2}\right),\tag{3.19}$$

the bar denoting the averaged area-mean over the radius, r_1. This gives the slope in terms of cylindrical coordinates (z along the axis, r perpendicular to it) as

$$\frac{dz}{dr} = \pm\left\{\left(\frac{v_u}{S_u}\right)^2 - 1\right\}^{\frac{1}{2}}\tag{3.20}$$

so that

$$Z = \pm\int_0^{r_1} \{4(\bar{v}/S_u)^2(1 - r^2/r_1^2)^2 - 1\}^{\frac{1}{2}}\,dr.\tag{3.20a}$$

This represents, with fair accuracy, the shape of a flame on a Bunsen burner (at least for a burner tube long enough for this flow regime to become fully established) except at the apex and near the flame base. At these points, although the above theory applies, the burning velocity is no longer a property of the composition and state of the reactant mixture alone.

At constant S_u, eqn (3.20) indicates that at the apex the flame slope changes abruptly from a particular positive value to numerically the same negative magnitude. This would imply infinite curvature on the axis and the underlying assumption of a unidimensional flame structure could not apply there; the increased transfer rates of heat and intermediates caused by the larger area of concave flame-front for a given volume of approaching reactants, would cause a local increase in burning velocity. The consequence is that apices of premixed flames tend to be rounded. Of more practical importance is what happens at the base of the flame, because this determines flame stability and hence the minimum and maximum rates at which reactants can be burnt under a particular set of conditions.

It was shown that whenever v_u falls below S_u, the flame must flash back and, since the flame surface is continuous, the whole flame is only stable as its least stable element. Now $v_u \to 0$ as $r \to r_1$ because of viscous drag at the burner walls. If S_u were constant, flames would *always* flash back. The reason why there is a finite range of stable burning is that S_u also tends to zero as r tends to r_1 because of heat losses to the walls. The issue of whether a flame will flash back or not is therefore determined by whether S_u or v_u tends to zero more rapidly as the wall is approached. The variation of burning velocity with proximity of a heat sink is again a property of the reactant mixture. The characteristics of the burner material are relatively unimportant because any solid is a very good heat sink in comparison with any gas and transport rates are controlled by the gaseous boundary layer. The limiting condition occurs when the flow velocity–distance curve near the burner edge becomes tangential to the burning velocity–distance graph. The gradient of the latter, near the heat sink,

$$g_f = -\left(\frac{dS_u}{dr}\right)_{r \to r_1} \tag{3.21}$$

is also a 'combustion property' of the reactant mixture and the condition for no flash back is that the flow velocity gradient must exceed this value. Thus, for stability,

$$-\left(\frac{dv_u}{dr}\right)_{r \to r_1} = g > g_f \tag{3.22}$$

Taking viscous flow as an example and differentiating eqn (3.19),

$$g = -\left(\frac{dv_u}{dr}\right)_{r \to r_1} = \left(\frac{4\bar{v}r}{r_1^2}\right)_{r \to r_1} = \frac{4\bar{v}}{r_1}. \tag{3.23}$$

Thus flash back is avoided so long as

$$\bar{v} > \frac{r_1 g_f}{4}. \tag{3.24}$$

For other flow regimes other equations must, of course, be used for calculating g. For $2000 < Rc < 3000$, for instance, g has been given as

$$\frac{0 \cdot 023\, \bar{v}^{1 \cdot 8} \rho^{0 \cdot 8}}{(2r_1)^{0 \cdot 2} \eta^{0 \cdot 8}}$$

In the case of complex geometries, data on skin friction drag coefficients may prove of use, the force per unit area being $= \eta g$. Fortunately the value of g_f remains a property of the reactant mixture, even in turbulent flow. The reason for this is that the relevant events occur so close to the wall that the pertinent region is that in the laminar sub-layer at the boundary.

Blow-off, the other limitation to the range of stable flames, is again controlled by a critical velocity gradient at the edge, g_b; and the two criteria are usually quoted together. In the case of blow-off, however, the relevant events occur above the burner mouth. The appropriate variation of S_u with distance from a cold boundary is now parallel, rather than perpendicular, to the burner axis. In the case of pre-mixed flames, the flame base overhangs the burner mouth and becomes almost horizontal there. This is due chiefly to the pressure drop across the flame which is caused by the acceleration of the expanding gas and has been mentioned previously. Although, as discussed, in the case of deflagration this is insignificant from the energy point of view (compared to gas enthalpies), it has the effect of driving reactants through any hole in the flame at velocities comparable to the burning velocity, and the 'dead space' that occurs above the burner lip due to quenching provides the opportunity for such a flow. The flame overhangs because it extends to burn up the reactants that are forced out along the burner edge, when they turn upward, away from the quenching surface. Considering now the variation of burning velocity with height above the burner rim, there is only one stable position ($v_u = S_u$) for the flame base, for each set of conditions. Forcing the flame lower increases heat loss, thereby decreasing S_u and inducing an upward drift due to the excess of flow velocity over burning velocity, until the flame is reinstated in its equilibrium position. The converse of this mechanism restores the flame if heat loss is lessened by raising its base.

We may now consider the effect of increasing flow velocity. For small increases, the flame base lifts to a new equilibrium position where the increased S_u balances the increased v_u. This, however, can continue only so long as the flame base remains within the zone in which wall proximity at all affects burning velocity. Beyond this distance, any further increase in flow velocity produces lift without a compensating increase in burning velocity and hence the flame blows off. The relevant events again occur so near the edge that the velocity gradient at the burner wall, g, largely defines the flow-field immediately above it. The detailed theory is not as simple as that for flash back because the maximum stable height is large enough in some flames for properties such as gas composition and temperature to alter somewhat between the burner port and flame edge, by diffusive processes. The critical velocity gradient for blow-off, g_b, is therefore not entirely a property of the reactants and it is advisable to quote the composition and state of the surrounding atmosphere with this criterion. An extreme case of this effect occurs when excess fuel from a pre-mixed flame burns as an outer cone of diffusion flame, as frequently happens on laboratory burners. In that case g_b, and hence the range of stable burning, is greatly extended, due to the presence of a ring of pilot flame—the base of the outer cone—close to the burner edge.

Apart from this, the treatment of the critical velocity for blow-off is exactly equivalent to that for flash back. The general condition for stable burning thus becomes

$$g_f \le g \le g_b \qquad (3.25)$$

and the problem of flame stability is usually solvable. Again, taking viscous flow as an example, the range of stable flow velocities may be written as

$$\tfrac{1}{4} r_1 g_f \le \bar{v} \le \tfrac{1}{4} r_1 g_b. \qquad (3.26)$$

All the remarks concerning g_f in different flow regimes, and in the presence of turbulence, apply equally to g_b. Numerical values of the two critical gradients have been measured for a range of fuels and summarized in the literature.[58-63]

There is another mode of flame stabilization which is not important in practice but will be mentioned here because it is sometimes induced by the application of electric fields.[64] So called 'lifted flames' occur at velocities above the blow-off limit and are of rather spectacular appearance as they remain stable with their bases at an appreciable height above the burner port. Following blow-off, a flame drifting downstream can be re-stabilized only by an increase in burning velocity, a decrease in flow velocity, or both. Both, in fact, tend to occur at a certain height if the initially laminar stream, or the boundary layer, undergoes transition to turbulent flow there. Burning velocity increases as discussed earlier (p. 51) and the increased radial diffusion of momentum diminishes v_u. Records obtained[60,65] by appropriate optical methods do indeed show that this theory explains the phenomenon of lifted flames.[60]

Burners producing freely stabilized flat flames (pre-mixed reactants)

This heading is intended to cover burners such as those devised by Egerton and Powling[5,66-68] and by Biedler and Hoelscher[3,69] in which flow velocity is balanced against a burning velocity essentially unaltered by heat losses to any solid body, so as to produce a flat flame perpendicular to the flow direction. This is in contradistinction to other flat flame burners (see next section) in which heat losses to a porous flame holder are the main method of flame stabilization.

The principle is illustrated by Fig. 3.7(a). The approach stream is first made uniform by an assembly of randomly distributed small obstacles, such as a bed of glass beads, which are followed by a stream-lining matrix. Alternatively, assemblies of fine grids or gauzes can be used to achieve this effect. Matrices constructed by winding a plane and an adjacent corrugated metal strip together into a spiral have been found particularly effective in aligning the previously randomized flow velocity vectors. At the low flow velocities at which these burners operate, the wakes downstream of the obstacles that effect flow rectification die out well ahead of the flame-front. The other feature of the burner which seems necessary for its successful operation is an obstacle to the flow downstream of the flame to cause a pressure drop or divergence of the streamlines. Powling and Egerton[66,67] used a gauze supported by the transparent 'chimney' that encloses the flame. Levy and Weinberg[25,71] obtained best results using a sheet of perforated asbestos, Biedler and Hoelscher[69] a plate baffle, and Friedman[72] found that a circular plate with a particular configuration of holes near its periphery

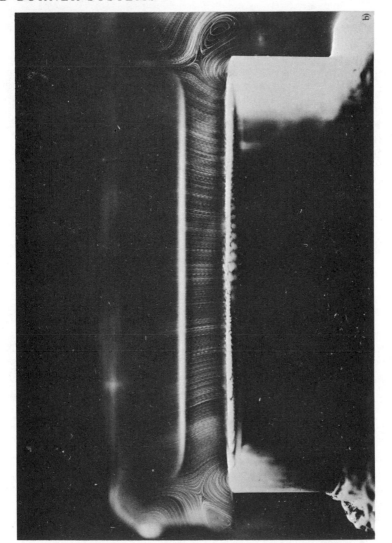

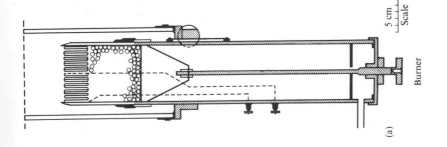

FIG. 3.7. Flat-flame burner for freely stabilized flames. From Weinberg.[70]

yielded the best flat flame. The underlying principle appears to be that of introducing a gradual decrease of stream velocity with distance by means of this divergence, so that slight movements of the flame will compensate for small variations in flow velocity, or fluctuations in fuel/air ratio, and hence in burning velocity. It will be noted that for this purpose it is essential that an increase in flow velocity (which causes the flame to drift downstream) brings it into a region of reduced flow velocity, and vice versa; convergent flow patterns do not permit such stabilization. The process is aided further by recirculation of hot products around the flame's periphery which stabilize the edge. Figure 3.7(b) shows such a free flat flame as well as its flow regime displayed by the tracks of small particles illuminated by a Tyndall beam that was interrupted at a frequency of 100/s. The torroidal vortex around the circumference is clearly visible.

At atmospheric pressure, the highest flow velocities at which such flame stabilization is possible are around 15 cm/s. This confines it to mixtures close to the limits of inflammability. The range of stability appears to be associated with flame thickness in the following way. All pre-mixed flames are, to some extent, 'self-flattening' and therefore resistant to flow perturbations fed into them. This is because the effect of any flame curvature is to increase burning velocity of concave sections—as seen from the reactant side—and vice versa, by the mechanism discussed previously. However, this effect can be appreciable only if the flame thickness is of the same order as the scale of the disturbance introduced. Now, flame thickness is roughly inversely proportional to pressure (it would be expected to be proportional to mean free path). Thus, at reduced pressures, the range of stability of flames on the flat flame burner is considerably extended and, at very low pressures, no special precautions are necessary to produce flat (and very thick) flames even for stoichiometric mixtures.[1]

The flat flame burner, in common with several other specialized burners was originally developed[73] to provide a flame that would be suitable for accurate burning velocity measurement, because of its simple geometry. Subsequently, however, it found very extensive application in the study of flame structure and that of electrical properties of flames. The latter include several studies of ion concentration profiles, for example refs. (30) and (74). As regards ion-current measurements, by using the matrix and top gauze as electrodes, the flame provides a uniformly distributed ion source between, and parallel to, plane electrodes, in an almost unidimensional system.

Flat flames stabilized by heat losses (pre-mixed reactants)

Figure 3.8 illustrates a different method of producing flat flames, originally developed by Botha and Spalding.[75] The flame burns close to the water-cooled porous disc—always within the distance within which burning

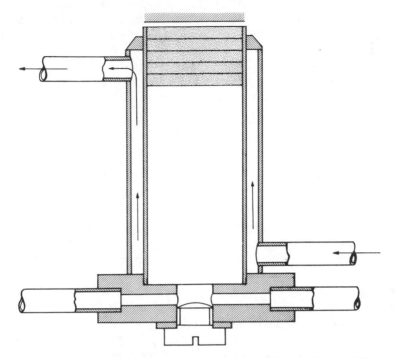

FIG. 3.8. Water-cooled porous-disc burner. After Botha and Spalding.[75]

velocity is affected by the proximity of a heat sink. The amount of heat lost to the plug by the flame depends on the flow velocity of the reactants and can be measured calorimetrically. As flow velocity is decreased, the flame attempts to flash back and, in moving nearer to the heat sink, finds a new position where the greater heat loss has reduced its burning velocity to equal the reduced flow velocity. The converse happens as flow velocity is increased, until the flame tends to lift off when the normal S_u is approached. Unless the flame is adapted to work also in the manner described in the preceding section, the condition of a stable, free-floating, flame cannot quite be attained,† but can be deduced by extrapolating to the flow velocity at which the heat loss would become zero.[75] The burner is best operated under conditions of reasonably small heat loss—if resemblance to any freely stabilizable flame is important—and the mechanism of such flames seems sensibly unaltered by the loss of heat, except for the decreased initial enthalpy, under these conditions.

† Because the temperature profiles of free flames tend to their initial values asymptotically, the essential distinction between the two types of stabilization has sometimes been obscured. Flame structure studies on free-floating flames show, however, that the temperature gradients fall to values too small to be detected, even by optical methods, several flame thicknesses above the burner top. This distance, moreover, can be varied by a factor of 2–3 times by trivial changes in the aerodynamics of the system.

Flames of this type have also been used for measuring burning velocities (by extrapolation to zero heat loss) and for studies of flame structure. They are particularly suitable for measurements of rates of ion generation by the saturation current method[76] (see Chapter 5), both because they allow a very close approach to unidimensionality and because they make possible the variation of flame temperature at constant mixture strength by varying flow velocity alone.

Spherical flames (pre-mixed reactants)

Burners in which the reactant inlet is a porous sphere and the spherical flame occurs in a region of symmetrical, radially expanding, flow have been developed by Fristrom and Westenberg.[3] They have the advantage that streamlines should be straight and the flow cross-sections known to be proportional to the square of the radial coordinate; no free flat flame is absolutely unidimensional, if only because of flow line divergence (see Fig. 3.7(b)), due to the pressure drop across the flame and to other causes. The authors advocate the use of spheres of sintered aluminium oxide for the inlet plug, as commercial sintered metal spheres tend to be insufficiently uniform. The flame is surrounded by a large Pyrex flask which provides the housing and exhaust system. This work was mostly carried out at reduced pressure and, under these conditions of increased flame-thickness, the symmetry does not depend critically on the number or distribution of exhaust pumping ports, nor is it much disturbed by buoyancy effects and the presence of an inlet tube.

Burners based on nozzles and on tubes of noncircular cross-section (pre-mixed reactants)

At atmospheric pressure, for mixtures of faster burning velocities, not close to limits of inflammability, different methods of flow rectification must be used. The obstacles used in burners described earlier produce wakes that would reach into the flame at the higher flow-velocities. At the same time, these flames are less capable of adjusting to such flow disturbances by the 'self-flattening' mechanism mentioned above because of their smaller thickness.

A simple method of producing a nearly uniform flow profile is by use of a nozzle. Although suitable nozzle shapes have been discussed in the literature[77,78] the shape is not critical, so long as the surface is smooth and the contraction ratio large enough to ensure that the increase in flow velocity is much larger than the amplitude of the initial velocity distribution; the ratio should not be less than $1:4$. For very lean mixtures, this device will produce button-shaped flames when the $v_u = S_u$ criterion is approached, and these resemble small free-floating flat flames, except for their down-turned edges. For faster mixtures, flash back occurs before this condition can be attained and the nozzle burner is generally used at higher flow velocities to produce flames whose shape approximates to right circular cones.

An extended flat face of the flame may be produced for these fast mixtures by terminating a nozzle in a rectangular slot. Since the machining of a nozzle that is to contract smoothly from a large circular to a small rectangular section is not an easy task, we have found it more convenient to mould this shape on the inside of a metal former, using plastic material that subsequently sets and will withstand moderately high temperatures. The inner surface is then polished, the flame itself acting as a sensitive indicator of progress in the process of polishing. The flames stabilized on such nozzle-slot burners are roof-shaped, like those on simple slot burners (merely tubes of rectangular cross-section) but the long sides of the 'roof' are flat. The same difference applies to V-shaped flames that can be stabilized on a wire bisecting the two short sides of the rectangular burner mouth.

High-intensity burners (pre-mixed reactants)

This section is concerned with burners and combustion chambers in which increases in combustion intensity are brought about by intermixing hot products with cold reactants, to achieve the effect discussed on p. 56; burners in which combustion intensity is increased because the reaction wave is associated with a shock-front, or because the enthalpy of the gas is increased in some other way, will be discussed in the following section.

The general method of increasing combustion intensity without adding to the energy of the system, as discussed earlier, is by recirculation of hot products and their mixing with cold reactants on a scale comparable with the dimensions of the combustion chamber. Such recirculation is conveniently induced by a vortex in the wake of some 'bluff' body such as a baffle, sphere, rod, V-shaped gutter, etc. Any such obstacle increases the pressure drop across the system—recirculation being bought at the expense of stream momentum—and in most applications only a small central part of the stream is so obstructed. The flame in the wake of such an obstacle, whose stabilization is favoured not only by the recirculation of hot products but also by the decreased linear velocity there, then acts as a pilot for burning in the remainder of the duct that it seeds with its own hot products. It is relevant to mention here that this type of stabilization has recently been promoted[79,80] on a laboratory scale, by using ionic wind effects to recirculate hot products and induce the desired vortex. One method, illustrated in Fig. 7.36 and Plate 1(iv), is detailed in Chapter 7. Essentially it is based on inducing a net unidirectional gas flow by using the difference in mobility between positive ions and electrons, the latter being discouraged from attaching by confining their trajectories to hot products.

An example of recirculation induced by burner contours occurs in the so-called 'tunnel burner', which consists essentially just of a tube joined co-axially on to a wider tube by a flange. Reactants enter from the narrower end and the torroidal vortex, which occurs downstream of the flange, rotates in this instance away from the axis at its downstream end. Under certain conditions of flow, the magnitudes of which are determined by the dimensions of the apparatus, the high-intensity 'tunnelling' regime sets in. In this the flame rapidly blows-off and flashes back over a range of positions determined by the vortex dimensions and the period of this fluctuation can synchronize with the organ-pipe frequency of the tube. The high combustion intensity of this mode of burning is accompanied by a very high noise level and, as might be expected, cannot be produced at all if the flange is replaced by a more gradual expansion.

As mentioned above, mixing is generally bought at the expense of pressure drop, which is costly. However, it is instructive to examine how far this principle can be extrapolated in the laboratory, i.e. up to what point increases in mixing rate result in increased combustion intensity. This introduces the concept of 'perfect' mixing, since there is no point in increasing mixing further, once conditions are homogeneous across the entire combustion chamber.

A region in which perfect mixing occurs may be defined as one in which conditions are uniform, although gases leave it at a reactedness finitely greater than that at which they enter. Such a model in which no spatial distributions exist over an appreciable volume is equivalent to attributing the properties of a mathematical element to a whole combustion chamber. The necessary discontinuity in composition and temperature occurs at the inlet, where the reactants are deemed to be brought to the reactedness within the chamber instantaneously by rapid inter-mixing with its contents. The reaction rate throughout the volume corresponds to the exit reactedness and the mean residence time is the time required for the desired conversion to take place. Although this is never fully attainable in practice, it is a rather useful idealization in which the rate of conversion becomes simply the single volumetric reaction rate corresponding to the conditions within the chamber, multiplied by the total volume. This model is to be regarded as an upper-limit approximation to the potentialities of a combustion chamber, the stability of the flame being determined entirely by the reaction rate. The condition of stability (cf. eqn (3.13)) may then be written

$$wQV = GA(H_{\text{out}} - H_{\text{in}}), \tag{3.13a}$$

where the reaction rate, w, corresponds to the exit reactedness and the inlet and outlet areas, A, have been assumed equal for simplicity alone. On the assumption of constant specific heat, the right-hand side of this equation reduces to a family of straight lines, whose slope is proportional to mass flow, G, on a diagram such as that shown in Fig. 3.9.

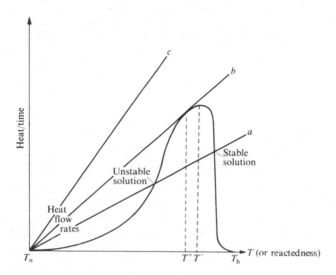

FIG. 3.9. Rates of heat generation and removal under perfect mixing. From Weinberg.[5]

For slow flows and hence small slopes, such as line (a), three intersections, corresponding to solutions of eqn (3.13a), occur. The first one, close to the origin, represents a cold-flow steady-state reactor. The reaction rate there is very small, but not quite zero, because $\exp(-E/RT)$ does not disappear for temperatures above zero absolute. The second intersection represents an unstable solution. Thus any slight decrease in temperature will cause the heat release rate to fall more rapidly than the exit enthalpy and cause a further temperature decrease, initiating a slide to extinction, i.e. to the first intersection. Conversely, the slightest increase in temperature can be seen to increase temperature further, as a result of increasing heat release more than heat removal, and thus to initiate transition to the third and stable intersection. The middle intersection therefore represents the ignition criterion of this idealized system, since attainment of a critical temperature will cause rapid transition to the flame condition. The latter is represented by the third intersection and this gives rise to an exit reactedness not far removed from the fully burnt state—another property that is a consequence of the shape of the reaction rate curve of all highly exothermic reactions in gases. It will be seen that another result of this shape is that quite large increases in flow velocity do not bring about any appreciable change in this product composition to begin with. Eventually, however, the state represented by line (b) sets in. This gives the limiting flow velocity beyond which blow-off must occur because any further increase in flow rate would lead to a net enthalpy outflow that cannot be balanced by any heat release rate available in the system (line (c)). Thus if the shape of the reaction rate curve is known, the maximum flow velocity for stable burning can be calculated. As shown in the diagram, this condition represents a reaction rate somewhat smaller than the maximum value discussed on p. 47. However, it is yet another consequence of the shape of reaction rate curve that the difference between reaction rates at T' and T'' (Fig. 3.9) is usually not very large.

Combustion chambers intended to approach the ideal of perfect mixing have been constructed, chiefly with the aim of determining effective rate laws. They consist usually[81–84] of two concentric spheres, combustion occurring in the shell between them—see Fig. 3.10. The small inner sphere is fed with reactants at high pressure through two tubes at the extremities of a diameter and it then admits the gases to the combustion space through several apertures that usually act as sonic orifices, the momentum of the jets being utilized for vigorous stirring of the chamber contents. The products leave through holes in the large outer sphere and are rapidly quenched, usually by water sprays.

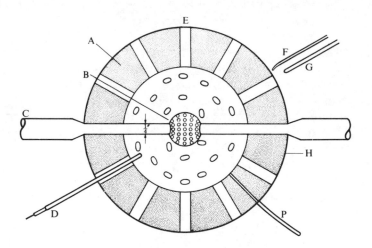

FIG. 3.10. Experimental approach to perfectly stirred reactor. A, insulating firebrick, i.d. 3 in; B, injector sphere, stainless steel, o.d. $\frac{3}{4}$ in, i.d. $\frac{5}{8}$ in, 68 holes 0·047 in diam.; C, inlet tube, stainless steel, o.d. $\frac{1}{2}$ in; D, gas sample probe, stainless steel; E, exit hole (one of 60), $\frac{1}{4}$ in diam.; F, spark wire; G, hydrogen ignition jet; H, nickel shell, o.d. 5 in, 0·038 in thick; P, pressure tap. After Longwell and Weiss.[81]

The manner in which such reactors have been used is to vary conditions, for example flow velocity, until blow-out occurs. The locus of the blow-out conditions is then used to calculate effective activation energies and reaction orders in terms of the overall kinetics. The information obtainable is thus somewhat similar to that deducible from flame structure studies and, while it is much less detailed, as well as being based on assuming the real existence of 'perfectly mixed' volume, it is less restricted as regards inlet reactant conditions. Also, the large heat release rates that are normally confined to the thin reaction zones of flames, demonstrably take place over the relatively large volume of a man-made combustion chamber (though one of excessive pressure losses) so that it has provided something of a practical as well as a theoretical ideal for gas turbine engineers and has been used to predict the performance of real flame tubes. Some relevant references that cover the principle of the method, some of its results, as well as discussions regarding the stability of such systems and the extent to which it can indeed be made to approximate to the ideal of perfect mixing, are appended.[81-85]

At the time of writing, devices aiming at perfect mixing have not yet been used for electrical studies (see however refs. 79 and 80). The stirred reactors, however, may ultimately prove useful in the study of ionization, perhaps particularly for measuring total rates of ionization by way of saturation currents (Chapter 5). In so far as they are perfectly stirred, they have the advantage over flames that the temperature at which chemi-ionization is studied could be defined less ambiguously and they may well provide the means for measuring rates of ion generation up to varying stages of completion of the reaction.

Burners for stabilized detonations and electrically augmented flames (pre-mixed reactants)

Although this is not usually their chief *raison d'être*, burners in this section are often capable of exceedingly high combustion intensities. This is because the reaction generally takes place at a very high temperature and without any corresponding dilution of reactants by products. This constitutes their principal distinction from burners of the preceding section. The manner in which enthalpy is added in some attempts to stabilize detonations—by means of a compressor that allows a stationary shock wave, followed by a reaction zone, to be stabilized somewhere downstream of it—may not appear fundamentally different from the role of the compressor in feeding reactants into the stirred reactor of the preceding section. However, in the latter, the extra enthalpy is all deemed to be consumed in promoting mixing and the reaction rate versus reactedness relationship is assumed to be that corresponding to the initial cold mixture, whereas in the systems to be discussed here, the additional enthalpy in no way involves dilution of the mixture with products and the reaction rate versus reactedness profile corresponds to a much higher initial temperature.

At the same time, the high combustion intensity is not a *necessary* feature of these devices and some systems will be discussed in which only the products are raised to an elevated temperature, for example by a discharge downstream of the reaction zone. This is a perfectly adequate method of increasing final enthalpy and ionization levels but, unless an appreciable fraction of the additional enthalpy diffuses upstream of the reaction zone, it allows no faster through-put than would the normal un-augmented flame ; combustion-intensity and flame stability are not greatly altered and therefore, when the flame blows off, so must the discharge. High combustion intensities are thus a usual, but not an essential, feature of burners discussed in this section.

Since detonation is the fastest mechanism of spontaneous propagation of a reaction wave through a mixture, the prospect of its stabilization in a duct has been regarded as highly desirable. The practical potentialities that prompted this kind of work[86–91] were originally mainly hypersonic ram jets and the possibility of using the experimental devices for high temperature kinetic studies. As it is necessary to attempt to match a detonation velocity, the 'burner' bears more resemblance to a supersonic wind tunnel than to the more familiar laboratory burners. Several experimental facilities have been constructed which differ in detail rather than in principle. At the upstream end, high-pressure air from a compressor is heated to an elevated temperature by passage through a heat exchanger. Since an explosion under conditions of such high pressure and temperature is to be avoided, the fuel, usually hydrogen, must be injected downstream of the compression and heating stages. This leads to the less calamitous, if more fundamental, risk of in-complete mixing and the occurrence of diffusion-controlled flames in the reaction zone, if the latter is not far enough downstream of the injector. The fuel is usually injected from a small needle along the axis, within the nozzle. If a convergent–divergent nozzle is used it is then possible to avoid combustion occurring in the divergent part, because of the rapid drop in temperature along it and the short residence time. In the system used by Nicholls and Dabora,[87,88] the nozzle is run under highly under-expanded conditions, so that further expansion takes place in the open jet, and burning occurs downstream of the Mach disc, where the Mach number falls and the static pressure increases (Fig. 3.11(a)[88]). When the ignition delay is very small, the phenomenon appears as shown in Fig. 3.11(b)—a superimposed shadow and direct photograph from a similar study.[92] Ignition downstream of oblique shocks has been used in some of the other references mentioned (for example ref. 89).

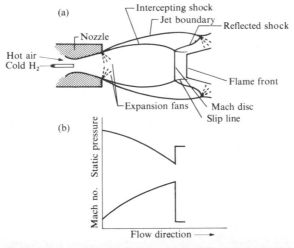

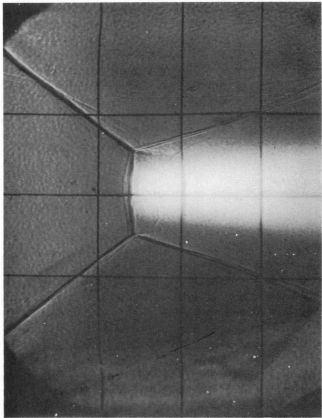

FIG. 3.11. Shock-induced combustion burner, (a) and (b) from Nicholls and Dabora;[88] (c) from Dr J. K. Richmond (private communication).

Much thought and discussion has been devoted to the question of whether the stabilized reaction wave is in fact a standing detonation, as distinct from a shock-initiated ignition. In a propagating detonation, of course, the shock wave is caused, or at least maintained, by the energy released during combustion; when the enthalpy recirculated becomes too small, the detonation will not continue to propagate unaided and a 'limit of detonability' is reached. Here, however, a shock will exist in the absence of any reaction and no feedback is necessary to maintain it. It has been indicated that, depending on conditions, either stabilized detonations or shock-initiated reactions can take place in burners of this type. The standing detonations that occur at higher temperatures are characterized by the shock wave being driven upstream by the combustion, to a stable position corresponding to a lower Mach number.

So far, the main fundamental use that has been made of these devices has been in kinetic studies. This has been carried out by determining reaction delay times, from velocities and the distances between the plane of pressure rise and the flame front. The agreement between H_2-O_2 kinetics obtained from other sources and these time lags seems good; as regards the difficulties in experimental procedure and detailed interpretation, the reader is referred to the original papers.

Whilst discussing stabilized detonations, reference must be made to a family of devices that fall somewhere in between burners stabilizing a stationary front, on the one hand, and vessels in which gas is stationary but the front is allowed to propagate, on the other. Detonation velocities are so high and the equipment required for reducing them to rest with respect to the burner so elaborate that there is some point in compromising between the two principles. Devices are being[93-96] developed in which the reactants are fed in at a velocity smaller than those used in the equipment described previously and the detonation wave is permitted to move in a controlled manner with respect to the apparatus. In one such device the propagation of the detonation front is constrained to a 'doughnut', the gases flowing through it radially, reactants being fed in through a slot in the inner wall. It will be seen that in this manner the maximum flow velocity is not required to balance the detonation speed but merely to replenish the reactant charge in the torroid, in the time the detonation travels around it once. Another approach is based on a flow system[96] similar to that of a cyclone separator. From the point of view of direct generation of electricity, devices of this kind would provide a slug of highly ionized gas travelling at high velocity in a circular path—the desirability of this will become apparent in Chapter 7.

Associating burners of this type—either for fully or partially stabilized detonations—with electric discharges using the principles discussed below, has been advocated[21] but, to our knowledge, not put into practice, at the time of writing. The point is that electrical energy, though thermodynamically highly organized and usually expensive, is so much easier to add to the gas stream than any other form that it is worth using if small amounts will greatly promote stabilization. Thus setting up a device in which the final gas enthalpy is made up of individually variable contributions due to adiabatic compression and the dissipation of chemical and electrical energy is likely to be experimentally much easier than one in which the last component is absent.

'Augmented flames'

Augmenting the enthalpy of flame product gases by electrical discharges is not a new concept. As early as 1924, Southgate[97] discussed a 'pyrelectric process' in which burner tubes, e.g. for furnaces, were provided with electrodes (referred to as 'pyrelectrodes'), along their axes. In recent years, the principle has been re-invented in several laboratories independently.[21,98,99] The experimental arrangements in these attempts differ considerably from that of Southgate and vary greatly from one another†; the above references cover at least five different methods as regards electrode configuration, position of discharge, and current-voltage characteristics, and the optimum form depends very much on the application envisaged.

The main difficulty in producing such a device lies in the tendency of discharges to contract into narrow arc channels of very high-energy concentration. This is a consequence of the rapid increase of electrical conductivity with temperature, for gases, from their near-zero conductivities at ordinary temperatures. If such a contracted discharge is associated with a flame, there results a minute volume of gas at an enthalpy that dwarfs the chemical contribution, the vast majority of the volume remaining at the normal combustion enthalpy. The higher temperature and ion concentration in the arc channel tends to confine the discharge to the same narrow path in the gas for a very long time. This can be readily demonstrated by passing a spark between two electrodes placed so as to contact a pre-mixed flame. Initially the discharge chooses a channel immediately downstream of the luminous flame surface, following the path of maximum ionization. Immediately afterwards, however, the spark channel drifts downstream at the gas velocity and becomes greatly elongated before it strikes back to the flame contour again, i.e. before its resistance increases beyond that of the flame. This characteristic behaviour of sparks has indeed been used[100] for flow visualization. The principle and the persistence of the spark channel are well illustrated in Fig. 3.12. This situation is less extreme for arcs of larger current and lower potential drops but the principle is similar. In due course, normal mixing processes will, of course, distribute the energy from a hot arc channel throughout the product gases, but this is inadequate for two reasons. One is

† See also ref. (129).

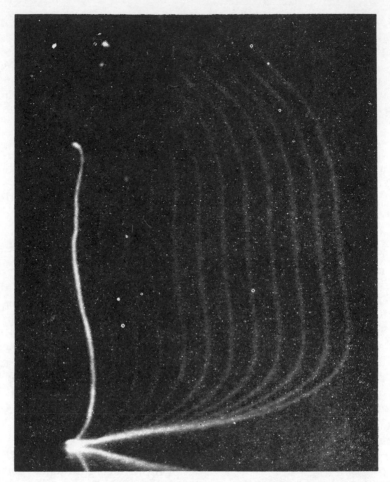

FIG. 3.12. Flow visualization based on persistence of spark channel. From Cochran.[100]

that the high temperature channel may be subject to appreciable energy losses
by radiation. However the main point is that one of the principal attractions
of augmented flames is their potentially very great resistance to blow-off, as
mentioned earlier, so that very large throughputs become feasible for small
burners. This attribute is lost if mixing becomes rate controlling.

Ensuring that an appreciable proportion of the gas is heated by the discharge is, in fact, very difficult.[101] This is illustrated in Table 3.1, which shows
how the fraction of the gas flow heated up by an arc varies with the fraction
of the stream area occupied by the arc channel, for various temperatures
(expressed as expansion ratios). It will be seen that, because of the great
reduction in density of the gas at the arc temperature, which in practice is
further combined with an increased viscosity, most of the flow tends to avoid

TABLE 3.1

By-passing of discharge (R, F, FX, M_1: expansion ratio, fraction of cross-section occupied by discharge, fraction of flow heated and Mach No. (cold), respectively)

F	FX					
	R = 10		R = 20		R = 30	
	$M_1 = 1$	$M_1 \ll 1$	$M_1 = 1$	$M_1 \ll 1$	$M_1 = 1$	$M_1 \ll 1$
0·10	0·023	0·014	0·015	0·009	0·012	0·006
0·20	0·056	0·036	0·038	0·023	0·031	0·018
0·30	0·098	0·066	0·069	0·044	0·056	0·035
0·40	0·15	0·11	0·11	0·072	0·088	0·058
0·50	0·22	0·16	0·16	0·11	0·13	0·091
0·60	0·30	0·23	0·23	0·17	0·19	0·14
0·70	0·42	0·33	0·33	0·25	0·28	0·21
0·80	0·55	0·46	0·47	0·37	0·41	0·32
0·85	0·64	0·55	0·55	0·46	0·50	0·41
0·90	0·74	0·67	0·67	0·58	0·64	0·53
0·95	0·86	0·81	0·81	0·75	0·78	0·71
1·00	1·00	1·00	1·00	1·00	1·00	1·00

the discharge channel. Regarding the latter as a hot solid body is often a better approximation than thinking of it as a perfectly permeable region.

As mentioned above, a variety of different principles has been used in practical attempts to spread discharges. In the early scheme of Karlovitz,[98] a relatively high-voltage (of the order of 10^4 V), low-current discharge is passed through the hot combustion products, which are in highly turbulent flow and frequently also seeded with readily ionizable salts. Both seeding and turbulence contribute to the spread of energy released in the discharge. A typical embodiment of this scheme is illustrated in Fig. 3.13. The advantages

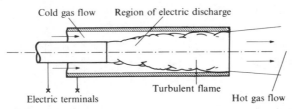

FIG. 3.13. Augmented flame burner according to Karlovitz's scheme. From Karlovitz.[98]

of being able to work at high-voltage, low-current characteristics are practical; they make possible the use of fairly conventional transformer equipment with available mains supplies. This is to distinguish it from, say, plasma-jet characteristics that are more typically of the order 10–10^2 V,

10^2–10^3 A and require much heavier leads and generators or large batteries of heavy duty cells; welding equipment is often suitable. On the debit side, it appears that in the former devices, the heat from the discharge is not well distributed until some distance downstream of the flame[99,102] and that the maintenance of the discharge is conditional upon the stabilization of a flame upstream of it. Thus if the flame blows off, so does the low resistance path and with it the discharge. Any increase in combustion intensity and improvement in flame stabilization is caused only by the hot electrodes and other parts of the apparatus which are heated incidentally. However, there is, no doubt, a wide range of potential applications for which improvement of combustion intensity is not important.

The early devices based on plasma jets[21] are illustrated in Fig. 3.14. It will be seen that when the reactants flow up the burner (Fig. 3.14(a)) it is necessary to shield the cathode by a thin sheath of flowing inert gas. This serves the dual purpose of minimizing chemical attack on the hot electrode and preventing ignition within the burner body. In Fig. 3.14(b), the combustible mixture is fed into the discharge from a ring of nozzles aligned so as to destroy the horizontal component of the jet momenta at the point of impingement. In this case also, the mass flow of the inert gas can be reduced to

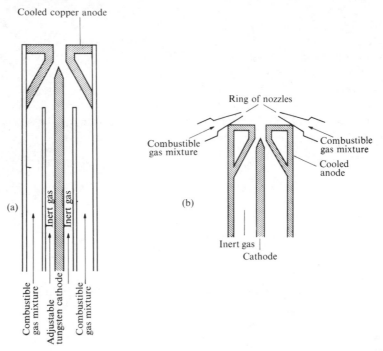

FIG. 3.14. Early augmented flames based on simple plasma jets. From Lawton, Payne, and Weinberg.[21]

an insignificant fraction of the total flow. The advantages of this scheme are greater resistance to arc instability (otherwise caused by effects such as the capture of free electrons by certain species and changes in thermal properties of the gas) and greatly improved mixing, caused by the high level of turbulence. The combustion intensity, however, tends to be mixing-controlled.

The various methods of spreading the discharge have been examined more fundamentally in a subsequent investigation.[99] One feature that has retarded understanding of these phenomena is image persistence on the retina of the eye. Thus even an 'ordinary' plasma jet, i.e. a nozzle anode provided with a coaxial cathode as in Fig. 3.14, gives the impression to the eye of producing a continuous and uniform 'flame'. Such distributed luminosity, however, may be merely the locus of the positions of a fluctuating discharge: a visual impression of uniformity is reliable only when the arc can be shown to be stationary. There is, of course, a wide range of rates of fluctuation fast enough to convey this impression to the eye and yet much too slow to contribute to the spread of heat (i.e. slow in terms of the scale based on the time of traverse by a pocket of gas).

High-speed photography indeed shows[99] that, at least at power densities relevant to augmented flames, a plasma jet is very far from heating the emerging gas uniformly. In the nozzle, an appreciable part of the volume flow is forced into close proximity to the more or less stationary column of discharge. Because of its greatly decreased density there, however, this is very much less than the corresponding mass fraction and a large part of the mass flow ascends close to a cool nozzle wall. Above the nozzle, ciné records show that an insignificant proportion of the gas is heated by the arc channel, in spite of the continuous, flame-like, appearance of this region to the eye. In a vertical section, the cascading of the arc appears to be caused by the gas flow itself. If the arc keeps to the same channel and drifts with it at the gas velocity until it is so extended that its resistance makes it prefer to strike back and re-commence the process, then the same region of gas is heated during one period. Moreover, in a horizontal section, any rotation or random jumping of the arc occurs at a frequency negligible on the time-scale based on gas flow. This is illustrated by Fig. 3.15, which covers a period of 1/16 s during which the gas flow would have advanced by over 30 cm, i.e. the gas in the jet is completely replaced between frames.

The following principles have been used, singly or in combination, in attempts to cause discharges to spread: (i) varying the flow pattern, either in an organized manner, or randomly, by inducing turbulence; (ii) seeding with readily ionizable materials; (iii) the use of radio-frequency discharges; (iv) inducing rapid oscillation or rotation of arcs by magnetic fields. Available space allows for only a cursory summary of the effects of each of these variables.

(i) As regards the organized flow pattern, the most successful configurations appear to be those that force gas into extended contact with the arc

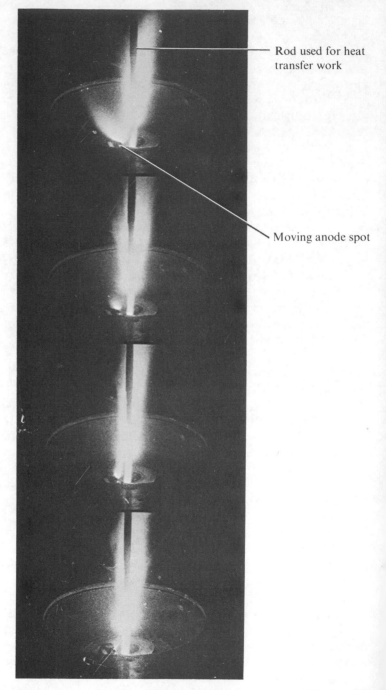

Rod used for heat transfer work

Moving anode spot

FIG. 3.15. Ciné sequence of arc movement in plasma jet. From Chen, Lawton, and Weinberg.[99]

column, and the least successful, those based on using the flow to move (for example, rotate) the discharge. An example of the latter is the vortex-spun arc.[103] Here a vortex, established by injecting gas tangentially at a high velocity, causes the arc to spin about the axial electrode at such rates that the eye sees a continuous disc-like discharge. With respect to the gas, however, the arc channel is stationary (except insofar as it *lags* behind gas movement). This is, of course, no criticism of the device as used for purposes other than augmenting flames—for example, it certainly minimizes anode erosion—but it does not increase the proportion of gas coming in contact with the discharge unless turbulence is induced.

In contradistinction, an example of using organized gas-flow to force reactants into extended contact with an arc channel is provided by the 'porous cavity' burner.[99] This is illustrated in Fig. 3.16. The reactants are

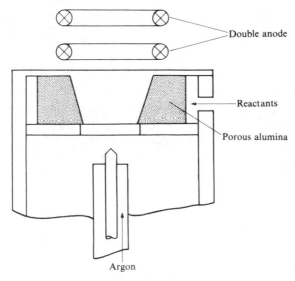

Double anode

Reactants

Porous alumina

Argon

FIG. 3.16. Arc burner based on divergent porous cavity. From Chen, Lawton, and Weinberg.[99]

fed in through the porous walls so that they cannot ascend through a cool boundary layer. This principle also makes it possible to minimize the distance between walls and arc channel both because of increased arc stability and because of the transpiration cooling of the walls. The cooling is regenerative, much of the heat being brought back by the gas flow. The high arc stability is brought about by having the least possible flow velocity parallel to the channel, for a given volume flow, particularly in the vicinity of the cathode which is recessed as shown. By tapering the sides of the cavity, the gradual admixture of fresh reactants to the arc can be made uniform along its length. This device is indeed capable of very high rates of through-put; it is thought likely that each part of the wall acts as a 'porous plug burner' (see p. 69), the flame being kept close to the wall even at the very high flow velocities employed because of the large burning velocity induced by the high final temperatures. The difficulties encountered are practical and are associated with keeping the burner in operation for extended periods of time. If porosity is not uniform, melting of the wall sets in at points of decreased permeability to flow and, since the consequence of this is a further reduction in transpiration, the instability grows and leads to rapid destruction of the porous inlay.

As regards spreading discharges by turbulence, the attainment of the objective depends critically on the scale. It is only at small scales of turbulence —comparable to the width of the arc channel—that the discharge experiences an increase in *diffusivity* that tends to disperse its boundaries. Fluctuating velocity distributions that extend over much larger distances must displace the arc channel bodily, much in the same way as does convection, except that the movement now is random and fluctuating. Experimental observations of this are once more complicated by the eye's inability to convey an instantaneous picture. Thus an individual arc channel distorted in a fluctuating manner by the random movement of turbules may look like a diffuse discharge and yet heat the same pockets of gas for a very long time. Bearing in mind that intensities of turbulence are generally small fractions of flow velocity, spread by frequent rupture of arc channels by turbules seems highly improbable. Thus scales of the magnitude required are so small as to be difficult to achieve and to maintain by normal methods (except as the incidental small-scale end of the scale spectrum). Turbulence is therefore most useful when employed in conjunction with methods that widen the discharge channel.

(ii) Seeding the reactants is an example of such a method. As mentioned earlier, the tendency of discharges to contract is a consequence of the feedback whereby such a contraction causes an increase in local temperature which, in turn, brings about an increase in ionization and hence in conductivity. It is therefore to be expected that seeding with an easily ionizable material will tend to counteract this instability by providing less resistance at lower temperatures and decreasing the rate of variation of conductivity with temperature. Small-scale experiments[99] have shown that this effect comes into operation only well downstream of the luminous zone of the flame, i.e. beyond the zone of attainment of maximum temperature. In applications where it is intended merely to heat up product gases and where seeding presents no obstacles, this is no disadvantage. The work of Marynowski, Karlovitz, and Hirt[102] has shown that the combination of turbulence with seeding succeeds in spreading the heat from the discharge uniformly, but only well downstream in an extended duct.

(iii) Radio-frequency discharges present the appearance of extended 'flame-shaped' regions (Fig. 3.17) of uniform luminosity extending over

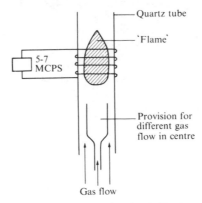

Fig. 3.17. Radio frequency heating in flowing gas.

much of the area of cross-section of the confining tube around which the exciting coil is placed. Although in recent years commercial devices for inducing r.f. plasma in gases at atmospheric pressure have become readily available, they are usually much more expensive, much more wasteful of input energy, and much less stable in operation than the devices discussed hitherto. Whilst the central luminous region appears to be heated uniformly, the peripheral annulus of cold gas once more carries a disproportionately large fraction of the mass flow, for reasons discussed previously.[101] In view of these factors, the use of r.f. discharges for augmenting flames would be justifiable and competitive with the much less elaborate devices only should they prove much more stable in the presence of an exothermic reaction. Using individual gases, the stability of r.f. discharges is greatest for species like argon which manifest neither an affinity for electrons nor the ability to absorb energy by dissociation. However, it seemed possible that the additional supply of radicals generated by the discharge in the reaction zone of a flame would so much increase the burning velocity, and hence resistance to blow-off, that a much increased throughput would become possible. Quite inelaborate experiments,[99] however, show at once that this is not the case. Not only does the addition of fuel or oxidant, when added singly to an argon flow, tend to extinguish the discharge, but a mixture in which an oxidation reaction is proceeding shows an even more remarkable effect. In experiments carried out when oxygen was already present, an addition of as little as 0·14 per cent by volume of ethylene was enough to extinguish an otherwise stable plasma. It was thought that this is brought about by essentially 'chemical' mechanisms such as electron capture by OH released during the reaction and perhaps involving increased charge recombination rates, since the effect was too drastic for a more physical explanation (in terms of thermal effects, or changes in circuit characteristics, due to the gas). The

gencral conclusion, as regards augmenting flames in this manner, must be
that the additional cost in capital equipment and power loss is probably
not justified for normal applications.

(iv) The use of a magnetic field to 'spread out' a discharge is not a recent
concept. The most familiar practical application is probably the Birkeland–
Eyde process in which the combination of nitrogen with oxygen is encouraged by passing the gases through an arc 'drawn out into a disc by
means of a magnetic field'. Here again it seems likely that a fluctuating
discharge is being observed and the question arises whether the phenomenon
'appears' as a disc to the gas at the flow velocities involved, as well as to the
eye.

A system relevant to augmented flames was studied[99] by placing, co-
axially, a small coil immediately above and around the mouth of a plasma
jet such as that shown in Fig. 3.14(a). The force resulting from the interaction
of the magnetic field with the radial component of the current causes the arc
to oscillate (a.c. magnetic field) or to spin (d.c.). The magnetic fields applied
were in the range 150–750 gauss and even quite small fields gave the discharge
the appearance of a continuous disc, greatly diminished wear of the anode,
and increased the stability of the arc. However, since appearances are so
unreliable, for reasons discussed previously, a Fairchild High Speed 16-mm
half-frame camera (maximum speed of 16 000 frames/s) was used, sometimes
in conjunction with a plane mirror, so that the camera was able to view
the movement of the arc in an end-on position. Figure 3.18 shows a series of
consecutive frames using (a) a.c. and (b) d.c. currents in the field coil. Because
of inductive lags in the circuit, the effect of a.c. fields is more spectacular but
far less conducive to uniform heating of the gas. All systematic work was
carried out using d.c. fields where the condition that all the gas be heated
uniformly, i.e. that every part of the stream, flowing upward at linear velocity
v, should experience the passage of a permeable discharge through it at least
once, is

$$\omega > 2\pi \, v/d, \qquad\qquad (3.27)$$

where d is the thickness of arc channel, defined in terms of the distance within
which gas will be raised to a high temperature by it, and ω is its angular
velocity. It was confirmed that the dependence of frequency, f, on current,
i, and magnetic field could be described by[104]

$$f = \text{constant} \times B^{0.6} i^{0.33}. \qquad\qquad (3.28)$$

Substitution of this into condition (3.27) makes it possible to calculate
maximum flow velocities for uniform heating according to the above model
(i.e. without relying on any subsequent spreading of heat at all). The results
depend on the diameter of the discharge and are in the region of 500 cm/s with
the relatively small fields used. Bearing in mind that the high-temperature

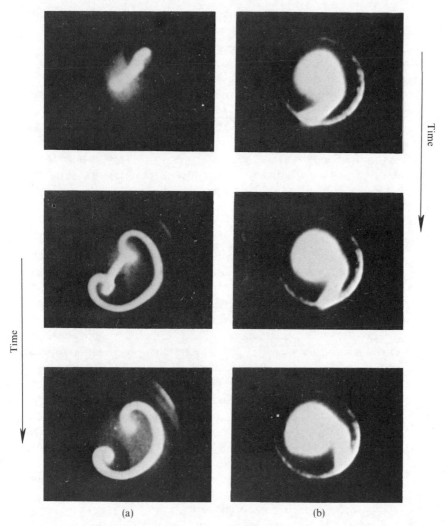

(a)   (b)

Fɪɢ. 3.18. High-speed ciné sequence of plasma jet arc in magnetic fields.

arc-channel may not be treated as a perfectly permeable region through which all the gas passes freely, the physical model of the gas 'seeing the discharge as a spiral staircase of pitch equal to the width of the hot zone' is an over-simplification. The saving grace of the principle, however, is that it is possible to heat a gas stream uniformly even just by passing it past a rapidly rotating hot spoke that is entirely solid and impermeable—to consider the opposite extreme. It has been shown that this device provides another successful method for producing an augmented flame of high stability and combustion intensity—Plate 1(v).

The large number of applications that have been envisaged for augmented flames include their use for welding, cutting, drilling (rocks, as well as metal), melting, smelting, sintering, spheroidizing, reducing ores, and synthesizing endothermic reaction products; their application as ignitors or 'pilot flames' in high intensity combustion equipment (particularly in propulsive systems) and as burners for particulate fuel suspensions to produce very high levels of radiation; their use as ion sources in methods of direct generation of electricity, and so forth. Table 3.2 indicates the extent of electric augmentation required to produce given temperatures and ion concentrations for a few common fuel-oxidant mixtures. It will be seen that in strongly dissociating gases the attainment of just 4000°K requires almost nine-tenths of the total energy input to be electrical. Beyond this, the phenomenon would become so nearly a pure discharge as to be outside our scope.

It is obviously impossible to consider all proposed applications in detail, but they can be subdivided according to whether they are based simply on the use of a hot, inert gas, whether they involve increased flame stability and combustion intensity, or whether they entail augmented ion concentrations and electrical conductivities also. It follows from the preceding discussion that all methods are suitable for generating hot product gases and elevated *thermal* ionization, whereas methods such as the magnetically spun plasma jet or the porous cavity have to be used when the application involves increased combustion intensity or flame stability. As regards applications based purely on the use of hot gas, the most important considerations are economic. The thermal augmentation required is often achievable by replacing the air of the reactant mixture, partly or entirely, by oxygen. The comparative costs of oxygen and electrical power that determine the choice on economic grounds vary considerably from one part of the world to another.

Many of these 'hot gas applications' involve heat transfer to a solid. This depends on dissociation, i.e. it is governed by the enthalpy as well as the temperature of the gas, because heat transfer from augmented flames is much enhanced by the recombination of dissociated species and release of the bond energies on, or near the surface. The theory of this effect on heat transfer coefficients has been calculated.[105,106] Measurements of heat transfer from augmented flames have been carried out by Davies[107] who found that, in addition to the above effect, heat transfer was greater than that of non-augmented flames of equal energy by factors of 2–4 times. Davies used augmented flames of the Karlovitz[98] design † and attributed observed increases in heat flux to non-uniform distribution of the added energy. It is likely that discharge-generated turbulence also played a part.

The level of ionization can also enter in heat transmission work, particularly when fields are used to increase heat transfer coefficients. It has been

† See also ref. (130).

Equilibrium products—augmented flames. From Lawton, Payne, and Weinberg.[21]

	0·00		0·01				0·10			
Partial pressure of potassium added as potassium carbonate (atmospheres)										
Temperature (°K)	4,000	4,500	3,000	3,500	4,000	4,500	3,000	3,500	4,000	4,500
Required ratio of electrical to chemical power — C+O$_2$	0·70	0·90	0·00	0·25	0·70	0·90	0·00	0·30	0·75	1·00
C+air	2·10	2·55	0·75	1·45	2·10	2·55	1·55	2·25	3·00	3·55
C$_2$H$_4$+3O$_2$	1·45				1·45				2·45	
C$_2$H$_4$+air	2·7	3·4			2·7	3·6			5·4	5·9
C$_2$H$_4$+15O$_2$	5·65				5·65				8·7	
10^{-14} × number of electrons per cm^3 — Based on { C$_2$H$_4$+air / nitric oxide } on air	0·09	0·4								
{ C+air } oxide	0·09	0·4								
Based on potassium	0·02	0·08	6·8	24	51	91	22	78	190	360
Conductivity (mhos/cm)			1·3	3·5	7·0	9·0	1·2	3·5	7·0	10·0
Fraction of potassium atoms ionized			0·03	0·11	0·27	0·56	0·01	0·04	0·10	0·22

shown[57] (see Chapter 7) that ionic winds can be used to increase heat transfer from flames to solid surfaces by setting up suitable circulation patterns and breaching boundary layers. In the highly-conducting augmented flame plasmas further work has shown[108] that such almost 'electrostatic' methods compare unfavourably with effects that can be induced using magnetic fields. Thus in the case of the magnetically-rotated plasma jet, heat transfer to surrounding walls is greatly increased by the turbulence and swirl generated by the rapidly rotating arc channel. This is an advantageous aspect of the discharge not being perfectly permeable to the gases it traverses and it opens up the possibility of stirring by something resembling a solid spoke at rates of the order of 10^5 r.p.m. Such a use of electromagnetic force is the exact opposite of MHD generation of electricity (Chapter 7) —instead of generating power by allowing the gas to do work on the charge carriers, fluid-mechanical effects are induced by using electrical power to set up forces and generate work on the gas. We think it probable that this principle will gain practical importance in future.

Heat transfer considerations are also important in applications involving high combustion intensities and in those aimed at synthesis of endothermic products. In the case of large electrical augmentation, all the species are split up beyond the discharge stage into radicals or atoms that are unstable at normal temperatures. The subsequent rate of heat release per unit volume (combustion intensity) as well as the yield of endothermic and other abnormal products depends entirely on how rapidly the gas is quenched. Electrical methods for varying rates of heat transfer are therefore of interest in this context also. Work is in progress on ignition and flame stabilization, in which the wide dissemination of heat and active species from an augmented flame to all the gas is the aim.

Finally, as regards the direct generation of electricity (see Chapter 7) it appears thermodynamically unsound to use a highly organized form of energy such as electrical power to augment the random thermal process, particularly in a conversion scheme. It is therefore thought that augmented flames are likely to prove useful in this context only if appreciable additional ionization, gained at the expense of a moderate electrical input, can be 'frozen' into the gas.

Burners for diffusion flames

In the case of reactants that are initially separate, flame shape is not determined by the interaction between a propagating reaction front and a velocity distribution in the in-flowing reactants but, as discussed previously, by the diffusion pattern of the reacting components. Although the simplest and most frequently used burner is again the cylindrical tube—the height of the diffusion flame on which is given on p. 58—several special burners have been developed for research purposes and some of these are well-suited to electrical studies.

The first burner to produce a flat diffusion flame[109] was intended chiefly as a tool for spectroscopic studies. It is illustrated in Fig. 3.19 and consists essentially of two adjacent tubes of rectangular cross-section sharing a long interface. Each tube carries the flow of one reactant and the assembly is sheathed by a flow of inert gas such as nitrogen. The results of spectroscopic structure studies on the approximately plane vertical flames have been summarized in detail by Gaydon and Wolfhard.[1]

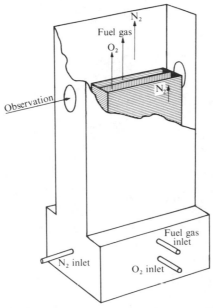

FIG. 3.19. Wolfhard–Parker burner. From Gaydon and Wolfhard.[1]

More recently, a burner for flat diffusion flames in a counterflow regime[43,44] has been developed from a method of establishing blow-out conditions as a measure of reaction rates.[40–42] As regards this method of studying effective flame kinetics, the concept of 'flame strength' has been discussed in the section on mechanisms of burning. The theory of an idealized form of this flame has been developed.[46] The original burner consisted essentially just of two co-axially opposed cylindrical tubes carrying the two reactant flows. Since a normal velocity distribution is established in each tube, blow-out tends to set in first on the flame axis, where the flow velocity is highest, i.e. the flame develops a hole in the centre.

The desire to make counter-flow diffusion flames flat arose out of their potential suitability for structure studies, particularly by optical methods. This burner[43,44] is illustrated in Fig. 3.20. Its operation depends on two

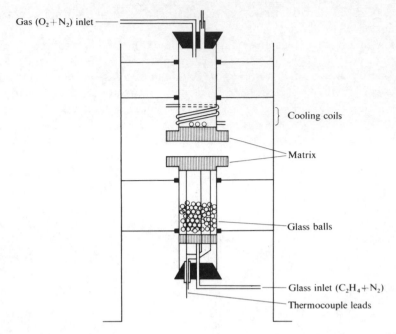

Gas ($O_2 + N_2$) inlet

Cooling coils

Matrix

Glass balls

Glass inlet ($C_2H_4 + N_2$)

Thermocouple leads

FIG. 3.20. Burner for flat, counter-flow, diffusion flame. From Pandya and Weinberg.[44]

principles, the first of which is the rectification of the velocity distribution, making velocity uniform across the burner mouths by means of beds of glass beads and matrices of corrugated and plane metal strip, much as in the flat flame burner for pre-mixed flames described previously. For best results in terms of flame flatness, however, it is desirable also to make the condition for stoichiometry coincide with that for aerodynamic symmetry, by dividing the nitrogen of air between the two streams in such a way as to make the flow velocities approximately equal when the rates of fuel and oxidant supply are in stoichiometric ratio. A typical flame and flow pattern is shown in Fig. 3.21 and the distributions of temperature, gas velocity, and heat release rate, as well as some composition and spectroscopic data are given in ref. (44).

Other burners for producing diffusion flames of simple geometry are feasible. Thus fuel emerging symmetrically into the atmosphere from a porous sphere will burn as an approximately spherical flame. This principle has usually been used with liquid fuels, the porous sphere simulating a drop, and it will be discussed further in the section of flames with heterogeneous fuel supply. Co-axial cylindrical tubes, one within the other, supplying fuel and oxidant into a draught-proof enclosure at suitably calculated rates can be used to produce flames that are sensibly cylindrical over a large part of their height.

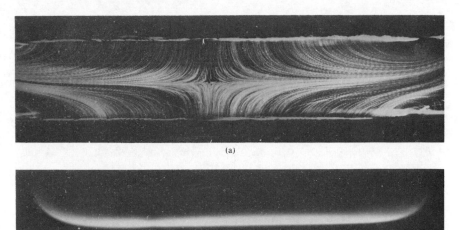

(a)

(b)

Fig. 3.21. Flat, counter-flow, diffusion flame and flow pattern. After Pandya and Weinberg.[44]

As regards suitability of these burners for electrical work, its features tend to favour the flat, counter-flow diffusion flame, particularly when the application of appreciable fields is involved. First, no part of it is pre-mixed—unlike in the case of rectangular burners where a small pre-mixed region persists above the quenching zone at the joint interface. Second, it can provide a good approximation to a unidimensional field distribution, if the burner mouths and matrices are used as the electrodes—preferably with guard-ring flanges. (The disadvantages of flames protruding beyond flanges include the risk of loss of some unburnt gas.) The unidimensional field geometry is particularly useful in measurements such as those of total rates of ionization by the method of saturation currents (Chapter 5) and is possible because of the absence of any solid surfaces around the flame periphery. In the case of the rectangular slot burners, lines of field tend to crowd into the burner itself and all cylindrical and spherical flames similarly suffer from asymmetrical effects due to fuel inlet tubes, burner bodies, etc. Third, counterflow flames are distorted least by ionic winds, when appreciable fields are applied, because the flow is directly opposed to the ionic wind direction in each electrode space (Chapter 4). In particular investigations, use has been made of some of their more specific features. Thus Place and Weinberg[55,56,111] made use of the relative position of their pyrolysis and main reaction zones in studies on the electrical aspects of flame carbon, whilst Heinsohn and Rezy[110] used their blow-out criterion to study the effect of fields on reaction rates. The former were based on the circumstance that when a field is applied between flanged burner mouths, ions from the flame pass through the pyrolysis zone and bring down carbon on the fuel-side electrode, away from oxygen, the polarity of this ion flux being determined solely by the field direction. The individual applications of these burners will be discussed in later chapters.

Heterogeneous systems

Even though flame reactions have been defined as occurring in the gas phase, one or both of the reactants, as originally supplied, may be liquid or solid. We want to avoid calling, for example a fuel droplet, which trails a stream of inflammable vapour as it travels through the air, a 'burner'. This is not because the flame process in the stream of trailing vapour is in any way different from that which would occur above the mouth of a cylindrical tube discharging gas at the same rate; the principal distinction is the absence of an independent control of the flow of gas, such as the gas valve with which all burners discussed hitherto can be provided. The rate of reactant supply in the cases discussed below is governed by heat and mass transfer to and from the droplet, particle, wick, etc. and is often determined, therefore, by feed-back from the combustion process itself. It is this aspect that provides an additional degree of freedom in the application of electric fields to such phenomena in efforts to control combustion processes electrically. As will be shown later, transport rates can often be altered quite drastically by electrical methods so that some control can be exercised in this way over the rate of supply of gaseous reactants in this kind of combustion process. In addition, electrical methods of dispersion of liquids and of solid powders, of guidance and confinement of the droplets and particles so dispersed, and of precipitation of products and residues, become applicable.

The subject can once more be separated into pre-mixed phenomena and those in which the reactants are initially separate. The latter, which are by far the more common, generally involve an initial phase difference between the two reactants—usually a liquid or solid fuel burning in air. Pre-mixed systems are confined to compositions designed as explosives or propellants in which an oxidizing agent is chemically combined with, or physically mixed into, the solid or liquid fuel.

Reactants separate

'Heterogeneous diffusion flames' cover the widest range of combustion phenomena, from the burning of a candle at one extreme, over coal-dust furnace flames, to forest fires and other conflagrations. In order to narrow this field sufficiently to allow discussion of types of burner, we must consider the general sequence of consecutive steps that constitute the burning process:
  (1) supply of (heterogeneous) fuel (for example, by flow of a liquid, by capillarity in a wick, or by excretion of tar by a burning piece of coal);
  (2) vaporization;

(3) mixing of reactants:

(4) reaction.

By definition of a diffusion flame, step (4) is so fast that it is never controlling and we therefore need not consider it here. Again, step (3) occurs by diffusion and determines the shape of the flame according to the principles discussed previously. Burner types are therefore characterized by (1) and (2); at least they differ from the burners of previous sections in these steps, which were absent hitherto. Usually it is step (2) that determines the rate of burning—the case where the fuel reservoir is inadequate is not normally an interesting one. We must nevertheless discuss some aspects of the first step because the form in which the fuel is supplied has an important bearing on the rate of vaporization: thus a suspension of coal dust burns very much faster than the parent lump of coal.

The burning of large aggregates (for example, piles of solid, pools of liquid fuel) has been studied from the point of view of prevention and extinction of accidental fires and is of no interest in the present context. On burners, on the contrary, the aim is generally to subdivide and to disperse the heterogeneous fuel in order to facilitate the burning.

As regards liquid fuels, historically the earliest burners are those using wicks. So long as capillarity does not fail, i.e. so long as supply of liquid to the tip of the wick does not become rate-controlling, the theory of the process is similar to that of the burning of large liquid drops; see below. Since, in this case, supply of fuel is determined by heat transfer from the flame, one would expect the rate of burning to be subject to influence by electric fields. The effect of such fields on combustion taking place on wick burners has been studied by Kinbara and Nakamura.[112,113] As will emerge in subsequent chapters, the effect of fields can here influence the process in a variety of ways; by the variation of carbon content of the flame and hence of radiation flux, and/or by varying aerodynamic flow and mixing patterns and hence changing the properties of the flame as well as the rate of supply of fuel. It is therefore not surprising that either extinction or increases in the rate of burning could be achieved by varying the field strength and configuration. Aeration of wick flames by inducting the stoichiometric requirement of air through a matrix of small channels surrounding individual strands of the wick, using the entrainment calculated for ionic winds (see Fig. 7.34 and Chapter 7) have been demonstrated by Lawton, Mayo, and Weinberg.[80]

The combustion of droplets and sprays of liquid fuel has received more study than any other phenomenon in this category because of its importance in some propulsive equipment. There is a size of droplet and particle below which vaporization and diffusion of vapour is so rapid that it occurs completely in the pre-heat zone and the phenomenon is then essentially a premixed flame. It differs from the homogeneous system only in the loss of latent heat to the fuel and, once corrected for this effect, manifests the same burning velocity. As the radius of the droplet increases, however, the rate of vaporization and its dependence on droplet diameter and environmental parameters becomes all-important because it determines the life time of the droplet (and hence the length of its burning trajectory, if the droplets are in motion). Many experimental studies have been carried out on droplets falling through atmospheres under controlled conditions.[114,115] For studying the rates of vaporization of large single droplets, a burner is available in the form of the wetted porous sphere.[116,117] Figure 3.22 illustrates the burner and the form of flame obtained.

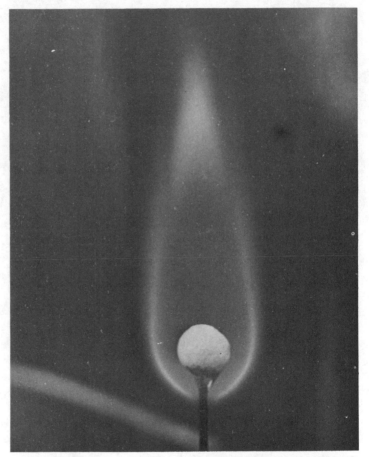

Fig. 3.22. Wetted porous sphere burner. From Wise, Lorell, and Wood.[117]

For individual stationary droplets, the lifetime is proportional to the square of the initial drop diameter. To see what this implies about the vaporization law, let m be the mass of the drop and let its rate of vaporization be proportional to the pth power of its radius. Then

$$-dm/dt = \beta_1 r^2 \, dr/dt = \beta_2 r^p, \tag{3.29}$$

where t is the time co-ordinate and the β's denote constants. Thus

$$r^{(2-p)} \, dr = -\beta_3 \, dt,$$

which integrates to

$$t_0 = \beta_4 r_0^{3-p}, \tag{3.30}$$

where r_0 denotes the initial radius and t_0 the lifetime. Equation (3.30) does not apply for $p = 3$, i.e. heat loss proportional to volume, the shrinkage then being exponential; this does not correspond to a practical case. For a particle vaporizing largely due to receiving radiation from a constant temperature enclosure, we would expect heat absorption to be very nearly proportional to its surface area and, therefore, $p = 2$. (dr/dt) is then constant and the lifetime proportional to initial radius. For the first-mentioned case, in the absence of appreciable convection or radiation, the experimental result

$$t_0 \propto r_0^2 \tag{3.31}$$

indicates

$$p \simeq 1 \tag{3.32}$$

and thus, when the 'droplet' is a wetted porous sphere in stationary air burning a fuel that does not produce much soot (and hence radiation), or when a travelling droplet is surrounded by air proceeding at the same velocity, the rate of vaporization is proportional to the radius, not the surface area. In the presence of forced convection, vaporization increases and p tends to $\frac{3}{2}$, so that

$$t_0 = \beta_5 r_0^{1.5}. \tag{3.33}$$

These two last results have been obtained both by theoretical analysis of the transfer processes and by experiment. They are not usually represented in this manner, but rather as graphs of relationships between dimensionless groups. In the context of the present discussion on burners, the wetted sphere, the train of droplets falling through an atmosphere under controlled conditions, and burners involving atomizers of liquid fuels, are suitable for research work.

Somewhat surprisingly, the effect of electrical fields on the burning process occurring on fuel-wetted porous spheres has not yet, to our knowledge, been studied. Diffusion flames on falling droplets have been subjected to transverse field by Nakamura.[113] A diagram of the experimental scheme is given in Fig. 7.29 : it was shown that extinction of the entire train of droplets beyond the electrodes could be achieved by application of a sufficient potential.

From the electrical point of view, a very interesting and potentially useful aspect of this part of the subject is the use of fields for atomizing and dispersing liquid fuel. A process that is sometimes referred to as 'electrostatic atomization' but does, of course, involve the passage of small currents, has been studied in several laboratories.[118-124] When a divergent field is applied to a jet of liquid trickling from a capillary, the attraction acting on the near end of the induced dipole causes the tip of the jet to be torn off at very high repetition rates. This results in atomization of the jet, the mean droplet size decreasing with increases in the applied potential. The phenomenon is discussed in some detail in Chapter 7 and illustrated in Fig. 7.1. Its main significance in the present context is that it provides the means to operate a burner that atomizes liquid fuels by using electric fields alone. It has been shown in unpublished work with Dr. K. Gugan in our laboratories that similar effects can be produced in powdered solids by subjecting fluidized beds to high potentials and divergent fields.

One of the potentially most useful aspects of this form of dispersion is that it can easily be arranged for the particles or droplets to retain their charge, which raises the possibility of governing the entire subsequent combustion process by means of fields replacing, at least in part, the tubes, baffles, and other hardware of the combustion engineer. For example, Fig. 7.2 shows the effect of applying alternating fields to charged dispersions of fuel droplets produced in this manner, the sinusoidal trajectories inducing vaporization in a shorter distance, with a view to increasing overall combustion intensity. This, together with influencing other phases of the combustion process such as entrainment of the stoichiometric requirement of air by field-induced ionic wind effects,[80] the ducting of charged particles without collisions with walls, once these become charged to the correct potential,[125] the increasing of combustion intensity by field-induced recirculation of hot products[80] and, finally, the electrostatic precipitation of any ash are discussed in Chapter 7. The limitations to such schemes, particularly those due to breakdown at the electrodes, are considered in Chapter 8.

Solid propellants

As regards pre-mixed systems in which reactants are not supplied in the gaseous phase, explosives are outside our present scope. The burning of solid propellants, however, is not so rapid that a steady state, constant-pressure deflagration concept could not be applied to them. Here again, some compositions burn in the manner of diffusion flames around individual grains, but the pre-mixed concept is applicable to one group of propellants. They have been left to the very end of this chapter because, whilst these are undoubtedly a combustion process, it is doubtful whether they are properly described by the term 'flame'. Our definition of flame processes (Fig. 3.1) implies that all the main heat releasing reaction occurs in the gas phase. This is not entirely true of propellant burning and propellants therefore have not been included in the table. Although some gaseous flame reactions do occur above the surface of a burning propellant, the whole process is very much influenced by the primary reaction zone that occurs first in the solid phase and then in the mixed phases of the 'fizz reaction zone'. Temperatures are likely to be over 1000°C by the point where the gas-phase reaction zone sets in—e.g. ref. (126).

The 'burner' here is, of course, the solid propellant itself and the rate of burning is determined by its composition, its geometry, and the ambient pressure. Thus, unless the latter is varied, the overall burning rate and its progress as the propellant is consumed can be 'programmed' into the system but no variation from the programme is possible during the process. Such 'programming' is usually imposed by casting the propellant in such a form that the burning area will vary as the flame front proceeds so as to give the desired variation of total mass consumption rate with time. One difficulty here is that the rate of burning is very pressure-dependent and generally increases with increasing pressure. This has the effect of amplifying any pressure oscillation that may arise incidentally in a rocket and leads to phenomena such as 'chugging', 'screaming', and other potentially destructive pressure fluctuations.

What brings the subject just within the scope of this monograph is that there is some prospect of controlling solid propellant burning electrically. Although the primary reaction zone appears to occur in the solid phase, the process is not uncoupled from the subsequent gaseous burning. It could, therefore, be influenced to some extent by variation in heat transfer rates from the gas to the surface and by other consequences which can be effected by applied fields and which are discussed in subsequent chapters. This intervention is particularly potent during the process of ignition and flame-spread across the propellant surface.[127] Such large changes in rates of flame spread have been induced by electric fields (factors of about 2000 times at atmospheric pressure (Chapter 7)) that further developments in this subject may be expected.

REFERENCES

1. GAYDON, A. G. and WOLFHARD, H. G. *Flames*, Chapman and Hall, London (1960).
2. LEWIS, B. and VON ELBE, G. *Combustion, flames and explosions of gases*. Academic Press, London and New York (1961).
3. FRISTROM, R. F. and WESTENBERG, A. A. *Flame structure*. McGraw-Hill, New York (1965).
4. MINKOFF, G. J. and TIPPER, C. F. H. *Chemistry of combustion reactions*, Butterworths, London (1962).

5. WEINBERG, F. J. *Optics of flames*. Butterworths, London (1963).
6. FRIEDMAN, R. and MACEK, A. *10th Int. Symp. Combust.*, p. 731. Combustion Institute (1965).
7. *JANAF Thermo-Chemical Tables*. The Thermal Lab., Dow Chemical Co., Midland, Michigan (1960).
8. ROSSINI, F. D. *et al. Circ. U.S. natn. Bur. Stand.* No. 46 (1947).
9. JUSTI, E. *Specifische Wärme, Enthalpie, Entropie, Dissoziation technischer Gase.* Springer, Berlin (1938).
10. General Electric Co. *Properties of combustion gases.* McGraw-Hill, New York (1955).
11. MAYER, Y. *Energie* 32, 65, 211 (1948).
12. DAMKÖHLER, G. P. and EDSE, R. *Z. Elektrochem.* 49, 178 (1943).
13. WOLFF, H. *Z. Elektrochem.* 54, 342 (1950).
14. WINTERNITZ, P. F. *3rd Int. Symp. Combust.*, p. 623. Williams and Wilkins, Baltimore (1949).
15. WEINBERG, F. J. *Proc. R. Soc,* A241, 132 (1957).
16. BRINKLEY, S. R. *High-speed aerodynamics and jet propulsion* (edited by R. W. Ladenburg, B. Lewis, R. N. Pease, and H. S. Taylor.), Vol. II. Combustion processes, Oxford University Press (1956).
17. MOFFATT, W. CRAIG. Magnetogas-dyn. *Lab. Rep.* 61–5, M.I.T. (1965).
18. GAYDON, A. G. and HURLE, I. R. *The shock tube in high-temperature chemical physics*, Reinhold, New York (1963).
19. GORDON, J. S. *WADC Tech. Report* 57–33 *Astia Doc.* 110735 (1957).
20. EGERTON, SIR ALFRED, GUGAN, K. and WEINBERG, F. J. *Combust. Flame* 7, 63 (1963).
21. LAWTON, J., PAYNE, K. G., and WEINBERG, F. J. *Nature, Lond.* 193, 736 (1962).
22. *High-speed aerodynamics and jet propulsion,* (edited by R. W. Ladenburg, B. Lewis, R. N. Pease, and H. S. Taylor), Vol. IX. *Physical measurements in gas dynamics and combustion.* Oxford University Press (1955).
23. CREITZ, E. C. *J. Res. natn. Bur. Stand.* 65A, No. 4, (July–Aug 1961); see also LEE, T. G. *J. phys. Chem., Ithaca* 57, 360 (1963).
24. LONGWELL, J. P. and WEISS, M. A. *Ind. Engng Chem. ind. (int.)* Edn 47, 1634 (1955).
25. LEVY, A. and WEINBERG, F. J. *Combust. Flame* 3, 229 (1959).
26. CHASE, J. D. and WEINBERG, F. J. *Proc. R. Soc.* A275, 411 (1963).
27. LINNETT, J. W. and HOARE, M. F. *3rd Int. Symp. Combust.* p. 195. Williams and Wilkins, Baltimore (1949).
28. CLARKE, A. E., ODGERS, J., STRINGER, F. W., and HARRISON, A. J. *10th Int. Symp. Combust.*, p. 1151. Combustion Institute (1965).
29. SPALDING, D. B. *Combust. Flame* 1, 287 (1957).
30. WORTBERG, G. *10th Int. Symp. Combust.*, p. 651. Combustion Institute (1965).
31. FOX, M. D. and WEINBERG, F. J. *Proc. R. Soc.* A268, 222 (1962).
32. BASU, S. and FAY, J. A. *7th Int. Symp. Combust.*, p. 277. Butterworths, London (1959).
33. JAARSMA, F. and FUHS, A. E. *Am. Rocket Soc. 14th annual meeting rep.*, p. 923 (1959).
34. EDWARDS, D. H. and LAWRENCE, T. R. *Proc. R. Soc.* A286, 415 (1965).
35. TUNG, C. C., KELLY, J. R., and TOONG, T. Y. *Proceedings 26th meeting AGARD propulsion and energetics panel* (1965).

36. CHINITZ, W., EISEN, C. L., and GROSS, R. A. *A.R.S. Jl* **29**, 573 (1959).
37. EISEN, C. L., GROSS, R. A., and RIVLIN, R. J., *Combust. Flame* **4**, 137 (1960).
38. HAND, C. W. and KISTIAKOWSKY, G. B. *J. chem. Phys.* **37**, 1239 (1962).
39. GLASS, G. P., KISTIAKOWSKY, G. B., MICHAEL, J. V., and NIKI, H. *10th Int. Symp. Combust.*, p. 513. Combustion Institute (1965).
40. POTTER, A. E. and BUTLER, J. N. *ARS Jl* **29**, 54 (1959).
41. POTTER, A. E., HEIMEL, S., and BUTLER, J. N. *8th Int. Symp. Combust.*, p. 1027. Williams and Wilkins (1962).
42. ANAGNOSTOU, E. and POTTER, A. E. *9th Int. Symp. Combust.* p. 1. Academic Press, New York (1963).
43. PANDYA, T. P. and WEINBERG, F. J. *9th Int. Symp. Combust.*, p. 587. Academic Press, New York (1963).
44. PANDYA, T. P. and WEINBERG, F. J. *Proc. R. Soc.* **A279**, 544 (1964).
45. ZELDOVICH, Y. B. *Zh. tekh. Fiz.* **19**, 1199 (1949).
46. SPALDING, D. B. *A.R.S. Jl* **31**, 763 (1961).
47. PORTER, G. *4th Int. Symp. Combust.*, p. 248. Williams and Wilkins, Baltimore (1953).
48. PORTER, G. *Combustion researches and reviews*, p. 108. Butterworths, London (1955).
49. SMITH, E. C. W. *Proc. R. Soc.* **A174**, 110 (1940).
50. TESNER, P. A. *7th Int. Symp. Combust.* p. 546. Butterworths, London (1959).
51. CULLIS, C. F. and PALMER, H. B. The formation of carbon from gases, *Advances in carbon*, Vol. 1, p. 265. Dekker, New York (1966).
52. GORDON, A. S. Combustion and propulsion, *5th AGARD colloquium high-temperature phenomena*, p. 111. Pergamon Press (1963).
53. THOMAS, A. *Combust. Flame* **6**, 46 (1962).
54. CALCOTE, H. F. *Combust. Flame* **1**, 385 (1957).
55. PLACE, E. R. and WEINBERG, F. J. *Proc. R. Soc.* **A289**, 192 (1965).
56. PLACE, E. R. and WEINBERG, F. J. *11th Int. Symp. Combust.*, p. 245. Combustion Institute (1967).
57. PAYNE, K. G. and WEINBERG, F. J. *Proc. R. Soc.* **A250**, 316 (1959).
58. LEWIS, B. and VON ELBE, G. *J. chem. Phys.* **11**, 75 (1943).
59. VON ELBE, G. and MENTZER, M. *J. chem. Phys.* **13**, 89 (1954).
60. WOHL, K., KAPP, N. M., and GAZLEY, C. *3rd Int. Symp. Combust.*, (a) p. 3, (b) p. 288. Williams and Wilkins, Baltimore (1949).
61. WILSON, C. W. *Ind. Engng Chem. ind. (int.)* Edn **44**, 2937 (1952).
62. PHILLIPS, V. D., BROTHERTON, T. D., and ANDERSON, R. C. *4th Int. Symp. Combust.*, p. 701. Williams and Wilkins, Baltimore (1953).
63. WALKER, P. L. and WRIGHT, C. C. *Fuel Lond.* **31**, 45 (1952).
64. CALCOTE, H. F. and PEASE, R. N. *Ind. Engng Chem. ind. (int.)* Edn **43**, 2726 (1951).
65. WEINBERG, F. J. and WILSON, J. R. *Combust. Flame* **10**, 89 (1966).
66. POWLING, J. A. *Fuel, Lond.* **28**, 25 (1949).
67. EGERTON, Sir ALFRED and THABET, S. K. *Proc. R. Soc.* **A211**, 45 (1952).
68. POWLING, J. A. The flat flame burner, *Experimental methods in combustion research* (edited by J. Suruge), sects. 2.2.1, 2.2.2. Pergamon Press (1961).
69. BIEDLER, W. T. and HOELSCHER, H. E. *Jet Propul.* **27**, 1257 (1957).
70. WEINBERG, F. J. Flame processes. *Enciclopaedia del petrolio e dei gas naturali.* Ente Nazionale Idrocarburi (1960).

71. LEVY, A. and WEINBERG, F. J. *7th Int. Symp. Combust.*, p. 296. Butterworths, London (1959).
72. BURKE, E. and FRIEDMAN, R. *J. chem. Phys.* **22**, 824 (1954).
73. EGERTON, Sir A. and POWLING, J. *Proc. R. Soc.* A**193**, 172, 190 (1948).
74. CALCOTE, H. F. *8th Int. Symp. Combust.*, p. 184. Williams and Wilkins, Baltimore (1962).
75. BOTHA, J. P. and SPALDING, D. B. *Proc. R. Soc.* A**225**, 71 (1954).
76. LAWTON, J. and WEINBERG, F. J. *Proc. R. Soc.* A**277**, 468 (1964).
77. MACHE, H. and HEBRA, A. *Öst. Akad. Wiss.* Abt IIa, **150**, 157 (1941).
78. HAHNEMANN, H. and EHRET, L. *Z. tech. Phys.* **27**, 228 (1943).
79. PARKER, T. A. and HEINSOHN, R. J. Proceedings of the 3rd conference on performance of high temperature systems, November 1964, Pasadena, California.
80. LAWTON, J., MAYO, P., and WEINBERG, F. J. *Proc. R. Soc.* A**303**, 275 (1968).
81. LONGWELL, J. P. and WEISS, M. A. *Ind. Engng Chem. ind. (int.)* Edn **47**, 1634 (1955).
82. HOTTEL, H. C., WILLIAMS, G. C., and BAKER, M. L. *6th Int. Symp. Combust.*, p. 398. Reinhold, New York (1956).
83. CLARKE, A. E., HARRISON, A., and ODGERS, J. *7th Int. Symp. Combust.*, p. 664. Butterworths, London (1959).
84. HOTTEL, H. C., WILLIAMS, G. C., NERHEIM, N. M., and SCHNEIDER, G. R., *10th Int. Symp. Combust.*, p. 111. The Combustion Institute (1965).
85. SCHMITZ, R. A. and GROSBOLL, M. P. *Combust. Flame* **9**, 337 (1965).
86. GROSS, R. A. *A.R.S. Jl* **29**, 63 (1959).
87. NICHOLLS, J. A., DABORA, E. K., and GEALER, R. L. *7th Int. Symp. Combust.*, p. 766. Butterworths, London (1959).
88. NICHOLLS, J. A., DABORA, E. K. *8th Int. Symp. Combust.*, p. 644. Williams & Wilkins, Baltimore (1962).
89. RUBINS, P. M. and RHODES, K. P. *AIAA Jl* **1**, 2778 (1963).
90. RUBINS, P. M., RHODES, K. P., and CHRISS, D. E. Arnold Engineering Development Center, *Technical Documentary Report AEDC-TDR*-62-78-(1962).
91. RUBINS, P. M. and CUNNINGHAM, T. H. M. Arnold Engineering Development Center, *Technical Documentary Report AEDC-TDR*-62-78-(1965).
92. RICHMOND, J. K. Boeing Scientific Research Labs, Seattle; Private communication.
93. VOITSEKHOVSKII, B. V. *Sov. Phys. Dokl.* **4** (6) (Trans. from *Dokl. Akad. Nauk. SSSR* **129** (6) Nov.–Dec. (1959) p. 1254.
94. NICHOLLS, J. A. and CULLEN, R. E. The feasibility of a rotating detonation wave rocket motor, *Propulsion Laboratory* (*Mech. Eng.*) *University of Michigan Report No. AD* 313550 (1962) and *AD* 414551 (1963).
95. (a) EDWARDS, B. D. Rotating wave device Mk. I, *Rolls Royce Report RR(OH)* 156.
    (b) PAYNE, K. G. Rotating wave device Mk. II, *Rolls Royce Report RR(OH)* 183.
96. HARRIS, D. and SWITHENBANK, J. Intentional combustion oscillations in propulsion systems, *Univ. of Sheffield, Dept. of Fuel Tech. and Chem. Eng. Report No. HIC* 45 (1963).
97. SOUTHGATE, G. T. *Chem. metall. Engng* **31**, 16 (1924).
98. KARLOVITZ, B. (a) *Pure appl. Chem.* **5**, 557 (1962). (b) *ASME reprint* 61-*WA*-251 (1962).

99. CHEN, D. C. C., LAWTON, J., and WEINBERG, F. J., *10th Int. Symp. Combust.*, p. 743. The Combustion Institute (1965).
100. COCHRAN, W. *Bull. Inst. Phys., Lond.* **17**, 214 (1966).
101. LAWTON, J. *Br. J. appl. Phys.* **18**, 1095 (1967).
102. MARYNOWSKI, C., KARLOVITZ, B., and HIRT, T. *Northern Natural Gas Co., Stanford Research Institute Report* (1965); *Ind. Engng Chem. (Proc. Des. Dev)* **6**, 375 (1967).
103. JOHN, R. R., and BADE, W. L. *ARS Jl* **34**, 4 (1961).
104. ADAMS, V. W., *Royal Aircraft Establishment Technical Note, Aero* 2896 (*ARC* 25085) June (1965).
105. HIRSCHFELDER, J. O. *J. chem. Phys.* **26**, 274, 282 (1951).
106. BUTLER, J. N. and BROKAW, R. S. *J. chem. Phys.* **26**, 1936 (1951).
107. DAVIES, R. M. *10th Int. Symp. Combust.*, p. 755. Combustion Institute (1965).
108. CHEN, D. C. C., LAWTON, J., and WEINBERG, F. J. In preparation.
109. WOLFHARD, H. G. and PARKER, W. G. *Proc. phys. Soc.* A**62**, 722 (1949).
110. HEINSOHN, J. H. and REZY, B. J. *ASME reprint* 65-*WA/Ener*. 3 (1966).
111. PLACE, E. R. and WEINBERG, F. J. *ARC* 25, 560 London (1964).
112. KINBARA, T. and NAKAMURA, J. *Scient. Pap. Coll. gen. Educ. Univ. Tokyo* **4**, 21 (1954).
113. NAKAMURA, J. *Combust. Flame* **3**, 277 (1959).
114. HALL, A. R. and DIEDERICHSEN, J. *4th Int. Symp. Combust.*, p. 837. Williams and Wilkins, Baltimore (1953).
115. BOLT, J. A. and SAAD, M. *6th Int. Symp. Combust.*, p. 717. Reinhold, New York (1957).
116. SPALDING, D. B., *4th Int. Symp. Combust.*, p. 847. Williams and Wilkins, Baltimore (1953).
117. WISE, H., LORELL, J., and WOOD., B. J. *5th. Int. Symp. Combust.*, p. 132. Reinhold, New York (1955).
118. ZELENY, J. *Proc. Camb. Phil. Soc. math. phys. Sci.* **18**, 71 (1915).
119. GRAF, P. E., Proc. A.P.I. Research conference on distillate fuel combustion, *A.P.I. publication 1701* (1962).
120. HOGAN, J. J. and HENDRICKS, C. D., *A.I.A.A. Jl* **3**, 296 (1965).
121. NAYYAR, N. K. and MURTY, G. S., *Proc. phys. Soc.* **75**, 369 (1960).
122. PESKIN, R. L. and LAWLER, J. P., Proc. A.P.I. Research Conference on distillate fuel combustion, *A.P.I. publication 1702* (1963).
123. VONNEGUT, B. and NEUBAUER, R. L. *J. Colloid Sci.* **7**, 616 (1962).
124. DROZIN, V. G. *J. Colloid Sci.* **10**, 158 (1955).
125. WEINBERG, F. J. *Combust. Flame* **10**, 267 (1966).
126. SOTTER, J. G., *Sheff. Univ. Fuel Soc. J.* **14** (1963).
127. MAYO, P. J., WATERMEIER, L. A., and WEINBERG, F. J. *Proc. R. Soc.* A**284**, 488 (1965).
128. MILLS, R. M. *Combust. Flame*, **12**, 513 (1968).
129. FELLS, I., GAWEN, J. C., and HARKER, J. H. *Combust. Flame*, **11**, 309 (1967).
130. FELLS, I. and HARKER, J. H. *Trans. Inst. Chem. Eng.* **46**, 236 (1968) and *Combust. Flame* **13**, 596 (1968).

4. Effects of Fields on Ions, Electrons, Charged Particles, and on Their Carrier Gas

THE purpose of the discussion below is to consider the forces acting on charged matter due to applied electrostatic and magnetic fields, and hence to establish how its motion will be affected by such fields. Both the electrostatic and electromagnetic systems of units (e.s.u. and e.m.u.) are based on such forces, the e.s.u. of charge being defined as that charge which will repel a similar charge with a force of 1 dyne, when situated 1 cm from it, in vacuum. The charge of an electron is $1 \cdot 8029 \times 10^{-10}$ e.s.u. The *intensity of the electrostatic field*, E, at any point in space is then defined as the force, F dynes, which would be exerted on unit positive charge there, if the field were not altered by its introduction. In view of the last qualification, a better definition is

$$E = \frac{\mathrm{d}F}{\mathrm{d}q},\tag{4.1}$$

where q is charge.

The field at a distance of 1 cm from unit charge in vacuum is, of course, unity, the inverse square law giving

$$F = \frac{q_1 q_2}{d^2} = q_2 E_1,\tag{4.2}$$

d being the distance between stations 1 and 2.

This is strictly correct only in vacuum. In any medium the field is modified by the effect it induces in the constituent molecules. The ratio $\left(\dfrac{\text{field in vacuum}}{\text{field in medium}}\right)$ is the *dielectric constant*, D. The field multiplied by the dielectric constant is termed the *electrostatic induction*. Since the field in the medium is (E/D), by the above definitions, the induction may be thought of as the 'field reduced to vacuum'.

These distinctions will not be found very important in the present context, because our interest is confined to gases. Except when dealing with the interaction of a small charge-carrier with an individual molecule during a collision, we shall be able to treat 'field' and 'induction' as synonymous, because of the low density of gases.

When charge is distributed in space, its effect on field distribution is most conveniently examined by means of Gauss's Law. In its general form, this states that the flux of induction across any closed surface, i.e. the integral over the whole surface area of the component of induction everywhere perpendicular to it, is equal to 4π times the sum of the charges contained within it. For the reasons stated, we shall generally be able to use the law in its 'vacuum' form

$$\int\int_A E_p \, dA = 4\pi \sum q, \tag{4.3}$$

where E_p is the component of *field* perpendicular to the area element dA and the summation on the right-hand side includes all space within the surface. To choose an example that will be useful in subsequent chapters, consider a system, unidimensional in x, which contains locally n_+ positive and n_- negative electronic charges, e, per unit volume. Applying eqn (4.3) to an element of area A and thickness δx,

$$\left(E + \frac{dE}{dx}\delta x\right)A - EA = 4\pi(n_+ - n_-)eA\delta x$$

or

$$\frac{dE}{dx} = 4\pi(n_+ - n_-)e. \tag{4.3a}$$

Since a field exerts a force on charged matter, energy must be supplied (or released) when such matter is moved in a field and each change in position corresponds to a change in its potential energy. The change in electrostatic *potential* between two positions is defined as the work done on a unit positive charge in transferring it from one station to the other. The absolute potential at a point is the work done in bringing unit charge to the point from infinity. It follows that field strength is, numerically, the differential with respect to distance of potential, V. If the potential difference between two points separated by an element of distance, δx, is δV, then the transfer of a particle of charge q from one point to another involves an amount of work

$$q \cdot \delta V = -qE\delta x,$$

i.e.

$$E = -\frac{dV}{dx}. \tag{4.4}$$

Field, being defined in terms of force, has direction as well as magnitude.

It is a vector, unlike potential which is a scalar quantity. The components of field are

$$E_x = -\frac{\partial V}{\partial x}, \qquad E_y = -\frac{\partial V}{\partial y}, \qquad E_z = -\frac{\partial V}{\partial z}. \tag{4.4a}$$

In vector notation,

$$E = -\operatorname{grad} V. \tag{4.4b}$$

The resultant field intensity of all the components in eqn (4.4a) thus lies in the direction along which the potential gradient is greatest, i.e. in a direction perpendicular to surfaces of constant potential. Lines of field are therefore always orthogonal to iso-potential lines.

We may now turn to the first topic of this chapter—the velocity attained by charge-carriers in fields. In vacuum there is nothing to offer resistance to the movement of such a particle. It is therefore free to accelerate under the influence of the electrostatic force, converting all its potential energy into kinetic. If its charge and mass be q and m, respectively, and it starts from rest, then movement down a potential difference ΔV induces velocity v given by

$$q\Delta V = \tfrac{1}{2}mv^2$$

or

$$v = \sqrt{\left(\frac{2q\Delta V}{m}\right)}. \tag{4.5}$$

Our main concern, however, is with charge-carriers in a largely uncharged gas. In such a case the particle is in the 'vacuum' of intermolecular space only during its free path between collisions (provided it is itself small compared with that path). For this distance it accelerates and draws energy from the field. In the steady state the additional energy is given up at the subsequent collision. For a larger particle, the mechanism is different, in that it 'sees molecular collisions as a viscosity of the gas'. Thus, in the Stokes regime, it experiences a retarding force (say $\beta_0 v$) proportional to its velocity. Its equilibrium drift velocity is then given by

$$qE = \beta_0 v. \tag{4.6}$$

In either case, in the steady state, all the energy is given up to the neutral gas and the particle acquires a constant velocity in a constant field. In eqn (4.6), this drift velocity is proportional to the field. It will be shown below that this is true also for molecular and atomic ions within wide limits of field strength. When this applies, the particle is said to have a constant *mobility*, K. Mobility is defined by

$$v = KE, \tag{4.7}$$

the velocity being taken as positive in the direction of the force (i.e. it is not usual to distinguish mobilities of opposite charges by signs).

A constant mobility is seen to imply a resistance to movement proportional to velocity, resulting in a drift velocity proportional to an applied force. This consequence is independent of whether or not the applied force is electrostatic in origin. It will therefore be convenient to consider the effect of an electromagnetic interaction before turning to a more detailed analysis of the resistance to movement of charge-carriers in gases.

We shall not have to revise the principles of magnetostatics, which are quite analogous to those of electrostatics, beyond noting that magnetic field, H, is defined as the force in dynes acting on a unit magnetic pole. The latter acts with a force of 1 dyne on a similar pole 1 cm distant in vacuum, and the inverse square law again applies. The difference between vacuum and gas is slight, as is that between magnetic induction and field flux. We are, however, more concerned with the magnetic field due to a current, i. This field, considered at a point a distance r away, is proportional to (i/r^2) and to the component of the length of the current's path $(\delta L \sin \theta)$ perpendicular to the direction of r. The direction of the field is perpendicular to both this component and the direction of r, and its magnitude is the basis of the electromagnetic definition of current. The current is defined to be 1 e.m.u. when the field is unity and so are r and $(\delta L \sin \theta)$. Thus,

$$\frac{i\delta L \sin \theta}{r^2} = H. \tag{4.8}$$

It will be more convenient to think in terms of a flow of current resolved along the direction of r—component $i_{\parallel}$—and at right angles thereto $-i_{\perp}$. The force acting on a magnetic pole of n units due to element δL, from which it is a distance r away, is now seen to be

$$F = \frac{i_{\perp}\delta L n}{r^2}, \tag{4.9}$$

the other component, $i_{\parallel}$, producing no effect. The force acts in a direction perpendicular to the plane containing $i_{\perp}$ and r. Now, an equal and opposite reaction must be acting on the element of current and this may be thought of as the effect of the magnetic field there due to the pole at r. This field being (n/r^2), the force per unit field is $(i_{\perp}\delta L)$. In general then, a magnetic field H acting on length δL of current i exerts a force

$$F = i_{\perp}H\delta L, \tag{4.10}$$

where $i_{\perp}$ is the component of current perpendicular to H and the direction of F is perpendicular to the plane containing both H and $i_{\perp}$.

Now, current is the rate of flow of charge, dq/dt. A single particle of charge q travelling a distance δL in time δt thus constitutes a current

$$q/\delta t = qv/\delta L$$

since its velocity

$$v = \frac{dL}{dt}.$$

It follows that a magnetic field will exert a force on a moving, charge-carrier, provided the particle velocity has a component, $v_\perp$, perpendicular to the direction of the field. Substituting above, this force is

$$F = Hqv_\perp \qquad (4.11)$$

and acts at right angles to both H and $v_\perp$.

This brings us to the point where we can treat the consequences of electrostatically and electromagnetically induced forces on charge-carriers in the same way. Two important distinctions must, however, be borne in mind. The first concerns units. Equation (4.11) is correct in e.m.u.s, as follows from its derivation, and the q it contains therefore differs numerically from those in equations preceding eqn (4.7). A conversion table (Table 4.1) is given below which is easy to memorize and makes possible the interconversion of all electrical parameters (bearing in mind that units of, for example, charge differ from those of current only by those of time, which remains unaltered, those of field and potential only by those of distance, and so forth).

TABLE 4.1

	e.s.u.	e.m.u.	Practical
current	c	1	10
potential	1	c	300

Here c is the speed of light in cm/s, which we may regard, in this context, purely as a conversion factor.

The second point is that the magnetic field will act only if the particle has a velocity component perpendicular to it and then the resultant force will be at right angles to this velocity. If the velocity induced by the electromagnetic force is at all appreciable, it must, at the next stage, be added vectorially to the original velocity, thereby altering in turn the direction and magnitude of the force. Consequently a quite complex trajectory can result even in the case of a region of constant field which, in the equivalent electrostatic case, would lead to a constant particle velocity and direction.

Bearing these points in mind, we may now consider the velocity (KE) induced in a charge-carrier by an electrostatic force Eq; the velocity per unit force is therefore (K/q). The force due to a magnetic field H, by eqn (4.11), is $Hqv_\perp/c$, where q is in e.s.u. The consequent induced velocity component is therefore $KHv_\perp/c$, at right angles to $v_\perp$ and H. The 'magnetic mobility' by analogy with eqn (4.7) is thus

$$K_m = Kv_\perp/c. \tag{4.12}$$

In general, a $v_\perp$ can arise due to three quite distinct causes: gas flow, the drift in an electrostatic field, and, if the charge-carriers are ions or electrons, the velocities of thermal motion. Velocities that can be induced by magnetic fields due to the first cause are usually small, unless a very fast gas flow is produced, for example, by a rocket nozzle. They are used in MHD generation of electricity to produce a relatively slow drift of charge (see Chapter 7) but the linear velocities of gaseous flow from normal burners, at the magnetic fields that can be sustained in gases, generally produce negligible drifts compared to those that can be induced by readily attainable electrostatic fields. Thermal motion gives rise to appreciable velocities which are, however, statistically random and therefore induce no net effect on the gas as a whole (though they do affect transport properties and breakdown fields, as discussed later in this chapter). Only in the presence of electric fields will magnetic fields appreciably affect ion trajectories, at normal flow rates, and we shall indeed encounter crossed electric and magnetic fields in practical use.

Dipoles

In regions of varying field intensity, a force acts also on polarizable molecules or particles, even if no net charge resides on them. Consider the case of field intensity E inducing opposite charges of magnitude q, a distance δx apart on a particle. The net force acting on such a particle is

$$F = q\left(E + \frac{\partial E}{\partial x}\delta x\right) - qE = \frac{\partial E}{\partial x}q\delta x. \tag{4.13}$$

The term ($q\delta x$) is the dipole moment. If the particle is a molecule, it is possible for it to have a permanent dipole moment. This occurs when the electrical centroid of the electron cloud does not coincide with that of the positive nuclei of the molecule, so that a dipole exists even in the absence of an applied field. Thus molecules such as CH_4 have none, whilst, for example, for CO and H_2O in the ground state the numerical values are of the order 0.12 and 1.8×10^{-18} e.s.u. respectively—these quantities depend on the state of excitation of the molecule. All this is, of course, in addition to polarization by an applied field. To a first approximation, the induced dipole moment is proportional to the field strength

$$q\delta x = \gamma E, \tag{4.14}$$

where γ is called the *polarizability*. This quantity is obviously related to the dielectric constant, D, since polarization determines the intensity within a dielectric when a field is applied. Thus in a gas consisting of polarizable molecules of mean polarizability $\bar{\gamma}$,

$$D = 1 + 4\pi n\bar{\gamma}, \tag{4.15}$$

where n is the number of molecules per unit volume. It is worth noting, in the case of molecules, that γ has the dimensions of volume and that it is in fact of the same order of magnitude as the molecular volume.

The force per unit field on such a polarizable particle, and hence its mobility, follows as before. An important distinction is that here F is proportional to $(\partial E/\partial x)$, as well as to E, since by eqn (4.13)

$$F = \gamma E \frac{\partial E}{\partial x}. \tag{4.16}$$

This can become important where strongly divergent fields occur in consequence of the geometry of the system (for example, near pointed electrodes). It follows from eqn (4.16) that, in general, dipoles will tend to concentrate at points of maximum field.

There is another force acting on neutral molecules and particles—that due to bombardment by ions—detailed in the second part of this chapter. It is a curious circumstance that in one-dimensional systems, the force on dipoles enters only when that due to ion bombardment is important and, usually, dominant. This comes about because by Gauss's law (eqn (4.3a)) dE/dx differs from zero only in the presence of either a unipolar cloud or, at least, an excess of one kind of charge; it is just then that the net mechanical force acting on neutrals due to collisions with ions drifting rapidly under the influence of the applied field becomes important.

MOBILITY OF CHARGE CARRIERS

Molecular ions and electrons exchange energy with neutral gas by individual encounters. However, in many cases of practical interest, for example, charged sprays or dust clouds, charged agglomerates of matter are involved which are of such a size that energy exchange is best treated in terms of the continuum properties of the fluid, viz. density and viscosity. In what follows, the term 'particles' will be reserved for such agglomerates.

The collision processes for particles, ions and electrons with neutral species are quite different. It is therefore convenient to treat the three types of charge carrier separately.

Molecular ions

If an ion of mass M_i and charge e moves through a gas under the action of an electric field E, it will accelerate between collisions at a rate of Ee/M_i, for a time τ. Thus, the average ion, which starts from rest after each collision, advances a distance $\frac{1}{2}(Ee/M_i)\tau^2$ in the direction of the field, between collisions. In other words, its mean drift velocity in the field direction is given by

$$v = \frac{e\tau}{2M_i}E. \tag{4.17}$$

The mobility deduced from this treatment—which ignores distributions of velocity and mean free paths and the persistence of velocity after collision—follows immediately from eqn (4.17)

$$K = \frac{v}{E} = \frac{e}{2M_i\bar{v}_i} = \frac{0 \cdot 75e}{nM_iQ\bar{c}_i} \tag{4.18}$$

where $\bar{v}_i$ is the ion collision frequency and Q the collision cross-section.

Langevin[1] carried out an exact analysis for a constant collision cross-section, for fields not large enough to interfere significantly with the thermal equilibrium between the gas and the ions. He found that

$$K = \frac{0 \cdot 75e}{nM_iQ\bar{c}_i}\left(1+\frac{M_i}{M}\right)^{0 \cdot 5} = \frac{0 \cdot 75e}{M_i\bar{v}_i}\left(1+\frac{M_i}{M}\right)^{0 \cdot 5} \tag{4.19}$$

for molecules of mass M in concentration n. The last equality follows from eqn (2.1).

When known values of collision cross-sections are used in eqn (4.19), the calculated values of mobility are too high by factors up to 5. The reason for this discrepancy is that no account was taken of the long-range dipole interaction between ions and molecules. Accordingly Langevin extended the theory to take into account energy exchange by both long range dipole forces and 'geometrical' collisions. The equation for mobility in this case is

$$K = \frac{A}{\sqrt{\{\rho(D-1)\}}}\left(1+\frac{M}{M_i}\right)^{0 \cdot 5}. \tag{4.20}$$

D is the dielectric constant and A depends upon the relative magnitudes of the effects of hard-sphere collisions and action at a distance. A is a function of the parameter Λ, where

$$\Lambda^2 = \frac{8PQ^2}{\pi(D-1)e^2}. \tag{4.21}$$

Since $(D-1)$ is proportional to density (see eqn (4.15)) A is independent of pressure, P, at constant temperature and, by eqn (4.20),

$$KP = \text{constant.} \tag{4.22}$$

When Λ is large, hard-sphere collisions dominate, and the product $A\Lambda$ tends to a limiting value of 0·75. On the other hand, for small values of Λ, polarization forces dominate and A tends to a limiting value of 0·5105—this is usually called the small ion limit. The appropriate expressions are:

$$\text{hard-sphere limit,} \quad K = \frac{0 \cdot 75e}{Q} \left(\frac{\pi}{8P\rho} \right)^{0 \cdot 5} \left(1 + \frac{M}{M_i} \right)^{0 \cdot 5}, \tag{4.23}$$

$$\text{small-ion limit,} \quad K = \frac{0 \cdot 5105}{\sqrt{\{\rho(D-1)\}}} \left(1 + \frac{M}{M_i} \right)^{0 \cdot 5}. \tag{4.24}$$

Equation (4.23) is equivalent to eqn (4.19). Although both give numerically inexact values when the geometrical collision cross-section is inserted, they are useful for calculating mobility when measured average collision cross-sections, which include the effect of polarization, are available.

The dependence of A upon Λ is shown in Table 4.2. In practice it is found that ionic mobilities follow the small ion equation quite well. Exceptions occur for a few gases with exceptionally low dielectric constants, such as helium.[2,3] The success of the small ion theory is rather fortunate because eqn (4.24) contains only macroscopic quantities, and it is not necessary to know collision cross-sections explicitly. Figure 4.1 shows the excellent agreement between the predictions of eqn (4.24) and experiment, for ions of various atomic weights in nitrogen.[2] Two important consequences follow from this equation. First, where $M_i \geq M$, the mobility for ions is very insensitive to the

TABLE 4.2

AΛ and A vs. Λ for Langevin mobility equation

Λ	A	$A\Lambda$
0·0	0·5105	0·0000
0·5	0·5886	0·2943
1·0	0·5483	0·5483
1·5	0·4402	0·6603
2·0	0·3514	0·7028
2·5	0·2886	0·7215
3·0	0·2436	0·7308
3·5	0·2104	0·7364
4·0	0·1849	0·7396

ionic mass, varying in the ratio $\sqrt{2}:1$ over the range $M \leq M_i < \infty$—this is clearly shown in Fig. 4.1 (molecular weight of N_2 is 28). Secondly, since $(D\text{-}1)$ is proportional to ρ,

$$K\rho = \text{constant, independent of temperature.} \qquad (4.25)$$

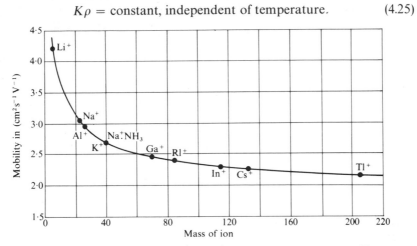

FIG. 4.1. Mobility of ions of different masses in N_2 at n.t.p. From Brata.[2]

Experiment has shown that the effect of temperature upon $K\rho$ is quite small, in agreement with eqn (4.25). Thus, for Li^+, Na^+, He^+, and Cs^+ in helium, the changes in $K\rho$ when the temperature is increased from 100 to 500°K (i.e. by a factor of 5) are $+30$, $+10$, $+30$, and -6 per cent respectively.[4] Still closer agreement is obtained, of course, if experiment is compared with the predictions of the full Langevin equation.[5]

In practice, when using any theory to estimate mobilities, it is necessary to ascertain whether clustering and charge transfer are likely to play an important role. These have already been discussed in Chapter 2. The effect of clustering is to reduce the mobility by increasing the cross-sectional area of the ion and is noticeable at low temperatures for small ions in polar gases (e.g. Li^+ in H_2O vapour) particularly in the absence of large electric fields.[6,7] On the other hand, charge transfer is important for ions in their parent gas (see Fig. 2.6). In this last process, a fast ion gives up its charge to a slower moving neutral, thus reducing the mobility.[8,9] These effects are often the cause of differences between experimental and calculated values.

So far the discussion has been limited to the case of moderate fields in which a linear dependence of drift velocity upon applied field is both predicted theoretically and found experimentally. When the fields are high, however, the random energy of the ions can be appreciably in excess of that of the gas. Under these conditions the Langevin treatment is invalid, because it assumes equal temperatures for ions and neutral gas molecules. An outline of the approach to the theory of mobility in strong fields is given below.

When an ion is accelerated in an electric field it gains kinetic energy, some of which is lost on collision, the rest being randomized. In the steady state, on average, an ion gives up on each collision the energy it gained in the previous mean free path. Thus if an ion loses a fraction δ of its total kinetic energy, U_i, on collision, the balance of energy gains and losses per unit time is

$$U_i \delta \frac{\bar{c}_i}{\lambda_i} = \tfrac{1}{2}\delta M_i \overline{c_i^2} \frac{\bar{c}_i}{\lambda_i} = Eev \qquad (4.26)$$

i.e. (energy loss per collision) × (collision frequency) = (work done by E on ion per unit time). Now

$$v = KE \propto \frac{\lambda_i}{\bar{c}_i} E, \qquad (4.27)$$

see, for example, eqn (4.19). $\bar{c}_i$ is a function of E and can be eliminated between eqns (4.26) and (4.27). Finally,

$$v \propto \delta^{\frac{1}{2}}(Ee\lambda_i)^{\frac{1}{2}}. \qquad (4.28)$$

The expression for δ in the case of solid elastic collisions between spheres, with mean kinetic energies U and U_i, is

$$\delta = \beta \frac{MM_i}{(M+M_i)^2}\left(1 - \frac{U}{U_i}\right) \qquad (4.29)$$

$$\simeq \beta \frac{MM_i}{(M+M_i)^2}, \quad \text{if } U_i \gg U. \qquad (4.30)$$

β is a constant which depends upon the velocity distribution; for a Maxwellian distribution it is equal to $8/3$.[10] Thus for high fields and solid elastic collisions, from eqns (4.30) and (4.28)

$$v \propto (E/P)^{\frac{1}{2}}, \qquad (4.31)$$

i.e.

$$K \propto (1/E)^{\frac{1}{2}}. \qquad (4.32)$$

This form of relationship was found experimentally by Munson and Tyndall,[11] and has been confirmed by numerous workers.

The calculation of numerical values for K in strong fields is difficult for two reasons. On the one hand, molecules are not hard spheres and therefore δ is uncertain. On the other, the constant of proportionality in eqn (4.27) is not known, although an approximation can be obtained by substituting $\bar{v}_i = \bar{c}_i/\lambda_i$ into eqn (4.19) and using the value of mobility thus obtained in eqn (4.27). A further discussion of this problem can be found in ref. (4).

In the special case of ions moving in their parent gas, when charge transfer predominates, Wannier[12] has shown that for high values of E/P

$$K = 1\cdot147(e\lambda_i/EM_i)^{\frac{1}{2}}. \qquad (4.33)$$

Surveys of recent data on ionic mobilities are given in refs. (4) and (9).

Mobility in mixtures of gases

From Langevin's general equation for mobility it is seen that $1/K$ is proportional to the gas density at constant temperature, i.e. each molecule individually contributes to the net value of $1/K$. Thus, in mixtures of gases,

$$\frac{1}{K} = \sum_j \frac{f_j}{K_j}, \qquad (4.34)$$

where f_j is the fraction of species j by volume and K_j the mobility of the ion in species j, when at the temperature and total pressure of the mixture. This result was first established experimentally by Blanc,[13] and is often called Blanc's Law. It is reliable so long as the ion is the same in the pure gases as it is in the mixture. Therefore if the mixture contains a constituent which causes clustering, for example water vapour, the experimental or theoretical values of K_j must be those for the clustered ion. Data for various mixtures are given by David and Munson.[14]

Before leaving the subject of ionic mobilities it is convenient to discuss ion diffusion.

Diffusion of ions

In a region of unipolar space charge, the current density is given by the sum of the diffusion- and field-induced currents, i.e.

$$j = Kn_i Ee - D_i \frac{dn_i}{dx} e, \qquad (4.35)$$

where D_i is the ion diffusion coefficient. The concentration gradient of the ions is directly proportional to their partial pressure gradient, i.e.

$$\frac{dn_i}{dx} = \frac{1}{kT} \frac{dP_i}{dx}. \qquad (4.36)$$

The effect of any pressure gradient can be represented as a net force acting on the ions. This force is equal to $-(1/n_i)(dP_i/dx)$ per ion. Moreover, from the definition of mobility (eqn (4.7))

$$v = K \text{ times (force on ion)}/e = K\left(Ee - \frac{1}{n_i}\frac{dP_i}{dx}\right)\frac{1}{e}. \qquad (4.37)$$

Therefore, the current density is given by

$$j = vn_i e = Kn_i Ee - K\frac{dP_i}{dx} = Kn_i Ee - KkT\frac{dn_i}{dx}. \qquad (4.38)$$

Comparing eqns (4.35) and (4.38) it is seen that

$$\frac{K}{D_i} = \frac{e}{kT}.$$ (4.39)

This relation is important in many aspects of ion flow. It follows directly from eqns (4.23) and (4.24) that, at constant pressure,

$$D_i \propto T^2, \text{ hard sphere,}$$ (4.40)

$$D_i \propto T^{\frac{3}{2}}, \text{ small ion.}$$ (4.41)

In general positive and negative ions have different diffusion coefficients, the extreme disparity occurring when the negative charge is carried by free electrons. Thus, even in the absence of applied electric fields, an originally neutral plasma will tend to lose its neutrality because of the different rates of diffusion of positive and negative charges. Charge separation of this sort does not occur unchecked, however, but is opposed by increasing electric fields. Eventually a stage is reached at which the faster charges are so retarded by the electric field set up by the charge separation that both signs of charge diffuse at the same rate. This process is known as ambipolar diffusion. Because of the very large fields set up by even quite small amounts of charge separation, it is usual in treating such diffusion problems to assume $n_+ = n_- = n$. Thus, since $j_+ = -j_-$, algebraically

$$-D_+\frac{dn}{dx} + K_+ nE = D_-\frac{dn}{dx} + K_- nE,$$ (4.42)

i.e.

$$nE = \left(\frac{D_+ - D_-}{K_+ + K_-}\right)\frac{dn}{dx}.$$ (4.43)

Substituting in the equation for current density,

$$j_+ = -j_- = -D_a\frac{dn}{dx}$$ (4.44)

where

$$D_a = \frac{K_+ D_- + K_- D_+}{K_+ + K_-}.$$ (4.45)

D_a is known as the ambipolar diffusion coefficient. For positive and negative ions $K_- \approx K_+$, i.e. $D_a \approx D_- \approx D_+$. On the other hand, if the negative charge is carried by an electron, $K_- \gg K_+$, and it follows that $D_a \approx 2D_+$. The ambipolar diffusion coefficient is useful in calculations involving the loss of ions from a plasma in the absence of external fields.

Electrons

Electrons are treated separately from ions because their collision processes are very much more complex, the theory of Langevin being inadequate to describe their behaviour in electric fields. When data is available on the variation of collision cross-section with electron energy, an accurate value of electron mobility can be computed by integration over the velocity distribution.[15]

$$K_e = \frac{4\pi e}{3M_e} \int_0^\infty \frac{c^3}{v(c)} \cdot \frac{df(c)}{dc} \cdot dc, \tag{4.46}$$

where $v(c)$ is the velocity-dependent collision frequency given by

$$v(c) = \frac{c}{\lambda} = \frac{c}{nQ(c)} \tag{4.47}$$

and $f(c)$ is the normalized electron velocity distribution. For a Maxwellian distribution

$$f(c) = \left(\frac{M_e}{2\pi k T_e}\right)^{\frac{3}{2}} \exp\left(-\frac{M_e c^2}{2k T_e}\right) \tag{4.48}$$

Pack and Phelps[16] have obtained an exact integration for a Maxwellian distribution whenever the collision frequency can be written in the following form

$$\frac{n}{v(U)} = \sum_{-b}^{m} a_i U^i, \qquad b < \tfrac{5}{2}. \tag{4.49}$$

The expression they obtained is

$$K_e = \frac{e}{nM_e} \sum_{-b}^{m} a_i \frac{\Gamma(\tfrac{5}{2}+i)}{\Gamma(\tfrac{5}{2})} \left(\frac{kT_e}{e}\right)^i. \tag{4.50}$$

For the particular case of a constant collision frequency and a Maxwellian distribution, the integration is straightforward, giving,

$$K_e = \frac{e}{M_e v}. \tag{4.51}$$

Frost[17] has made a compilation of electron collision frequency data for a number of species found in combustion systems over the energy range 0·05 to 1 eV. He expressed his data in terms of $v(U)/n$, the collision frequency per molecule, as a function of electron energy in volts, as shown in Table 4.3, and deduced the following simple approximate expression for the effective collision frequency between electrons and ions,

$$\frac{v(U)}{n_+} = 0·476B\frac{kT_e}{eU} \propto \frac{1}{U}, \tag{4.52}$$

where

$$B = \frac{7 \cdot 75 \times 10^{-6}}{(kT_e/e)^{\frac{3}{2}}} \ln \left\{ \frac{1 \cdot 55 \times 10^{7}}{n_e^{\frac{1}{2}}} \left(\frac{kT_e}{e} \right)^{\frac{3}{2}} \right\}. \tag{4.53}$$

TABLE 4.3

Electron collision frequency as a function of electron energy. From Frost.[17]

Gas	$10^8 \, \nu(U)/n, \, \text{s}^{-1} \, \text{cm}^{-3}$
H_2O	$10U^{-\frac{1}{2}}$
CO_2	$1 \cdot 7U^{-\frac{1}{2}} + 2 \cdot 1U^{\frac{1}{2}}$
CO	$9 \cdot 1U$
O_2	$2 \cdot 75U^{\frac{1}{2}}$
O	$5 \cdot 5U^{\frac{1}{2}}$
H_2	$4 \cdot 5U^{\frac{1}{2}} + 6 \cdot 2U$
H	$42U^{\frac{1}{2}} - 14U$
OH	$8 \cdot 1U^{-\frac{1}{2}}$
N_2	$12U$
He	$3 \cdot 14U^{\frac{1}{2}}$
Ne	$1 \cdot 15U$
K, Cs	160

A survey of modern determinations of electron drift velocities in various gases over wide ranges of applied fields can be found in reference.[9]

Sodha, Palumbo, and Daley[18] have considered collisions between electrons and charged macroscopic particles such as would occur, for example, in dust suspensions used to seed MHD generators (see Chapter 7). For n_p positively charged particles per cm^3, with each particle of radius r_p carrying N charges,

$$\frac{\nu(c)}{n_p} = c \left\{ \pi b_1^2 + \frac{4\pi N^2 e^4}{M_e c^4} \ln \left(\frac{\sin \chi_1/2}{\sin \chi_2/2} \right) \right\}, \tag{4.54a}$$

where

$$\chi_i/2 = \cot^{-1} \left(\frac{M_e c^2}{N e^2} \cdot b_i \right),$$

$$b_1 = r_p \Big/ \sqrt{\left(1 + \frac{2N e^2}{r_p M_e c^2} \right)},$$

$$b_2 = 7 \cdot 45 \times 10^2 \Big/ \sqrt{\left(\frac{kT_e}{n_e e} \right)}.$$

(The case of negatively charged particles has also been treated. The resulting expressions are even more lengthy and the reader is referred to the original paper.) The average collision frequency is given by[19]

$$\frac{\bar{v}}{n_p} = \pi r_p^2 \left(\frac{3kT_e}{M_e}\right)^{\frac{1}{2}} \left\{ \left(\frac{8}{3\pi}\right)^{\frac{1}{2}} (1 + \alpha N) + \frac{4N^2\alpha^2}{9} \ln\left(\frac{\sin \chi_1/2}{\sin \chi_2/2}\right) \right\}, \quad (4.54b)$$

where

$$\alpha = e^2/r_p kT_e,$$

$$\cot(\chi_1/2) = \frac{3}{N\alpha}(1 + \tfrac{2}{3}N\alpha)^{\frac{1}{2}},$$

$$\cot(\chi_2/2) = \left(\frac{3}{fN^3\alpha^3}\right)^{\frac{1}{2}},$$

and f is the fraction of the total volume occupied by particles.

Of rather more practical importance than the electron mobility is the electron conductivity, σ_e, in an electrically neutral ($n_+ = n_-$) ionized gas. It is defined as the electron current density divided by the local electric field,

$$\sigma_e = \frac{j_e}{E} = K_e n_e e. \quad (4.55)$$

Thus in order to calculate σ_e it is necessary to know the electron concentration. In an equilibrium system, i.e. in the absence of chemi-ionization and applied electric fields so large as to put the electrons significantly out of equilibrium with the neutral gas, this can be found from the Saha equation, discussed in Chapter 2. The case of strong electric fields is dealt with later in this section. Positive ions have mobilities at least two orders of magnitude less than electrons and therefore contribute negligibly to the total conductivity of a neutral plasma which for all practical purposes can, therefore, be taken as simply σ_e.

Electron mobility and conductivity in gas mixtures

In a mixture consisting of electrons, neutral molecules, ions, and particles, the total collision frequency, $v_t(U)$, for electrons of energy U is the sum of the collision frequencies with the individual constituents, thus,

$$v_t(U) = \sum \left\{\frac{v(U)}{n}\right\}_j n_j = n\sum \left\{\frac{v(U)}{n}\right\}_j f_j, \quad (4.56)$$

where f_j is the number fraction of species j. Frost,[17] substituting the values of $v(U)/n$ given in Table 4.3 into eqn (4.50), has calculated electron mobilities for a number of mixtures of caesium and potassium with inert gases and combustion products.

When just an estimate of electron mobility is required, or when only average collision frequencies are available, eqn (4.51) can be used: thus

$$K_e \approx \frac{e}{M_e \bar{v}_t} = \frac{e}{n M_e \sum_j (\bar{v}/n)_j f_j}. \tag{4.57}$$

It is easy to show that this equation is equivalent to Blanc's Law.

In order to obtain a sufficiently high conductivity, for example for magneto-hydrodynamic power generation (Chapter 7), it is often necessary to seed combustion products with easily ionized materials, such as the alkali metals. It might seem at first sight that the conductivity would increase continuously with increasing amounts of seed at a constant temperature because of the increasing electron concentration. However, because of the large electron collision cross-sections of the alkali metals, seeding beyond an optimum value leads to a decrease in conductivity.[20–22] It is useful, by way of illustration, to consider the case in which the seed metal is present only as free atoms and ions, and there is no electron attachment, as for example in the case of potassium in argon. In terms of average collision cross-sections, the electron collision frequency is, from eqn (2.1),

$$\bar{v} = \bar{c} n (f_N Q_N + f_K Q_K + f_+ Q_+). \tag{4.58}$$

The subscripts N, K, and + refer respectively to argon, potassium, and potassium ions. In the region of practical interest, the fraction of the metal ionized is much less than 0.1 and ψ, the ratio of moles of seed, n_s, to moles of carrier gas, n_N, is less than 0.05. It follows from Chapter 2, that

$$n_+ = n_e = x n_s = x \psi n_N = n_N \psi^{\frac{1}{2}} \sqrt{\left\{ \frac{F(T)}{P} \frac{\omega_+}{\omega_N} \right\}}, \tag{4.59}$$

$$n_s = \psi n_N, \tag{4.60}$$

and, since $f_N \simeq 1$, $n_N \simeq n$.

Substitution in eqn (4.57) yields

$$\sigma = \frac{e^2}{\bar{c} M_e} \psi^{\frac{1}{2}} \sqrt{\left\{ \frac{F(T)}{P} \frac{\omega_+}{\omega_N} \right\}} \Big/ \left[Q_N + \psi Q_K + Q_+ \psi^{\frac{1}{2}} \sqrt{\left\{ \frac{F(T)}{P} \frac{\omega_+}{\omega_N} \right\}} \right]. \tag{4.61}$$

On differentiation with respect to ψ at constant temperature and pressure, it is found that the maximum value of σ occurs when $\psi = \psi_{max} = Q_N/Q_K$, i.e. when $n Q_N = n \psi Q_K$, in other words when the total collision cross-section for the carrier gas is equal to that of the seed material. The contribution of electron-ion collision is often small, and if it can be ignored,

$$\sigma/\sigma_{max} = (\psi/\psi_{max})^{\frac{1}{2}}/(1 + \psi/\psi_{max}). \tag{4.62}$$

This expression is independent of temperature to the extent that the ratio

Q_N/Q_K is Figure 4.2 is the comparison with experiment over a range of temperatures.[22] The equation was normalized at the experimentally determined value of $\psi_{max} = \frac{1}{760}$.

When appreciable salt formation and electron attachment occur, they greatly influence the conductivity—this has been mentioned in Chapter 2; a further discussion is given in Chapter 6.

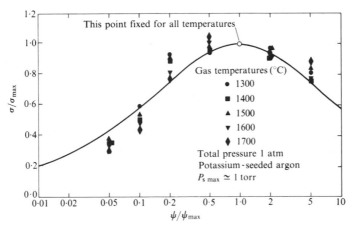

FIG. 4.2. σ/σ_{max} vs. ψ/ψ_{max} for potassium seeding in argon.[22]

Electron temperature, mobility, and conductivity at high (E/P)

In the presence of electric fields, the electron temperature is raised above that of the neutral gas. Although the qualitative description is the same as that for ions, the effect of electric fields upon electrons is far more marked, owing to their small mass. Since $M_e \ll M$, eqn (4.29) reduces to

$$\delta = \beta\frac{M_e}{M}\left(1 - \frac{U}{U_e}\right), \tag{4.63}$$

where $\beta = \frac{8}{3}$ for elastic collisions between spheres with a Maxwellian velocity distribution. Thus the value of δ for elastic collisions between electrons and molecules is of the order of 10^{-4}; this is to be compared with values of around unity for ions.

In molecular, as opposed to atomic, gases the values of δ for electrons is invariably greater than that calculated from setting $\beta = \frac{8}{3}$ in eqn (4.63). This is because of the ease with which inelastic collisions can occur with molecular gases, even at low energy. Massey and Burhop[5] have tabulated data for air, $H_2, N_2, O_2, CO, NO, CO_2, N_2O$, HCl, and NH_3 for electron energies in the range 0·1 to 6 eV. Over this range δ lies between 6×10^{-4} and 0·15.

Carrying out an energy balance between rates of gain and loss, as before,

$$Eev = \beta \frac{M_e}{M}(U_e - U)\bar{v}, \tag{4.64}$$

where β is an experimentally determined constant. Also

$$v = K_e E = \frac{eE}{M_e \bar{v}}. \tag{4.65}$$

Combining these two equations and using the relationship $U = \frac{3}{2}kT$,

$$T_e - T = \frac{2E^2 e^2 M}{3\beta k \bar{v}^2 M_e^2}. \tag{4.66}$$

This equation can be used to derive an expression for electron drift velocity in high fields. When $U_e \gg U$, i.e., $T_e \gg T$

$$\tfrac{1}{2}M_e \overline{c_e^2} = \tfrac{3}{2}kT_e \approx \left(\frac{Ee}{\bar{v}M_e}\right)^2 \frac{M}{\beta}, \tag{4.67}$$

furthermore, $\bar{v} = \bar{c}_e/\lambda_e$ and $\bar{c}_e = 0{\cdot}92\sqrt{(\overline{c_e^2})}$; therefore by elimination of $\bar{c}_e$ between eqns (4.66) and (4.68)

$$v \propto \beta^{\frac{1}{4}}(Ee\lambda_e)^{\frac{1}{2}} \propto \beta^{\frac{1}{4}}(E/P)^{\frac{1}{2}}. \tag{4.68}$$

The mobility, v/E, therefore decreases as $1/E^{\frac{1}{2}}$. It is also interesting to note that, when inelastic collisions set in, β increases and with it the drift velocity. The curve marked 'experimental' in Fig. 4.3 shows the square root dependence at lower values of E/P and the more rapid increase that occurs once inelastic collisions take place.

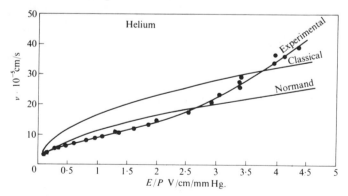

FIG. 4.3. Electron drift velocity at high E/P in helium. From Neilsen.[23]

Strictly speaking, it is correct to express the random energy of a gas in terms of temperature only when equilibrium prevails, i.e. when there is a

Maxwellian distribution. However, the presence of an electric field continually displaces the system from equilibrium. In order for the electron velocity distribution to be Maxwellian in an applied field, the rate of energy transfer between electrons must be faster than that between electrons and other species,[24] i.e.

$$n_e Q_e \gg \sum_j \beta_j \frac{M_e}{M_j} n_j Q_j, \tag{4.69}$$

where Q_e is the electron–electron collision cross-section. Thus, the lower the electron concentration, the greater the departure from the Maxwellian distribution. Although these departures from equilibrium will have an influence on diffusion coefficients and mobility, their most serious effects arise in calculations of rates of reactions that involve electrons, for example rates of ionization. The reason is that reaction rates are dependent not upon the average electrons but only upon those in the high-energy tail of the distribution. It is instructive to compare the Maxwellian distribution to that derived by Druyvesteyn,[25] which applies when electron–electron interaction is negligible, all collisions are elastic and have a constant cross-section, Q, and $U_e \gg U$. Druyvesteyn's distribution is given by

$$dn_e(U_e) = 1 \cdot 04 \, n_e \frac{U_e^{\frac{1}{2}}}{\bar{U}_e^{\frac{3}{2}}} \, e^{-0 \cdot 548 (U_e/\bar{U}_e)^2} \, dU_e, \tag{4.70}$$

where

$$\bar{U}_e = \frac{0 \cdot 427 \, Ee}{nQ} \left(\frac{M}{M_e} \right)^{\frac{1}{2}} \tag{4.71}$$

as compared with Maxwell's distribution, which is given by

$$dn_e(U_e) = 2 \cdot 05 \, n_e \frac{U_e^{\frac{1}{2}}}{\bar{U}_e^{\frac{3}{2}}} \, e^{-1 \cdot 5(U_e/\bar{U}_e)} \, dU_e. \tag{4.72}$$

The $(U_e/\bar{U}_e)^2$ term in the exponential of the Druyvesteyn law indicates that the high-energy population is far less than for the Maxwell law at the same average energy.

An interesting comparison of various distributions is given by Loeb;[4] treatments of specific distributions are given by Holstein[26] and Barbiere[27] for monatomic, and by Lewis,[28] Engelhardt, Phelps, and Risk,[29] and Carleton and Megill[30] for diatomic, gases.

In calculating conductivities in ionized gases subjected to applied electric fields, it is necessary to calculate the effect of the enhanced electron temperatures upon the level of ionization. If the criterion for a Maxwellian distribution is fulfilled, ion concentrations can be calculated from the Saha equation using the calculated value of electron temperature.[31] On the other hand, if the electrons do not have a Maxwellian distribution, the electron concentration in the steady state must be calculated from first principles by equating rates of ionization to rates of recombination (see Chapter 2).

Mobility in a magnetic field

When a magnetic field of strength H is applied to an ionized gas, the charged species experience a force as a result of their random motion transverse to the field, as discussed above. This force, which for charge e is of magnitude evH/c (see eqn (4.11)), is directed at right angles to both v and H, v being the velocity in the plane perpendicular to the field and c the velocity of light *in vacuo*. Thus v is continually changed in direction but not in magnitude. The force is also of constant magnitude, and, being always directed at right angles to the instantaneous velocity, leads to circular motion, i.e. the charged species are made to rotate around lines of force. From Newton's second law,

$$M\frac{v^2}{r} = eHv/c, \tag{4.73}$$

i.e.

$$\frac{v}{r} = \omega_b = \frac{eH}{Mc}. \tag{4.74}$$

The angular velocity of this rotation, ω_b, is known as the cyclotron frequency and is characteristic of the charge to mass ratio in a given magnetic field.

If a magnetic field is applied along the z-axis and an electric field along the x-axis, the resulting electron motion can be found by equating the force to the rate of loss of momentum, it being assumed that the electron loses all its momentum in the direction of the force on collision. Thus

$$x\text{-direction:} \quad Ee + v_y eH/c = M_e \bar{v} v_x, \tag{4.75a}$$

$$y\text{-direction:} \quad -v_x eH/c = M_e \bar{v} v_y. \tag{4.75b}$$

Solving these equations for v_x, $K_\perp$, the mobility at right angles to magnetic field, is obtained:

$$K_\perp = \frac{v_x}{E} = \frac{e}{M_e \bar{v}} \frac{1}{(1 + \omega_b^2/\bar{v}^2)} = \frac{K}{1 + \omega_b^2/\bar{v}^2}. \tag{4.76}$$

The conductivity is reduced by the same factor. It is interesting to note that there is also a movement in the y direction, at right angles to both applied fields; solving for v_y

$$v_y = \frac{Ee}{M_e \bar{v}} \frac{\omega_b/v}{(1 + \omega_b^2/\bar{v}^2)}. \tag{4.77}$$

This matter is discussed further in Chapter 7, in the context of magnetohydrodynamic power generation.

Mobilities of droplets and particles

The presence of charged particulate matter is especially important in the context of electrical properties of flames, for several reasons. As regards practical applications, many of the most promising, discussed in various sections of this monograph, involve particulate suspensions of reactants, products or intermediates. The control of flame carbon and of liquid fuel sprays as well as precipitation of ash are obvious examples. The theory of maximum effects attainable by applying fields to flames (Chapter 8) indicates the reason for this. It will be seen that for certain effects, for example, maximum rates of mass transport by 'electrolysis', the magnitudes attainable continue to improve as the size of the charge carriers increases. In addition, there is the more spontaneous and universally occurring formation of charged particulate matter—ion clustering, attachment of charges to microscopic and submicroscopic particles present in unfiltered and undried air generally, and electron emission by particles of low work-functions at elevated temperatures. It is for these reasons that a theoretical study of the response of droplets and particles to electric fields in the presence of flame ions has been carried out.[32] The theory depends very much on the size of the particle and on whether the particle is charged to a constant extent or is free to continue acquiring charge along its trajectory in the field.

Dependence on size is determined by how the particle 'sees' the molecules with which it collides. As discussed above, in the case of the very smallest ions, the target area for impact with molecules is greater than the geometrical target area because of dipole effects induced in neighbouring molecules by the divergence of the field due to the ion's charge. This is followed by a range of sizes in which such forms of attraction are negligible, while the interaction with neutral gas can still be treated in terms of individual collisions. Larger particles experience such collisions only in the form of a gas viscosity, leading into the Stokes regime. Lastly, particles of sufficient size and/or velocity attain Reynolds numbers large enough for the density of the gas to take the place of viscosity as the property determining drag. The theory can then be developed in terms of Newton's Law and it has been shown[32] that this regime extends to sizes large enough to include all cases likely to be encountered in practice.

The small sizes having been dealt with under ionic mobilities, this section is concerned with particles so large in comparison with the mean free path (i.e. of micron size and above) that they no longer sense the gas as individual collisions but acquire the velocities of bodies moving through fluids under the influence of an impressed force. Moreover, the particles are large enough to be multiply charged, so that the driving force, neglecting gravitational effect, is

$$F = EeN, \qquad (4.78)$$

where N is the number of electron charges (e) carried. The 'drag' force which opposes the above, and is equal to it under equilibrium conditions, is a function of velocity that depends on the Reynolds number of the particle in the manner shown in Fig. 4.4. At low values of Re (up to ~ 3, based on

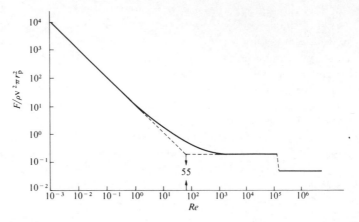

FIG. 4.4. Drag coefficient against Reynolds number. From Gugan, Lawton, and Weinberg.[32]

the diameter) Stokes' Law applies to within 20 per cent. Thus, in equilibrium,

$$F = 6\pi\eta r_p v \qquad Re \leq 3. \tag{4.79}$$

For $700 \leq Re \leq 2 \times 10^5$, Newton's Law may be used, i.e.

$$F = 0.22\pi r_p^2 \rho v^2. \tag{4.80}$$

Beyond this, the dependence changes once more, but it can be shown that larger values of Re are irrelevant in the present context. Many of the sizes of interest unfortunately fall in the transition region $3 \leq Re \leq 700$. Although empirical laws will be applied to this regime, it is more instructive to consider first the limiting values given by Stokes' and Newton's Laws. Figure 4.4 is a graph, on logarithmic scales, of drag coefficient for spherical particles against Reynolds number, wherein the dashed lines are the asymptotes to the Stokes and Newton regimes, extended into the transition region. They intersect at about $Re = 55$. There the error due to using either law is a maximum. It falls off on either side provided eqn (4.80) is used for $Re > 55$ and eqn (4.79) for smaller Re.

It was shown in Chapter 2 that particles in the presence of ions, particularly when a field is applied, continue to acquire charge by diffusion and bombardment charging, the latter being dominant for particles of micron size and above. This would be expected to occur, for example when particles travel in a field, in the presence of ions, from a flame to an electrode; and the calculation of mobility must take this into account. The case of a constant charge applies when the particles become charged in a region of relatively plentiful space charge (such as a region of corona discharge or a flame) and thereafter are considered in a space where no smaller charge carriers exist or, if they do, where the theoretical equilibrium charge that the particles would acquire is no greater than the initial value (this will apply only if field intensity does not increase with distance).

In the case where EeN is independent of distance x, equating the electrical force to drag, we obtain

$$EeN = 6\pi\eta r_p v \qquad (4.81)$$

in the Stokes regime, giving a mobility

$$K = v/E = eN/6\pi\eta r_p. \qquad (4.82)$$

In the Newton regime,

$$EeN = 0.22\pi r_p^2 \rho v^2, \qquad (4.83)$$

so that the velocity is proportional to the square root of the field strength and the 'mobility' depends on field strength

$$K = v/E = (eN/0.22\pi\rho E)^{0.5}/r_p. \qquad (4.84)$$

In both cases K is inversely proportional to the particle radius.

Let us now consider the particle sizes for which each regime is relevant. The 'Stokes mobility' may be used down to very small sizes and can be extrapolated to diameters of the order of a mean free path by the use of a semi-empirical correction due to Cunningham[4]

$$K = \frac{Ne(1 + \lambda/r_p)}{6\pi\eta r_p}, \qquad (4.85)$$

where λ is the mean free path. For smaller sizes there is a region of uncertainty until the regime of the 'classical' Langevin equation (eqn (4.20)) is reached.

A process of interpolation has been proposed[4] but, since λ at s.t.p. is of the order 10^{-1} μm while collision radii of molecules are of the order 10^{-4} μm, this regime will not be of great practical interest here.

The upper limit of the Stokes regime is determined by the Re of the particle, a criterion that is most readily expressed in terms of the (constant) particle charge, since Re proves to be independent of diameter, in this regime.

$$Re = v(2r_p)\rho/\eta = EeN\rho/3\pi\eta^2 \qquad (4.86)$$

by eqn (4.81). If we take 3 as the maximum then, approximately, for air at s.t.p.

$$eN \leq (7 \times 10^{-4})/E. \qquad (4.87)$$

Taking the maximum field short of breakdown as 30 kV/cm = 100 e.s.V./cm, the minimum value of charge at which Stokes' Law can be suspect is about 7×10^{-6} e.s.u., which is, approximately, 15 000 electron charges. It must be borne in mind, however, that this value is inversely proportional to the field and rises, for example, to $4 \cdot 5 \times 10^5$ electron charges at 10^3 V/cm. Under charging conditions permitting particles long times at field strengths close to breakdown conditions, particles of a few microns diameter can in theory attain such charges.

The Reynolds number, computed in the Newton's Law regime, is again independent of radius for a given charge

$$Re = \frac{(2r_p)\rho}{\eta} \cdot r_p^{-1} \left(\frac{EeN}{0 \cdot 22\pi\rho}\right)^{0 \cdot 5} = \frac{2}{\eta}\left(\frac{EeN\rho}{0 \cdot 22\pi}\right)^{0 \cdot 5} \qquad (4.88)$$

by eqn (4.83). On taking the lowest acceptable Re as 700, the limiting charge becomes

$$eN > \tfrac{1}{4}\{(700)^2\, 0 \cdot 22\pi\}(\eta^2/E\rho) \simeq 2/E \qquad (4.89)$$

for air at s.t.p. at the maximum (breakdown) field. The minimum charge for Newton's Law to apply is therefore about 2×10^{-2} e.s.u., which corresponds to approximately 5×10^7 electron charges. The smallest particle that can acquire such a charge (by charging to equilibrium in a breakdown field) is one of about 80 μm.

By setting the limits of Re at 3 and 700, errors have been kept to within about 20 per cent. From the academic point of view it is perhaps more satisfying to extend both limits to meet at Re 55 and leave the discussion there. This will always yield an order-of-magnitude estimate of mobility. When more accurate values are required for practical work, one of the empirical laws for the transition region $3 < Re < 10^3$ must be used. A relatively simple one, which is, in fact, the best straight line on the log–log graph of Fig. 4.4 is

$$2F/\rho v^2 \pi r_p^2 = 18 \cdot 5/Re^{0 \cdot 6}. \qquad (4.90)$$

This gives

$$K = v/E = 0 \cdot 121(eN)^{0 \cdot 715}/r_p(E\rho)^{0 \cdot 286}\eta^{0 \cdot 428}. \qquad (4.91)$$

Mechanisms of charge acquisition by droplets and particles have been discussed in Chapter 2 and it was shown (eqns (2.51) and (2.52)) that rates of charging, and equilibrium charges attained, depend on local electrical parameters such as field strength and ion concentration. In considering the mobility of particles not constrained to a constant charge, the new feature is that, as these particles travel along lines of force, they pass through regions of varying E and n, so that their rates of charging and equilibrium charges vary with time, and this must be taken into account. The most interesting case in the present context is again that of the unipolar regions between the zone of ion generation and the electrodes, in which fields of large magnitudes are applied. As regards diffusion charging, the variation of charge with time is given by eqn (2.54b), the variation of N with x being calculable by the theory of Chapter 8. However, for the model under consideration, bombardment charging (eqn (2.51)) tends to be much more important than diffusion charging, as was discussed in Chapter 2. Although the equilibrium charge due to diffusion charging is potentially higher, it takes so much longer to attain that the former is the dominant mechanism except at the lowest of particle sizes and/or field strengths. The numerical comparisons on p. 31 show that even for quite small fields, only 'bombardment' charging needs to be considered for particles of radii greater than 0·5 μm and this is borne out by the rates of approach to equilibrium calculated below.

In a unipolar cloud, E increases continuously with distance according to Gauss' law (eqn (4.3a)). Hence the equilibrium charge due to bombardment increases with distance (eqn (2.52)). Let us consider first the case of particles that acquire their equilibrium charge at every point in the field; this represents the upper limit idealization—that of full charging—complementary to that of a fixed small charge. It is not *a priori* obvious that this is a valid approximation in any real system, because the time spent in each region may not allow the particle to pick up a sufficient number of charges there, or because a sufficient number of ions is not available in the vicinity of each particle. However, it turns out[32] that, although this approximation can never be exactly fulfilled, it is convenient and quite adequate for virtually all flames. In order to simplify the mathematics, this will be demonstrated only after the laws relevant to full charging have been established. (Ne) now becomes equal to $(3Er_{\mathrm{p}}^2)$ and the equations for particle velocity (eqns (4.81) and (4.83)) become

$$v = EeN/6\pi\eta r_{\mathrm{p}} = E^2 r_{\mathrm{p}}/2\pi\eta \qquad (4.92)$$

in the Stokes' regime and

$$v = \left(\frac{EeN}{0\cdot22\pi r_{\mathrm{p}}^2 \rho}\right)^{0\cdot5} = \left(\frac{3}{0\cdot22\pi}\right)^{0\cdot5} \frac{E}{\rho^{0\cdot5}} = \frac{2\cdot085E}{\rho^{0\cdot5}} \qquad (4.93)$$

in that covered by Newton's Law.

For particles freely acquiring charge as they progress, the mobility equations are thus quite different now. It is in the Stokes' regime that a 'mobility' dependent on field strength appears:

$$K = Er_{\mathrm{p}}/2\pi\eta. \qquad (4.94)$$

In the Newton regime, we not only have a true mobility, but this is independent of radius and becomes a constant at any one gas density. For air at s.t.p. its value is 58 cm^{2}s^{-1} e.s.V^{-1} and this may be compared with values for ions under similar conditions, which are typically about 10^3. It is remarkable that under these circumstances large particles (theoretically the regime extends to particles several millimetres in radius) show velocities as great as $\frac{1}{15}$ of those of molecular ions, regardless of their size, provided they can always attain their equilibrium charge.

Substituting from eqns (4.92) and (4.93), Reynolds numbers in the two cases are

$$Re = \rho(Er_p)^2/\pi\eta^2 \tag{4.95}$$

in the Stokes case, and

$$Re = (4\cdot16\rho^{0\cdot5}/\eta)(Er_p) \tag{4.96}$$

for the Newton regime.

The relationship is thus most readily expressed in terms of (Er_p). For air at s.t.p., this product is approximately $1\cdot55 \times 10^{-2}, 6\cdot64 \times 10^{-2}$, and $0\cdot85$ e.s.u. at Re 3, 55, and 700, respectively, using eqn (4.95) for the first result, eqn (4.96) for the last, and either for the middle one (the point of intersection at which the two regimes are interpolated to meet). The upper limit of the Newton's Law regime occurs at about $Re\ 2 \times 10^5$, where Er_p becomes 242 e.s.u. At the break-down field of 100 e.s.V./cm, this would represent a particle of $r_p = 2\cdot42$ cm. This is well beyond the range of conceivable possibilities. If trajectories (as distinct from mobilities) are calculated, gravitational forces must be taken into account long before this point. The calculation is only intended to show that the conclusions drawn for the Newton's Law regime will cover the largest particles encountered in practice. These conclusions are summarized, in terms of mobility, in Fig. 4.5.

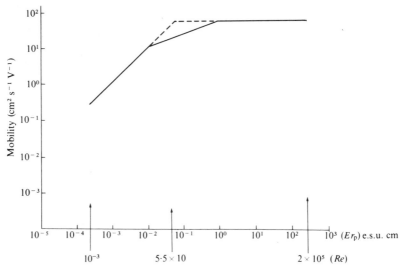

FIG. 4.5. Theoretical mobility against (field × radius) and against Reynolds number for fully charged particle. From Gugan, Lawton, and Weinberg.[32]

The generalization cannot be taken much beyond this stage. At low field strengths, up to a few hundred V/cm, Stokes' Law will cover most of the particle and droplet sizes of interest. At high field strengths, up to breakdown, particles of more than a few microns radius must be considered in terms of Newton's Law or some transition formula. Again, using eqn (4.90) for practical purposes, the mobility in the intermediate regime can be shown to be

$$K = v/E = 0.27(Er_p)^{0.43}/\rho^{0.28}\eta^{0.43}, \tag{4.97}$$

once more a function of (Er_p). For air at s.t.p., this becomes approximately

$$K = 73(Er_p)^{0.43}. \tag{4.98}$$

The full line in Fig. 4.5 between Re 3 and 700 is a plot of this relationship. At Re 55 it is just under half of the result obtained by interpolation of Stokes' and Newton's Laws.

It now remains to be considered to what extent the residence times of particles and the ion concentrations in every zone are in fact sufficient for charging to approach equilibrium. This has been discussed in detail in ref. (32) and we will consider here a particular case only—that of relatively low space charges—in order to illustrate the approach.

To this end we shall define f as the fraction that the actual particle charge is of the equilibrium value N (i.e. particle charge $= 3r_p^2 Ef$) and consider its variation in x, the distance coordinate between an ion source, such as a plane flame, and either electrode. An example of a practical system that approximates to the theoretical model is the flat counter-flow diffusion flame (Chapter 3), the field-induced velocity of the charged particles at the field strengths envisaged in the electrode spaces making the gas flow velocities unimportant. The particle mobility, K_p, will consequently be decreased to

$$K_p = f K_{p,f=1} \quad \text{and} \quad K_p = f^{0.5} K_{p,f=1} \tag{4.99}$$

in the Stokes and Newton regimes, respectively.

We shall consider the Newton regime first, since the greatest deviations of f from unity must occur there. This is because particles travel faster and therefore have less time to be charged to their equilibrium value as this increases during the passage of the particle through the ascending field. There are two parameters which do not vary with x. First, the flow rate of particles per unit area, G, must be conserved. Thus, writing K_{p1} for K_p at $f = 1$,

$$G = \text{const.} = n_p K_{p1} f^{0.5} E, \tag{4.100}$$

so that the number of particles per unit volume,

$$n_p = G/K_{p1}Ef^{0.5}. \tag{4.101}$$

Second, the total current density, j_T, made up of the current densities due to ion flow and particle flow, $(j_i + j_p)$, must be conserved. It follows that

$$dj_i/dx = -dj_p/dx \qquad (4.102)$$

and, since

$$j_p = G \cdot 3r_p^2 Ef \qquad (4.103)$$

and

$$j_i = n_i K_i Ee \qquad (4.104)$$

that

$$j_i = j_T - 3Gr_p^2 Ef \qquad (4.105)$$

and

$$n_i e = (j_T - 3Gr_p^2 Ef)/K_i E. \qquad (4.106)$$

All the relevant variables in x have now been expressed in terms of two: E and f. These are determined by two differential equations. The first is Gauss's Law, which here becomes

$$dE/dx = 4\pi(n_i e + 3r_p^2 Efn_p). \qquad (4.107)$$

On substituting from eqns (4.106) and (4.107) this takes the form

$$dE/dx = \varepsilon_1/E - \varepsilon_2 f + \varepsilon_3 f^{0.5}, \qquad (4.108)$$

where

$$\varepsilon_1 = 4\pi j_T/K_i, \ \varepsilon_2 = 12\pi Gr_p^2/K_i, \text{ and } \varepsilon_3 = 12\pi r_p^2 G/K_{p1}.$$

The second equation required must express the variation of f with x in terms of the rate of charge attachment to particles, which has been considered above. In the section on particle charging (Chapter 2) it was shown that each particle has a 'catchment area' within which it collects all the ions that approach it and that this area decreases with the charge already accumulated. At fraction f of the equilibrium charge, the area is $3\pi r_p^2(1-f)^2$, as can be seen from eqn (2.53), which gives the rate of charge acquisition by a single particle. It follows that, if we consider an element containing n_p particles per unit volume

$$-dj_i/dx = j_i n_p 3\pi r_p^2 (1-f)^2 = dj_p/dx. \qquad (4.109)$$

Substituting from eqns (4.103), (4.104), (4.101), and (4.105) this takes the form

$$\frac{d(Ef)}{dx} = \{(\varepsilon_4/Ef^{0.5}) - \varepsilon_5 f^{0.5}\}(1-f)^2 \qquad (4.110)$$

where

$$\varepsilon_4 = \pi j_T/K_{p1} \text{ and } \varepsilon_5 = 3\pi Gr_p^2/K_{p1}.$$

Solving eqns (4.108) and (4.109) as simultaneous differential equations gives the distributions of E and f in x. Fortunately, it proves unnecessary to consider the perfectly general case. At first sight, there appear two quite separate reasons why f should fall short of unity. The first, which would apply even to a single particle, is that a particle moving too fast through a rapidly increasing E may never have time to charge up to the ever-increasing equilibrium value. The second presupposes a particle concentration so high that there are not enough ions to go round, so that j_i falls to very low values near the electrode. It can be shown,[32] that in normal combustion systems only the former is limiting. As the particle flux, G, tends to zero, ε_2 and ε_3 in eqn (4.108) become negligible as compared with ε_1, i.e. the space charge due to particles becomes small by comparison with that due to ions. Under these circumstances eqn (4.108) can be integrated directly (see Chapter 8) to yield the field distribution as

$$E^2 - E_0^2 = 2\varepsilon_1 x. \tag{4.111}$$

Here E_0 is the value of E at $x = 0$. In eqn (4.110) ε_5 similarly tends to zero and that equation becomes

$$E\frac{df}{dx} + f\frac{dE}{dx} = E\frac{df}{dx} + f\frac{\varepsilon_1}{E} = \frac{\varepsilon_4(1-f)^2}{Ef^{0.5}}; \tag{4.112}$$

multiplying by E and substituting from eqn (4.111) gives

$$(2\varepsilon_1 x + E_0^2)(df/dx) = \{\varepsilon_4(1-f)^2/f^{0.5}\} - \varepsilon_1 f. \tag{4.113}$$

If f is close to unity, the first term on the right-hand side tends to zero and f falls rapidly, until $(df/dx) = 0$. If, on the other hand, f is less than its value at $(df/dx) = 0$, $(df/dx) > 0$ and f increases. The implication, physically, is that if f for some reason becomes too small, the particles slow down so much that the charging rate gains on the rise in equilibrium charge, and vice versa, so that under all conditions f tends to a constant value given by $(df/dx) = 0$. Introducing this condition into eqn (4.113) gives the equations for the steady-state value of f as

$$\varepsilon_4(1-f)^2 = \varepsilon_1 f^{\frac{3}{2}} \tag{4.114}$$

or, putting $\varepsilon_4/\varepsilon_1$ in terms of its constituent parameters

$$(K_i/4K_{p1})(1-f)^2 = f^{\frac{3}{2}}. \tag{4.115}$$

Now $K_i/K_{p1} \simeq 15$ and the solution of

$$3.75 = f^{\frac{3}{2}}(1-f)^{-2} \quad \text{is} \quad f = 0.63.$$

Thus, so long as the particle concentration is sufficiently smaller than that of the ions, f will tend to 0·63 regardless of its initial value and

$$K_p \rightarrow 0.63^{0.5}K_{p1} \simeq 0.8\, K_{p1} \tag{4.116}$$

in the Newton regime. It has been shown[32] that this result is an excellent approximation to all probable practical systems, first because even at high combustion intensity the particles do not normally contribute greatly to the total space charge, and second, because in those special cases where they might (for example, the combustion of finely powdered metal dusts in oxygen, all the particles to be manipulated by fields) the effect on f is very gradual. This is indicated in Table (4.4) obtained by numerical integration for specific cases.

TABLE 4.4

Results in dense particle space charges

Particle space charge as percentage of that of ions	4	19	130	320
Percentage decrease in mobility due to particle space charge	1·3	2·6	12·5	21·3
Percentage decrease in f, due to particle space charge	3	6·4	22·3	36·5

Other charging mechanisms (Chapter 2) compete with or complement the above, if they are at all appreciable. Thus diffusion charging always complements, on the above model. Because of the unipolar clouds involved, the approach to equilibrium is more rapid in those cases where diffusion charging is not negligible. Thermionic charging of particles of low work-function (p. 32) is a rather special case because it is unsymmetrical in that only positive charge can be conferred and in that it applies only to a few substances that produce particles of sufficiently low work-function. Also it occurs only in zones of high temperature and since the distribution of these with respect to that of field is so much a property of the system, the case cannot be generalized. Thus in the above-mentioned diffusion flame system, temperatures are high only where the field is low, so that particles merely start with a positive charge calculable on eqn (2.45) and quickly revert to their equilibrium charge when they commence their trajectory to the electrodes. On the other hand, in a pre-mixed flat flame (Chapter 3) in regions of hot products, electron emission can either increase or decrease the lag behind equilibrium charge, depending on the polarity of the applied field. In the zones close to the cold electrodes, bombardment charging will always reassert itself and the mobility theory will revert to the above model.

ELECTRICAL BREAKDOWN IN GASES

The practical effects that electric fields have on combustion processes are ultimately limited by the onset of electrical breakdown of the gas. In this section the subject of breakdown is treated in outline, emphasis being given only to those aspects that are directly relevant to the monograph.

Consider plane electrodes, separated by a distance a, across which a potential, V, sufficient to cause secondary ionization, is applied. An electron emitted from the cathode, by irradiation or some other means, will be accelerated by the electric field and, on average, will undergo, say, α ionizing collisions per centimetre of path in the direction of the applied field. Thus if there is an electron current i at distance x from the cathode, the increase in electron current along path length δx is given by,

$$\delta i = i\alpha \, dx, \tag{4.117}$$

i.e.

$$\ln(i_1/i_0) = \alpha a, \tag{4.118}$$

where i_0 and i_1 are the electron currents emitted by the cathode and collected by the anode, respectively. There is also a positive current $i_0(e^{\alpha a} - 1)$ reaching the cathode which results from the positive ions formed by the multiple electron collisions in the gas. The positive ions as well as incident photons and excited species can cause further emission at the cathode; see Chapter 2. Thus if there is a probability γ of electron emission per ion, it can be shown that

$$i_1/i_0 = e^{\alpha a}/\{1 - \gamma(e^{\alpha a} - 1)\}. \tag{4.119}$$

A relationship of this form accounts well for the variation of current with electrode separation found in practice at constant field strength. It is apparent from eqn (4.119) that the current increases indefinitely when

$$1 = \gamma(e^{\alpha a} - 1). \tag{4.120}$$

This indicates that the gap has broken down and has passed over to a self-sustained discharge, the current in which is limited only by the applied potential and the resistance of the circuit.

From theoretical arguments, the dependence of α upon the applied field can be shown to be given approximately by an expression of the form

$$\frac{\alpha}{P} = \beta_1 e^{-\beta_2(E/P)}, \tag{4.121}$$

where β_1 and β_2 are constants characteristic of the gas. The significance of the term E/P is that it is proportional to $Ee\lambda_e$, which is the energy gained by an electron moving a mean free path in the direction of the field. Substituting (4.121) into (4.120) for conditions at breakdown, it follows that

$$V_B = \beta_2 a P / \{C + \ln(Pa)\}, \tag{4.122}$$

where

$$C = \ln\left\{\beta_1/\ln\left(1 + \frac{1}{\gamma}\right)\right\}. \tag{4.123}$$

This expression agrees qualitatively with experiment and predicts a minimum breakdown potential, which is found in practice—see Fig. 4.6—but this occurs at pressures far too low to be of interest here ($Pa = 1$ to $10\,\text{mmHg cm}$). At higher pressures, $Pa > 10^2\,\text{mmHg cm}$, the logarithmic term varies very

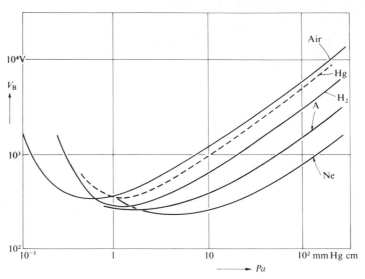

FIG. 4.6. Minimum breakdown potential for iron cathodes as a function of Pa. From von Engel.[8]

slowly, i.e. V_B increases almost linearly with Pa and E/P becomes almost constant. Figure 4.7 shows a graph of V_B against Pa for air at relatively high pressure. The near constancy of E/P ($40\,\text{v/cm.mm Hg}$ for air) at high

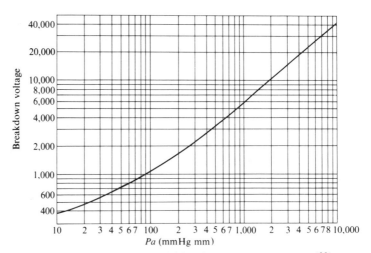

FIG. 4.7. Breakdown potential for air at 20°C. From Cobine.[33]

pressures indicate that there breakdown tends to occur at a particular electron temperature and is independent of the electrode spacing, unlike in the low pressure case.

The breakdown potentials, relative to air, of a number of gases are shown in Table 4.5.[8,33,34]

TABLE 4.5

Breakdown potential relative to air

Air	1	H_2S	0·90
O_2	0·85	SO_2	0·3
N_2	1·15	Cl_2	2·5
CO_2	0·95	CSF_8	5
H_2	0·65	CCl_4	5·7
NH_3	1	CCl_2F_2	3·2

The role of pressure in the various expressions is that of a scaling factor for the mean free path at constant temperature. Density could equally well have been used. Thus at high pressure

$$E_B \propto \rho. \tag{4.124}$$

This expression is very useful for extrapolation of data to different temperatures, there being little information on breakdown at temperatures encountered in combustion systems.

The nature of the cathode material would be expected to affect the breakdown potential inasmuch as it determines γ. A study has been made using hydrogen[35] in which it was found that different cathode materials gave significantly different breakdown potentials only at low pressures. At values of Pa in excess of 10 mmHg cm, differences become unimportant. Such behaviour would be expected from eqn (4.122), for when $Pa \gg \beta_1(\ln(1 + 1/\gamma))$ the influence of γ upon V_B is very small. At the pressures that are encountered in combustion systems, the effect of the nature of the cathode material is negligible. However, irregularities on the surfaces caused, for example, by accumulation of carbon or ash and by abrasions, are important because they lead to crowding of lines of force on protuberances, which can result in premature breakdown, see Chapter 8.

So far only plane electrodes have been considered. Another case of practical interest is that of cylindrical geometry. This case arises in a number of ways in combustion systems. Flames often have cylindrical geometry, wire gauzes are often used as electrodes, and it is sometimes desirable to make the lines of electric force diverge in order to obtain maximum effects. Peek[36] has shown that for co-axial cylindrical electrodes the following empirical relationship applies:

$$E'_B = E_B\left(1 + \frac{\beta_3}{\sqrt{r}}\right). \tag{4.125}$$

E'_B is the breakdown field at the surface of the inner electrode of radius r cm, E_B is the breakdown field for the plane case, and β_3 is an empirical constant that is equal to 0·3 for air at s.t.p.

In combustion systems the distribution of electric field is non-uniform, reaching a maximum at the electrodes in the vicinity of which, therefore, breakdown occurs first (chapter 8). Clearly, correspondence between combustion systems and those just discussed can only be expected in the range of values of Pa where breakdown is sensibly independent of electrode spacing and dependent primarily upon electron temperature, i.e. upon the local value of E/ρ.

The pressures at which flames are usually burnt are such that the value of a necessary to make the product Pa greater than, say, 10^2 mmHg cm, is only a small fraction of the flame/electrode separation. Therefore, the variation in field strength in a region extending a distance a from the electrode surface is usually quite small. Under these circumstances, the theory for a constant field can be applied over small regions close to the electrodes, in which the onset of breakdown can be predicted by the criteria of eqns (4.124) and (4.125), the field in the equations being that at the electrode. The validity of these equations can be checked in particular cases by calculating the change in field strength associated with the corresponding value of a.

The maximum practical effects that electric fields can have on combustion systems are limited by breakdown at one or other of the electrodes. Therefore, in view of the fact that the gases tend to be hot and the electrodes dirty, it is desirable to consider means of mitigating the adverse effects that these factors have upon breakdown. This will be detailed in Chapter 8; only the fundamentals are considered here. The use of geometry, pressure, and temperature follow from the aforegoing. Another possibility is to sheathe the electrodes in a flow of gas of high dielectric strength, some values of which are given in Table 4.4. A principle not considered so far is to apply a magnetic field perpendicular to the electric field;[37] this is discussed next.

For pressures relevant to combustion, the product Pa is such that E/P at breakdown is sensibly constant, i.e. breakdown occurs at a critical electron temperature that is independent of Pa (see eqn (4.68)). Consider two cases of electrons moving in a gas at incipient breakdown, in one of which a magnetic field is applied whilst in the other the electric field acts alone. Since the electron temperatures (and therefore the rates of energy lost by collision) are to be the same in both cases, the rate at which electrons gain energy from the field must also be the same (see eqn (4.64)). Hence

$$(E_M)_B ev_M = E_B ev, \tag{4.126}$$

where subscript M implies the presence of a magnetic field.

From eqns (4.126) and (4.76),

$$\frac{(E_M)_B^2}{E_B^2} = \frac{K}{K_M} = 1 + \frac{\omega_b^2}{\bar{v}^2}. \tag{4.127}$$

Assuming a Maxwellian distribution this expression may be rewritten, using CGS units, as

$$(E_M)_B^2 = E_B^2 \left(1 + \frac{e^2 H^2}{c^2 M_e} \frac{\lambda_e^2 \pi}{8kT_e} \right) \tag{4.128}$$

$$= E_B^2 \{ 1 + \beta_4 (H/P)^2 \}, \tag{4.129}$$

where β_4 is a constant and c the speed of light in cm/s.

For nitrogen at 0°C, taking a value for T_e at breakdown of $2 \cdot 7 \times 10^{4 \circ} K$ and data on collision cross-sections from ref. (8), the expression becomes

$$(E_M)_B^2 = E_B^2 \{ 1 + 2 \cdot 7 \times 10^{-5} (H/P)^2 \}, \tag{4.130}$$

where H is in gauss and P in mmHg. This effect is thus of academic interest, but is unlikely to be important in practice at atmospheric pressure and above, maximum attainable steady magnetic fields being only of the order of 10^5 G. However, it could become important at sub-atmospheric pressures.

FORCES ON NEUTRAL GAS: 'IONIC WIND'

So far, we have considered only those particles which carry a net, or an induced, charge. However, a constant mobility implies that while the charge-carrier moves down a potential gradient and thus draws energy, its own kinetic energy does not increase, so that the energy drawn must be handed on to the neutral gas, unless it is dissipated in some other way.

The detailed picture, as discussed above, is that the field modifies the velocities and directions of the 'thermal' trajectories of the ions during each mean free path in such a way as to increase the velocity component in the field direction. The additional momentum acquired by the ion is $(Ee\tau)$, where τ is the mean free time between collisions. This is lost completely, on average, in the collision which terminates the mean free path. Thus, the additional momentum in the field direction is conserved and results in a body force on the gas which must be identical in magnitude to that which would be expected if ions were not free to move with respect to the gas. The force per unit volume is therefore

$$F = Ee(n_+ - n_-). \tag{4.131a}$$

In the flame–electrode spaces, the equation generally reduces to

$$\pm F = Een_{\pm} \tag{4.131b}$$

The reason is that at the appreciable field strengths at which these forces become noticeable (but before the onset of secondary ionization), positive and negative charge carriers co-exist virtually only in the very thin flame reaction zone where they are created. However, the negative charge carrier is normally an electron in these high-temperature regions and it will be shown in Chapter 8 that therefore it must either be present in very small concentration or that the field intensity must be low. Electrode regions where no charges are generated, and in which ions of only one polarity drift, occur even when electrodes are immersed in the flame, because of quenching. In cases where body-forces and ionic winds are of interest, how-ever, the usual configuration is that of electrodes (or, at least, one electrode) well separated from the flame. Under these circumstances, the pressure distri-butions set up in the unipolar electrode regions by the forces of eqn (4.131b) entirely dominate the aerodynamics of the system. Now, in these zones the current density is given by

$$j = Een_{\pm}K_{\pm} \tag{4.104}$$

so that

$$F = \pm \frac{j}{K_{\pm}} \tag{4.131c}$$

by substitution into (4.131b). The sign notation is cumbersome and it will be more convenient to bear in mind that, whilst the current is unidirectional, the body force acts in the direction of local charge movement. Since current is conserved in a unidimensional system, F is inversely proportional to mobility. This is why the effect of electrons is small while charged clusters, droplets and particles have a pronounced 'drag' effect.

These arguments apply only so long as the field is well below E_B, the value at which secondary ionization sets in. When this condition is approached in an electrode space, the secondary charges produced in the ionizing col-lisions, which occur because the energy acquired in a mean free path is now large enough, rapidly reduce the net body force. Once electrons arise in a region of positive space charge, they carry much of the current, cause effective mobility to increase steeply, and tend to produce further secondary ionization because their paths between inelastic collisions allow them to acquire large energies at quite moderate field strengths. The effect of all this is a rapid deterioration from the well-behaved system that obeys the above theory to the disappearance of body force altogether, just below the potential at which secondary ionization is first detected—see Fig. 4.10. This transition is of no particular interest in itself, except as a boundary, limiting the use of wind effects, and all subsequent calculations of attainable magnitudes are tied to it.

The static pressure distribution in the normal system is readily calculable. If the gas could be held stationary, the pressure difference between two points, x_1 and x_2 would be

$$\pm \int_{P_1}^{P_2} dP = \int_{x_1}^{x_2} \frac{j}{K_\pm} \, dx \qquad (4.132a)$$

which reduces to

$$\pm P_2 = \frac{j}{K_\pm} (x_2 - x_1) + P_1 \qquad (4.132b)$$

for an ion of constant mobility in a unidimensional system (constant j) and to

$$\Delta P = j\left(\frac{a_+}{K_+} - \frac{a_-}{K_-}\right) \qquad (4.132c)$$

excess pressure on the positive electrode, where a_+ and a_- are the lengths of the electrode spaces.

Normally the gas is free to flow or circulate and a complex flow pattern can be set up by the above driving force distribution, depending on the aerodynamics of the system. The simplest case is that of inviscid flow in the absence of buoyancy, in one dimension, x, with no resistance to gas movement. The momentum flux at distance x from the ion source being ρv_x^2, the increase in it across an element of thickness δx and unit cross-sectional area is

$$\rho \frac{\partial v_x^2}{\partial x} \delta x = \frac{j}{k} \delta x - \frac{\partial P}{\partial x} \delta x, \qquad (4.133a)$$

the last term disappearing in the absence of pressure gradients. This integrates to

$$(v_x^2 - v_0^2) = \int_{x_0}^{x} \frac{j}{K} \, dx = \frac{j}{K} (x - x_0) \qquad (4.133b)$$

if mobility is constant. When $v_0 = 0$ at $x_0 = 0$, i.e. in the symmetrical case where $(a/K_+) = (a_-/K_-)$,

$$v = \pm \left(\frac{jx}{K_\pm}\right)^{\frac{1}{2}} \qquad (4.133c)$$

and its terminal value, as it flows without obstruction through a perfectly porous electrode, is

$$v_a = \pm \left(\frac{ja_\pm}{K_\pm}\right)^{\frac{1}{2}}. \qquad (4.133d)$$

Wind velocities in flame systems resembling the unidimensional model have been measured experimentally and the results compared with theoretical predictions. In early work[38,39] 'bat's wing' flames of coal gas, hydrogen, and ethylene were employed between, and parallel to, plane electrodes, 10 cm square. One of the electrodes carried a glass tube as a pressure probe, a metal gauze acting as the electrode within the probe mouth—Fig. 4.8.

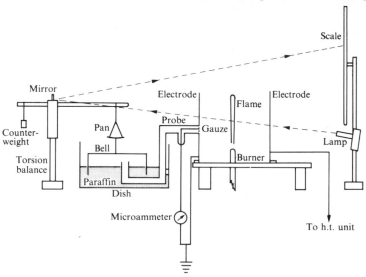

Fig. 4.8. Measurement of Pitot pressures due to small ionic winds. From Payne and Weinberg.[39]

The pressure at the probe (of the order of 10^{-6} atm) was measured by weighing a mica 'bell' on a torsion balance, the mouth of the bell being closed by a paraffin bath. The enclosed body of air communicated with the probe via glass tubing. A small plane mirror on the torsion wire of the balance acted as the fulcrum of an optical lever system. Thus any apparent change in the weight of the bell due to a change of pressure inside it was shown by the movement of the light spot of the optical lever system.

This manometer was calibrated by adding weights to a small balance pan hanging from the same point of suspension on the balance arm as the bell. Its pressure sensitivity was greater than 3×10^{-2} dyn/cm^2/mm deflection. The electrode containing the probe was kept earthed, the gauze electrode in the probe being earthed through a microammeter which measured the current falling within the probe. Potentials from 3 to 30 kV, of both polarities, were applied to the other (movable) electrode by an EHT generator across variable distances of the order of 10 cm. Some typical results are shown in Figs. 4.9 and 4.10. The former confirms the proportionality of Pitot pressure (itself proportional to v^2) to current, for various electrode-flame separations.

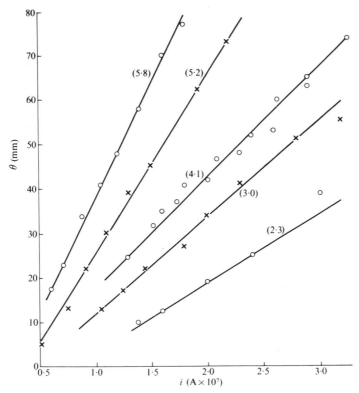

FIG. 4.9. Deflection θ versus probe current i for various flame-electrode spacings (coal gas, electrode separation = 8·2 cm; probe negative). From Payne and Weinberg.[39]

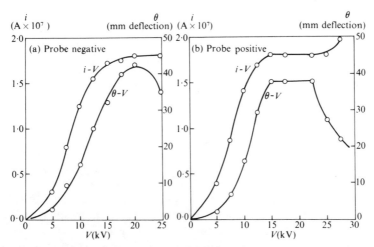

FIG. 4.10. Probe current i versus voltage V, and deflection θ versus voltage for hydrogen bat's-wing flames. From Payne and Weinberg.[39]

The latter shows what happens at first saturation, and then secondary ionization, are encountered. When j becomes constant, the deflection θ which represents the Pitot pressure at first levels off as predicted by the above theory. At about $22.5\,kV$ the effect of occasional ionization in energetic collisions becomes significant, and the curve plunges steeply. Beyond about $25\,kV$ this becomes a frequent event and j recommences to rise steeply, as shown by (b) which illustrates measurements in the flux of negative charge. The limit of usefulness discussed earlier is well demonstrated—any current increase beyond saturation here, in fact, brings about a decrease in the effect investigated.

Numerical values of gas velocity at the electrode were compared with a theory like that of eqns (4.133) (in fact the theory was based on energy rather than momentum conservation, but the discrepancy—a factor of 1.12 on velocity—is trivial) in order to determine what 'drag' law would be required in practical calculations to modify the inviscid theory. It transpired[39] that, in the case of hydrogen and coal gas flames, the theory was quite well obeyed without allowing for any resistance to flow. The mobility of the negative charge carrier indicated that it was an ion in the anode space.

Ethylene flames showed some interesting deviations from the general pattern. Unlike the other flames, they were powerfully deflected toward the negative electrode and were highly luminous. It will be shown later that their positive charge carriers are likely to include quite large carbon (or rather soot) agglomerates of up to many thousands of atomic masses. Only in this case were the Pitot pressures on the two electrodes considerably different, an effect caused by the positive 'ions' being very much larger. It was shown that although the theory allowing for no resistance to flow was still obeyed, it now accounted for net velocities relative to the overall circulation induced by an overwhelming effect in one of the electrode spaces. This was demonstrated by assuming the mobility of the negative charge carrier to be no different from that in other flames and subtracting its drag from the overall unidirectional circulation.

In subsequent studies,[37] a single flame surface burning on a modified porous plug burner (see Chapter 3), mounted horizontally, was used. A plane sheath of flowing cold gas helped to shield the test space from hot combustion products—see Fig. 4.11. The primary objective of this investigation was to compare maximum wind velocities attainable in practice with theoretical predictions (see below and in Chapter 8) and the gas flow rates were made large enough to permit use of particle tracking methods. In these, the trajectories of small particulate inclusions are photographed as broken tracks in a strongly convergent light beam, which is interrupted at a known frequency. It is, of course, essential to introduce the particles in a region where they cannot acquire charge from the ions whilst subject to a field.

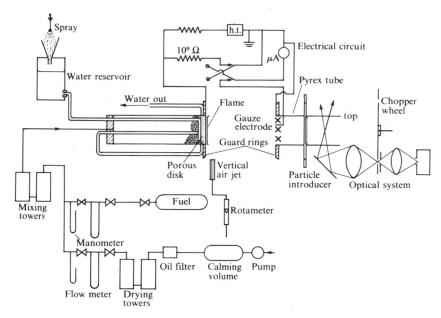

FIG. 4.11. Measuring ionic wind velocities by particle tracking. From Lawton, Mayo, and Weinberg.[37]

The theory for inviscid flow in the absence of buoyancy and other retarding forces has been established also for other geometries.[40] Thus the cylindrical configuration corresponds to some practical cases and it will be shown later that the ultimate limitations can indeed be extended by arranging for the lines of field to diverge. In the presence of appreciable space charge, divergence tends to occur even in the plane parallel electrode system unless it is deliberately prevented. For this reason it is worth noting that in the case of a cylindrical flame (radius r_0) burning coaxially between two cylindrical electrodes, the wind velocity at the electrode (radius r, positive if measured outward from flame and vice versa) is given by

$$v = \pm \left\{ \frac{i_c}{2\pi K \rho}(1 - r_0/r) \right\}^{0.5}, \qquad (4.134)$$

where i_c is the current per unit length (positive outward). This is based on a model corresponding to that discussed above for the planar system. The corresponding spherically symmetrical case involves, in addition, the assumption that gas entrained by an ionic wind, as well as the reactants, can be supplied (or products removed) through tubes traversing the outer electrode, without destroying the symmetry of the system. Under these conditions, the velocity of the ionic wind, presumed to be flowing without impediment through porous electrodes, is

$$v = \left\{ \frac{i_s}{4\pi K \rho} \left(\frac{r - r_0}{r^2} \right) \right\}^{0.5} \tag{4.135}$$

where i_s is the total current.

If we now allow for the fact that the ion source is of finite extent, even the simple model of a flat flame parallel to planar electrodes gives rise to a much more complex flow pattern than may be apparent at first sight.

Under the influence of body force F per unit volume, the gas must accelerate and, as the flow is virtually incompressible at the velocities involved, it must entrain more gas at right angles to the current density vector. The volume entrained per second, $\delta \dot{V}$, in a distance δx by a stream tube of cross-sectional area (δA) is given by

$$\delta \dot{V} = \delta A \left(v + \frac{\partial v}{\partial x} \delta x \right) - v \delta A \tag{4.136}$$

so that, as a fraction of the volume-flow along the tube,

$$\frac{\delta \dot{V}}{\dot{V}} = \frac{\delta v}{v} = \frac{\frac{1}{2}\sqrt{(j/K\rho x)}\delta x}{\sqrt{(jx/K\rho)}} = \frac{\delta x}{2x} \tag{4.137}$$

by eqn (4.133c). This initially very large, but gradually decreasing, fraction of the volume flow must be entrained across other stream tubes. Thus the simple unidimensional theory assumes not only that there is no effect on velocity (i.e. $v = 0$ at $x = 0$, the ion source) due to ion drift on the other side of the flame, but also that every tube of ion-current can entrain its required volume across all the other intervening tubes, without conflict with the one-dimensional model.

It is relatively easy to deduce the flow configuration for various simplified models and deduce a qualitative picture of what happens in practical cases. A simple example[37] is that of a small ion source of radius R, in a two-dimensional system that is symmetrical about the x-axis—Fig. 4.12. The point of R being small is that velocity can be assumed not to vary across the ion stream; an alternative formulation of this simplification is that there is perfect radial mixing.

Equation (4.136) then becomes

$$2\pi R v_r \delta x = \pi R^2 \frac{\partial v_x}{\partial x} \delta x, \tag{4.138}$$

where v_r is the velocity of the radially entrained gas. Thus

$$v_r = \frac{R}{2} \frac{\partial v_x}{\partial x}. \tag{4.138a}$$

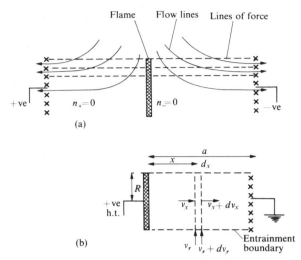

FIG. 4.12. Symmetrical system with gauze electrodes and free entrainment. From Lawton, Mayo, and Weinberg.[37]

But by eqn (4.133c)

$$\frac{\partial v_x}{\partial x} = \frac{1}{2}\left(\frac{j}{\rho K x}\right)^{\frac{1}{2}}$$

so that

$$v_r = \frac{R}{4}\left(\frac{j}{\rho K x}\right)^{\frac{1}{2}}. \tag{4.139}$$

From this expression, the distribution of mass entrainment along the ion stream can be deduced. The fraction of the total mass entrained up to x is proportional to the square root of the fraction that x is of the electrode distance, so that half of the total mass flow is entrained already by the first quarter of the distance.

If the velocity components are allowed to vary with r (or y in the cartesian system) as well as with x—as they must if the source is of appreciable extent—then the above method of calculating entrainment velocities must be applied to each element. Thus the equation to be solved with eqn (4.133a) is

$$\frac{1}{r}\frac{\partial}{\partial r}(rv_r) = \frac{\partial v_x}{\partial x} \tag{4.140a}$$

in a cylindrical system, or

$$\frac{\partial v_y}{\partial y} = \frac{\partial v_x}{\partial x} \tag{4.140b}$$

in rectangular coordinates.

Writing these in a more general form, the continuity equation, which for incompressible fluids is

$$\text{div } v = 0, \tag{4.140}$$

must be solved together with the equations of motion to satisfy the appropriate boundary conditions, in order to obtain the equation for the flow lines. Using eqn (4.133a) and the general Eulerian equation of motion for steady state, irrotational flow gives

$$\frac{\nabla v^2}{2} = \frac{j}{\rho K} - \frac{1}{\rho}\nabla P. \tag{4.133}$$

The solutions obtained for the special cases discussed in this section are for systems symmetrical about the flame, as regards force fields, i.e. (a/K) is the same in the two electrode spaces so that $v = 0$ at $x = 0$. Furthermore, none of this takes into account the effects of viscosity, buoyancy, and lack of permeability of the electrode. It is not difficult to modify the above differential equations to take into account the effects of viscosity. However, in view of the significant temperature—not to mention composition—dependence of viscosity, such equations would have to be solved simultaneously with those of temperature distribution, which would make the problem entirely intractable. Figures 4.13(a), (b), and (c) illustrate qualitatively the effects on flow pattern of impermeability of electrodes (which gives rise to recirculation of the entrained gas) and of asymmetry due to one of the electrode spaces, or one of the charge carriers, being larger than the other. This gives rise to a net flow velocity at the ion source and leads to deflection (in addition to the usual distortion) of the flame. Figure 4.13(c) may be compared with the photograph of a soot-bearing diffusion flame in such a field— Fig. 7.25(a).

In general, any lack of symmetry is clearly brought out by the deflections and distortions of flames in electric fields. The case of geometrical asymmetry reinforcing a difference in mobility is illustrated in Plate 1(ii). Optical[41] records showing the boundaries between hot and cold gas are even more informative and illustrate[42,43] both splitting of the hot gas stream and any overall deflection—see Fig. 4.14, 7.26, and 7.27.

These wind effects are of potential practical value and have been used, for example, to modify flame shape and stability, rates of flame propagation, entrainment of air into fuel, combustion intensity, heat transfer from flames to solid surfaces, and rates of flame spread in solid propellants. Their realm of usefulness may be summarized as all those systems in which a relatively slow (10^2–10^3 cm/s for a single stage) but accurately directed gas flow, induced without baffles and such like obstacles, would be of value. An account of such applications is given in Chapter 7. Occasionally, wind effects are a nuisance in other electrical measurements, for example when they cause flame distortion in the measurement of saturation currents.

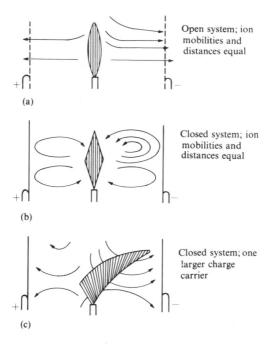

(a) Open system; ion mobilities and distances equal

(b) Closed system; ion mobilities and distances equal

(c) Closed system; one larger charge carrier

FIG. 4.13. Flow patterns and flame shape. From Lawton, Mayo, and Weinberg.[37]

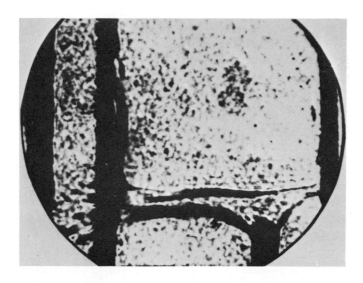

FIG. 4.14. Schlieren photograph of C_2H_4 diffusion flame in a field. Negative electrode (left-hand side) was a gauze. From Watermeier.[43]

So far we have considered only free entrainment. Particularly from the practical point of view (see Chapter 7) this is unnecessarily restrictive. One reason is that the above models give the largest gas velocity at the electrodes where it is likely to be least useful. It is usually at the combustion process itself (i.e. at the ion source) that wind effects could most usefully be applied. Another reason is that we would expect much larger velocities to result from entrainment in the vicinity of a small aperture which is the sole inlet supplying air to a large ion stream. The use of electrical body forces in an unconfined system is analogous to the use of convective forces for improving combustion without employing the principle of a chimney or flue. For this reason various cases of confined entrainment have been analysed.[37] A very obvious system that corresponds to the geometry of Fig. 4.11 is sketched in Fig. 4.15 where the open area for entrainment at the ion source is A_e and the

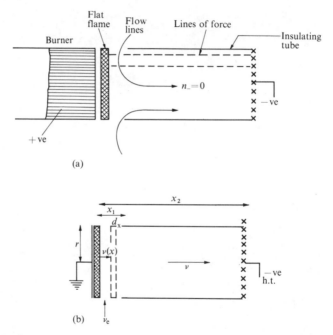

(a)

(b)

FIG. 4.15. Flat flame with gauze electrode and confined entrainment. From Lawton, Mayo, and Weinberg.[37]

velocity of the gas traversing it, v_e. With A the area of cross-section of the ion stream, of radius r, see Fig. 4.15(b), the entrainment velocity is given by

$$v_e^2 = \frac{j}{K\rho\left(1+2\dfrac{A_e^2}{A^2}-\dfrac{2A_e}{A}\right)} \left\{2x_2 - \frac{rA_e}{2A}\right\}. \qquad (4.141)$$

Even larger velocities are attainable in similar systems of cylindrical geometry.[37]

The practical applications to the recirculation of hot-flame products, in order to increase combustion intensity, and to the aeration of diffusion flames by inducting predetermined amounts of air, which are discussed further in Chapter 7, are based on calculations of this form.

The maximum attainable values for these effects are limited by the onset of secondary ionization, as discussed previously. All of them depend directly on the current (e.g. eqns (4.131)–(4.133)) that can be drawn and maxima limiting ion currents are analysed in detail in Chapter 8. Using results from there to illustrate the attainable magnitudes relevant to this section, the maximum body forces, static pressures, and wind velocities at the electrodes the theory predicts for the unidimensional system are

$$F_{max} = \frac{E_B^2}{8\pi a}, \tag{4.142}$$

$$\Delta P_{max} = \frac{E_B^2}{8\pi}, \tag{4.143}$$

$$v_{max} = \frac{E_B}{2\sqrt{(2\pi\rho)}}, \tag{4.144}$$

At s.t.p., for a breakdown strength of $30\,kV/cm$ (air, and clean, smooth electrodes) these values become respectively, $800\,dyn/cm^3$, $400\,dyn/cm^2$, and $550\,cm/s$. The maximum body force is based on an electrode separation of $0.5\,cm$ and is about 800 times greater than the maximum body force (ρg) in natural convection, in air at s.t.p. Maximum rates of convected heat flow due to fields applied to flames are in the range of 10–$40\,cal/cm^2/s$.

As regards velocities achievable by confined entrainment, appreciably larger magnitudes can be attained in specified regions. The one-dimensional case of Fig. 4.15 gives the following relationships:

$$\text{tube velocity, } v, = \frac{E_B}{2\sqrt{(\pi\rho)}}, \tag{4.145}$$

$$\text{entrainment velocity, } v_e, = \frac{E_B}{\sqrt{(2\pi\rho)}}, \tag{4.146}$$

$$\text{mass flux} = \frac{E_B A}{2}\left(\frac{\rho}{\pi}\right), \tag{4.147}$$

$$\text{momentum flux} = \frac{5E_B^2 A}{32\pi}. \tag{4.148}$$

The point of eqns (4.147) and (4.148) is that the numerical values depend on the parameter to be optimized and this in turn, is determined by the practical application envisaged. For instance, in order to obtain the maximum amount of aeration of a flame, the mass flow must be maximized whilst, if the object of the entrainment is to increase mixing (as when it is desired to increase combustion intensity) the momentum of the entrained stream would be optimized. The maximum values for these cases, on the basis of the above conditions are 778 cm/s and 1100 cm/s for the tube and the entrainment velocity respectively, $1 \cdot 0$ g/cm^2/s for the mass flux, and 900 g/s^2/cm for that of momentum.

There are various reasons why the above idealized maxima cannot be attained in practice; conversely, there are various modifications of the above systems which improve upon them. Among the reasons for the breakdown occurring at a lower field strength at the electrode are: that hot gases, with their lower breakdown strengths, tend to become entrained into the electrode spaces; that gauzes or other electrodes that are permeable to flow cause convergence of lines of field upon the elements of which they are made; that such convergence and hence earlier breakdown is also brought about by carbon particles or other solid specks deposited on the electrodes under the influence of the field. Methods of delaying breakdown include the use of geometries in which lines of force diverge towards the (negative) electrode, the application of magnetic fields, and of higher pressures. All these factors will be examined in detail in Chapter 8.

The theoretically predicted maxima have been compared with those attainable in the laboratory, using ionic wind velocity as an indicator by Lawton, Mayo and Weinberg.[37] Limitations due to field convergence and pressure drops arising in the case of gauze electrodes, the entrainment of flame products, and deposition of flame-carbon were examined in detail and very good agreement with the above theory was obtained.

REFERENCES

1. LANGEVIN, P. *Annls Chim. Phys.* **5**, 245 (1905).
2. BRATA, L. *Proc. R. Soc.* **A141**, 454 (1933).
3. POWELL, C. F. and BRATA, L. *Proc. R. Soc.* **A138**, 117 (1932).
4. LOEB, L. B. *Basic processes of gaseous electronics*, University of California Press, Berkeley, California, (1961).
5. MASSEY, H. S. W. and BURHOP, E. H. S. *Electronic and ionic impact phenomena*, Clarendon Press, Oxford (1952).
6. MUNSON, R. J. and TYNDALL, A. M. *Proc. R. Soc.* **A172**, 28 (1939).
7. MUNSON, R. J. and HOSELITZ, K. *Proc. R. Soc.* **A172**, 43 (1939).
8. VON ENGEL, A. *Ionized gases*. Clarendon Press, Oxford (1965).
9. BROWN, S. C. *Basic data of plasma physics*. Wiley (1959).

10. CRAVATH, A. M. *Phys. Rev.* **36**, 248 (1930).
11. MUNSON, R. J. and TYNDALL, A. M. *Proc. R. Soc.* **A177**, 187 (1940).
12. WANNIER, G. H. *Bell Syst. tech. J.* **32**, 170 (1953).
13. BLANC, A. *C.r.hebd. Séanc. Acad. Sci., Paris* **147**, 39 (1908).
14. DAVID, H. G. and MUNSON, R. J. *Proc. R. Soc.* **A177**, 192 (1940).
15. ALLIS, W. P. *Handb. Phys.* **21**, 413 (1956).
16. PACK, J. L. and PHELPS, A. V. *Phys. Rev.* **121**, 798, (1961).
17. FROST, S. L. *J. appl. Phys.* **32** (10), 2029 (1961).
18. SODHA, M. S., PALUMBO, C. J., and DALEY, J. T. *Br. J. appl. Phys.* **14**, 916 (1963).
19. SODHA, M. S., KAW, P. K., and SRIVASTAVA, H. K. *Br. J. appl. Phys.* **16**, 721 (1965).
20. ROSTAS, F. *International symposium on MHD electrical power generation*, ENEA, Paris, 1964, p. 91.
21. ZIMIN, N. *International symposium on MHD electrical power generation*, ENEA, Paris, 1964, p. 318.
22. HARRIS, L. R. *General Electric Research Laboratory Report No.* 63-*RL*-3334G (1963).
23. NIELSEN, R. A. *Phys. Rev.* **50**, 950 (1936).
24. BEN DANIEL, D. J. and TAMER, S. *Physics Fluids* **5**, 500 (1962).
25. DRUYVESTEYN, M. J. *Physica* **10**, 69 (1930).
26. HOLSTEIN, T. *Phys. Rev.* **70**, 367 (1946).
27. BARBIERE, D. *Phys. Rev.* **84**, 653 (1951).
28. LEWIS, T. J. *Proc. R. Soc.* **A244**, 166 (1958).
29. ENGELHARDT, A. G., PHELPS, A.V., and RISK, C. G. *Phys. Rev.* **135**, 1566 (1964).
30. CARLETON, N. P. and MEGILL, L. R. *Phys. Rev.* **126**, 2089 (1962).
31. KERREBROOK, J. *2nd Symposium on Engineering aspects of magnetohydrodynamics*, Columbia, 1962, p. 327.
32. GUGAN, K., LAWTON, J., and WEINBERG, F. J. *10th Int. Symp. Combust.*, p. 709. The Combustion Institute (1965).
33. COBINE, J. B. *Gaseous conductors.* Dover (1958).
34. THOMSON, J. J. and THOMSON G. P. *Conduction of electricity through gases*, Vol. II, p. 506. Cambridge University Press (1928).
35. LLEWELLYN-JONES, F. *Ionization and breakdown in gases*, p. 65. Methuen (1957).
36. PEEK, F. W. *Dielectric phenomena in high-voltage engineering.* (1929).
37. LAWTON, J., MAYO, P. J., and WEINBERG, F. J. *Proc. R. Soc.* **A303**, 275 (1968).
38. PAYNE, K. G. and WEINBERG, F. J. *Proc. R. Soc.* **A250**, 1316 (1959).
39. PAYNE, K. G. and WEINBERG, F. J. *8th Int. Symp. Combust.*, p. 207. Williams and Wilkins, Baltimore (1962).
40. LAWTON, J. and WEINBERG, F. J. *Proc. R. Soc.* **A277**, 468 (1964).
41. WEINBERG, F. J. *Optics of flames.* Butterworths (1963).
42. SAUNDERS, M. J. and SMITH, A. G. *J. appl. Phys.* **27**, 115 (1956).
43. WATERMEIER, L., D.I.C. Thesis, Imperial College, London, (1965).

5. Experimental Methods for the Study of Flame Ionization

IT IS convenient to think in terms of some particular combustion system by way of providing a framework within which the methods of this chapter can be described. Perhaps the simplest one may wish to study is the flat, laminar, pre-mixed flame. Measured distributions of temperature, heat release, and ion concentration perpendicular to the plane of such a flame are illustrated in Fig. 6.1. Although only total ionization is recorded, various ion species are involved and their relative proportions are unlikely to remain the same across the flame.

In the vicinity of the reaction zone, non-equilibrium ionization predominates—a feature that is particularly prominent for a hydrocarbon flame of low final temperature, as shown in Fig. 6.1. This is discussed in detail in chapter 6 and we need only note here that it is thought to be due to the energetic molecular fragments which propagate the main reaction and occasionally participate in reaction steps that release an electron. The thickness of the reaction zone, at atmospheric pressure, varies from a maximum of about 3 mm, for near limit flames, to very small fractions of a millimetre for stoichiometric mixtures. It increases with falling pressure, however, and this can be used to spread it over distances of the order of 1 cm. Ionization reverts to equilibrium values in the zones beyond that of maximum temperature.

In the case of normal hydrocarbon flames, the non-equilibrium ionization peaks, which are thought to be caused by chemi-ionization, give by far the highest concentrations of free charges—so much so that it is quite usual[1,2] to calculate recombination rates from ion concentration measurements downstream, without any reference to equilibrium ionization. Some exceptions to this are pure hydrogen, CO, and H_2S flames, in which non-equilibrium ionization appears to be negligible[3] and flames in which the product ionization has been greatly augmented by seeding with easily ionizable materials or by increasing product temperature. When complications due to geometry are also taken into account, this then gives a general representation of the kinds of ion source we are dealing with.

The experimental methods are designed to measure the concentrations (or rates of generation), the identity, and/or the position of origin of the flame ions. They can be subdivided into 'internal' and 'external' methods according to whether they attempt the measurement *in situ* (for example, attenuation of electromagnetic waves) or whether they draw the ions out by means of a field for examination elsewhere (for example, in a mass spectrometer). This subdivision is one of convenience and is less fundamental than might appear at first sight. Thus any probe immersed in the flame gases is certainly *intended* to measure unperturbed flame conditions, but in the relatively cold boundary layer that surrounds it, ions experience conditions similar to those that ions deliberately drawn out of the flame would be exposed to. The subdivision is therefore based to some extent on the assumptions that various schools of research have felt justified to make. The scheme is summarized in Table 5.1.

TABLE 5.1

Ionization parameter	'Internal'	'External'
Concentration or rate of generation	Conductivity. e.m. radiation (including attenuation, refractive index, inductive coupling, cyclotron resonance, capacitance measurements). Probes (including Langmuir, double)	Saturation current density
Identity	Mobility *in situ* (e.g. Hall effect)	Mass spectrometry, mobility in ambient gas
Position	All above, limited in resolution by size of probe regions or wavelength of e.m. radiation	'Ion photography' and all above, limited in resolution by geometry of lines of force and by diffusion

Generalizing somewhat, *in situ* methods would, in principle, be preferable for all flame measurements, if they could be made non-interfering and did not lack discrimination, sensitivity, and resolving power. As regards the use of electro-magnetic interactions, which alone are free from interference with the delicate flame processes, the resolving power is limited by their effective wavelength. For adequate sensitivity in the case of plasmas as tenuous and path lengths as short as those available in flames, this must be large, giving poor resolution.

Withdrawal of ions by fields can also be effected without interference with flames. By the time the ions reach the station at which they are examined, however, they may have changed identity by attachment or charge transfer. Only the flux of *charge* (current density) is conserved, subject to suitable field configuration.

Finally, solid surfaces are likely to interact with the ions, if used externally, and with the flame processes also, if used internally.

'INTERNAL' METHODS

The measurement of mobility

The techniques used for *in situ* measurements of mobility are based upon one of three principles: balancing the motion of the ion in an applied field against a gas stream,[4,5] distorting the equipotential surfaces in a conducting region by means of an applied magnetic field[4,5] (i.e. the Hall effect) and measuring the thickness of the space charge layer surrounding a wire immersed in the ionized gas.[6] An example of each is given in turn.

The apparatus used by Wilson[4] to measure the mobility of sodium ions in flame gases is illustrated in Fig. 5.1. The potential applied between the

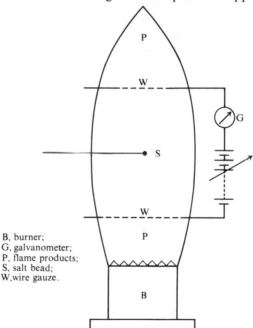

B, burner;
G, galvanometer;
P, flame products;
S, salt bead;
W, wire gauze.

FIG. 5.1. *In situ* measurement of mobility by drawing ion against gas flow. From Wilson.[4]

wire gauzes W was such that positive ions were drawn against the gas stream. Thus when a bead of salt is introduced positive sodium ions cannot flow to the lower gauze until the condition

$$K_+ E > U \tag{5.1}$$

is fulfilled, U being the gas velocity. If E_0 is the field at which a current is first detected, the positive ion mobility is equal to U/E_0. Using this method, Wilson obtained a value of 1 cm^2/V/s for the mobility of the sodium ion in the flame gases. Furthermore, he found that the negative charge present has a mobility too high to be measured by this means.

The apparatus Wilson[4] used to measure the mobility of the negative charges is illustrated in Fig. 5.2. A flame was stabilized on a long quartz slot

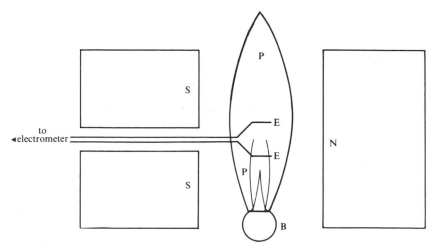

B, slot burner; E, rotatable quartz shielded electrodes; P, flame products; N,S, poles of electromagnet

FIG. 5.2. Measurement of mobility using the Hall effect. From Wilson.[4]

burner between the poles of an electromagnet through which two probes from an electrometer could be inserted into the flame gases, an electric field being applied at right angles to the plane of the diagram. The probes were rotated until they lay in an iso-potential surface. Then a magnetic field, H, was applied and the probes rotated through an angle θ until they lay in an iso-potential surface once more. The mobility of the charge carrier is given by

$$K = \frac{\tan \theta}{H} \times 10^8 \text{ cm}^2/\text{s/V}. \qquad (5.2)$$

The values of K found by Wilson lay between 2450 and 3900 cm^2/s/V, indicating the presence of significant numbers of free electrons in flames. A fuller discussion of the Hall effect, upon which this method is based, can be found in Chapter 7 in the section dealing with magnetohydrodynamics.

The apparatus used by Kinbara and Ikegami[6] for sheath thickness measurements is shown in Fig. 5.3. A potential of either sign could be applied to a cylindrical electrode of radius r_1 and length l, immersed in the flame

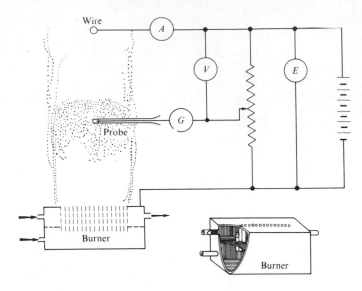

FIG. 5.3. The measurement of mobility from sheath thickness. From Kinbara and Ikegami.[6]

gases, the burner being earthed, and a probe used to investigate the resulting distribution of potential. The analysis of the results was based upon the assumption that the space around the wire can be divided into two distinct regions. In the region near its surface it was supposed that there is a non-ohmic zone of space charge $(n_+ \neq n_-)$ that has a well-defined boundary, on the other side of which the gas is neutral $(n_+ = n_-)$ and the potential, V, follows Ohm's Law. For a cylindrical ohmic conductor

$$V \propto \ln r, \tag{5.3}$$

where r is the radial distance from the axis of the wire. Accordingly the edge of the space charge region can be located as the radius, r_2, at which the graph of V against $\ln r$ ceases to be linear. The mobility of the charge carrier moving towards the electrode can then be deduced from the expression

$$K = \frac{i\{r_2^2 \ln (r_2/r_1) - \frac{1}{2}(r_2^2 - r_1^2)\}^2}{lV_0^2(r_2^2 - r_1^2)}, \tag{5.4}$$

where i is the total current and V_0 the potential difference between the electrode and the sheath edge.

Using this technique, K_e and K_+ were measured to be 3000 to 6000 cm^2/V/s and 1 cm^2/V/s, respectively. The validity of the division of the current-carrying gas into two distinct regions, in the manner described, is best judged for particular cases by reference to the discussion on Langmuir probes presented later in this chapter.

The measurement of conductivity, charge concentration, and electron temperature

The various methods are conveniently classified according to whether d.c. or a.c. electric fields are employed. D.c. methods involve the use of Langmuir probes and measurements of conductivity, whereas a.c. techniques employ the influence that ionized gases have upon the propagation of electromagnetic waves and upon the selectivity and resonant frequency of circuits of which they are made a part.

D.c. conductivity measurements

Whilst the d.c. conductivity of an ionized gas is itself very important from a practical point of view, for example in MHD power generation, it can also be used as means of deducing more fundamental parameters. In Chapter 4 it was shown that the conductivity, σ, of a plasma comprising equal numbers of positive ions and electrons is given by

$$\sigma = (K_e n_e + K_+ n_+)e \simeq K_e n_e e. \tag{5.5}$$

The approximation is valid also in the presence of negative ions—even if 90 per cent of the negative charge resides on ions—in view of the very high mobility of electrons. Thus if either electron mobility or concentration is known, *a priori*, the other can be deduced from a measurement of conductivity.

An apparatus used by Dibelius *et al.*[7] for conductivity measurements is shown in Fig. 5.4. The current/potential characteristic was measured

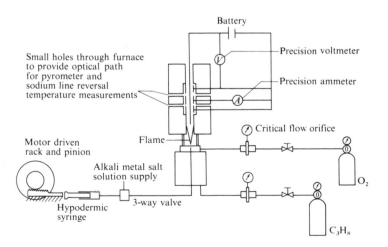

FIG. 5.4. Cylindrical conductivity cell. From Dibelius *et al.*[7]

between a central graphite cathode and an outer ring with guard ring electrodes mounted above and below. Thermionic emission from the cathode was sufficient to prevent the formation of a positive ion sheath at its surface which might otherwise have invalidated the results. The conductivity was determined from the relation

$$\sigma = \frac{\ln(r_2/r_1)}{2\pi l(V/i - R_0)} \frac{\text{mhos}}{\text{cm}},$$ (5.5a)

where r_1 and r_2 are the radii of the inner and outer electrode, respectively, l is the length of the outer electrode, V and i are the potential and current in volts and amperes, and R_0 is the resistance of the circuit exclusive of the flame.

For the success of this method, the cathode must freely emit electrons and not cool the gases significantly at its surface. Moreover, the field geometry must be well defined by the guard rings. Constant geometry conductivity cells of various designs have been used by a number of workers, for example, Pain and Smy[8] and Poncelet, Berendsen, and van Tiggelen.[9]

In order to avoid the difficulties associated with the formation of space charge sheaths and cool layers at the surfaces of the electrodes, van Tiggelen[10] used a variable geometry technique, similar to that frequently employed in the study of electrical discharges[11], to investigate the ionization in the reaction zone of a premixed flame. The apparatus is shown in Fig. 5.5.

A roof-top shaped, premixed flame (see Chapter 3) was stabilized on a quartz slit at either end of which were movable air-cooled electrodes. The experiment consisted of measuring the potential drop between the electrodes as a function of electrode separation at constant current, i. In this way the potential drops in the disturbed regions near the electrodes are kept constant and the field in the flame, E_0, can be evaluated as the gradient of the graph of potential against separation. The resistance per unit length of reaction zone is then given by E_0/i. On the supposition that ionization is limited to a reaction zone, the thickness of which can be estimated by his theory of flame propagation,[12] van Tiggelen calculated the mean conductivity of the flame. Further, by assuming that electron conduction dominates (see Chapter 6) and taking a typical value of electron mobility, he was able to obtain a mean value of the electron concentration.

Conductivity measurements of the kind described above have the merit of experimental simplicity. However, the interpretation is made difficult by the tendency of the current to crowd into regions of highest conductivity in a rather indeterminate manner. At best one obtains a mean over a relatively large volume, and only if conditions are uniform can the measured value of conductivity be associated with the local magnitude.

If the electrodes are made sufficiently small, it becomes possible to determine a wider range of plasma parameters, for then the potential drops in the

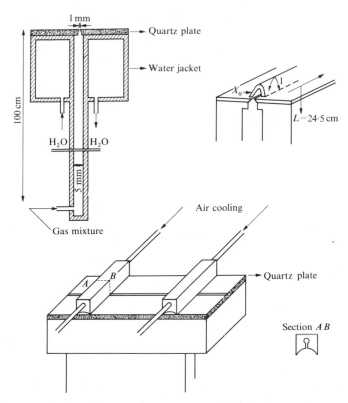

FIG. 5.5. Rectangular slit burner; shape of flame; measuring electrodes. From van Tiggelen.[10]

vicinity of the electrode surfaces may be interpreted by Langmuir probe theory, to give local values of electron temperature and ion and electron concentrations. This important technique will now be discussed in detail

Langmuir probes

Ever since the pioneering work of Mott-Smith and Langmuir,[13] Langmuir probes have been used extensively to measure concentrations of ions and electrons, and electron temperatures, in discharges. Recently they have been applied to flame ionization studies to an increasing extent.

The experimental conditions encountered in combustion are so wide that no one probe theory suffices to cover them all. It is therefore necessary to consider several theories in an attempt to cover the required range of conditions. Although there have been a number of reviews of probe theory, Chen's[14] being particularly useful, there are none that both cover the range of experimental conditions encountered in combustion and take account of some of the more recent developments that are of particular importance in this field. For these reasons the subsequent discussion on probes is rather detailed. However, the emphasis will be laid on the interpretation of data rather than on the theory *per se*.

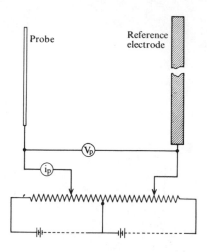

FIG. 5.6. Single probe.

Figure 5.6 shows a typical arrangement. The probe, usually a thin cylindrical wire or a small sphere, can be biased either positively or negatively with respect to a reference electrode. In this way a current-potential characteristic, as in Fig. 5.7, can be obtained. Over the region AB the probe is biased strongly negative with respect to the ionized gas and repels electrons. An insignificant number of electrons reach the surface under these circumstances, the current to the probe being due almost solely to positive ions. As the potential is made less negative the current falls, owing to a decrease in the flux of positive ions reaching the probe. In the region BC the electron current becomes appreciable and, at point C, is equal to the positive ion current. The potential V_f at which this occurs is called the floating potential, since it is the surface potential that an insulated body immersed in the plasma would attain in the steady state, i.e. when $i_p = 0$. In the region CD more electrons than positive ions reach the probe, causing the current to reverse direction. At the point D the probe is at plasma potential, which is taken as zero, and receives the random, thermal ion, and electron currents. At the point E the positive ion current is reduced to an insignificant level by the repulsive potential and the increase in net current in the region EF is due to an increase in the number of electrons reaching the probe under the action of the attractive potential.

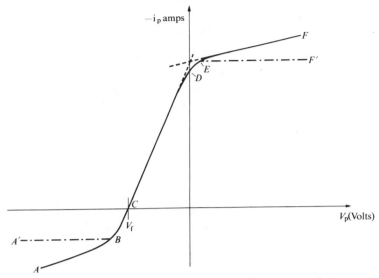

FIG. 5.7. Current-potential characteristic of Langmuir probe.

The interpretation of these characteristics to obtain reliable information
on ion and electron concentrations and electron temperatures over a wide
range of pressure and charge density has posed severe theoretical problems.
Fortunately, a considerable amount of work has been carried out in recent
years that makes it possible to interpret data obtained by the probes over a
wide range of conditions. Before considering various techniques of interpretation, a simplified treatment will be given in order to provide insight into
the processes occurring.

Consider a plane electrode biased strongly negative with respect to a
plasma. The repulsion of electrons greatly reduces their concentration in
the vicinity of the surface and leads to the formation of a positive ion sheath,
the effect of which is to shield the plasma from the field caused by the negative
charges on the probe's surface. Thus the potential, V_p, between the probe
and the plasma tends to be concentrated within the sheath region. In the
absence of collisions, the ions arrive at the probe surface at a rate equal to the
random thermal current density, $(j_+)_r$:

$$j_+ = (j_+)_r = n_0 e \left(\frac{eV_+}{2\pi M_+} \right)^{\frac{1}{2}}, \tag{5.6}$$

where V_+ is the positive ion energy in electrostatic electronvolts (1 e.s.eV =
300 eV) and $n_0 (= n_+ = n_e)$ is the concentration of charged species in the

unperturbed plasma. Electrons, on the other hand, are repelled, the random current being reduced by the Boltzmann factor, $\exp\left(-\dfrac{|V_p|}{V_e}\right)$, i.e.

$$j_e = n_0 e \left(\frac{eV_e}{2\pi M_e}\right)^{\frac{1}{2}} \exp\left(-\frac{|V_p|}{V_e}\right). \tag{5.7}$$

The net current density reaching the probe is given by

$$j_P = n_0 e \left\{ \left(\frac{eV_+}{2\pi M_+}\right)^{\frac{1}{2}} - \left(\frac{eV_e}{2\pi M_e}\right)^{\frac{1}{2}} \exp\left(-\frac{|V_p|}{V_e}\right) \right\}. \tag{5.8}$$

At very negative potentials the current density tends to $(j_+)_r$; this is indicated by the dotted line $A'B$ in Fig. 5.7. Assuming the ion temperature to be equal to that of the gas, the ion concentration can be found from eqn (5.6). At floating potential, V_f, $j_p = 0$. I.e., from eqn (5.8),

$$V_f = \frac{V_e}{2} \ln\left(\frac{V_e}{V_+} \cdot \frac{M_+}{M_e}\right). \tag{5.9}$$

For example, if $T_+ = T_e = 2500°K$ and H_3^+O is the positive ion, $V_f \simeq 1\cdot1$ V. Taking logarithms in eqn (5.7),

$$\ln j_e = \ln\{j_p - (j_+)_r\} = \text{constant} - \frac{|V_p|}{V_e}, \tag{5.10}$$

$(j_+)_r$ being known from the positive ion saturation region of the characteristic. Thus, by plotting $\ln j_e$ against $|V_p|$, the electron temperature can be found from the gradient of the line. In the case of electrons it is not safe to assume that the temperature is equal to that of the gas, since electrons lose only small amounts of energy on collision and can therefore reach elevated temperatures in the presence of an energy source, for example, an electric field or vibrationally excited molecules; see Chapter 2.

The quantity V_p is the potential of the probe relative to the local plasma, whereas the measured potential V is that of the probe relative to the reference electrode. Thus, if there is a potential gradient in the plasma, V does not equal V_p, and it is necessary to identify the local plasma potential in order to deduce V_p. This is usually done by extending EF backwards and CD forwards (Fig. 5.7) and taking the point of intersection to establish the plasma potential, which is then chosen as the zero for the potential axis.

At strongly positive potentials, the roles of ions and electrons are simply reversed, and the current density tends to the random thermal value for electrons, viz. $n_0 e \sqrt{(eV_e/2\pi M_e)}$, from which the electron concentration can be deduced using the measured value of electron temperature. It will be noticed that, in view of the very small mass of the electron, the electron saturation current is two to three orders of magnitude greater than that of the ions. For this reason it is important that the reference electrode should be very much larger than the probe. If it is not, positive ion saturation will set in at its surface before electron saturation occurs at the probe.

The above discussion is intended merely to introduce the subject and is inadequate for the detailed interpretation of data because of its many implicit idealizations. In the subsequent discussion, theoretical aspects will be limited to essentials only, the main emphasis being laid upon the use of the various probe theories for the reduction of experimental data.

One of the most important idealizations to question is that of the plane probe. If a probe is sufficiently large to be effectively plane (i.e. two dimensional effects caused by expansion of the sheath can be ignored) the drain of current from the plasma is usually excessive. On the other hand, if it is made small enough to minimize disturbances to the plasma, effects due to curvature become important. For these reasons, theories have been developed for spherical and cylindrical probes.

The distribution of charge around a probe at a strongly negative potential is shown in Fig. 5.8. Since the probe collects charge from the plasma, there

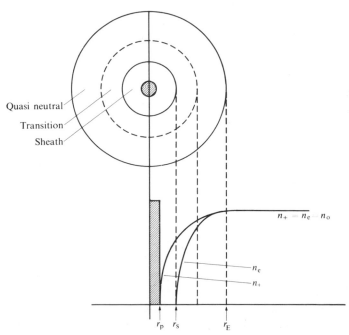

FIG. 5.8. Distribution of charge round a negative probe.

must be net diffusion towards it, even at relatively great distances. The result is that a quasi-neutral diffusion region is formed ($n_e \approx n_+$) in which concentrations are less than in the undisturbed plasma. Nearer the probe there is a transition region where appreciable charge separation begins to occur. Closer still there is a sheath in which electrons are present in only very low concentrations. As the potential is made more negative, the sheath expands and more positive ions are collected, i.e. in non-planar systems there is no saturation current owing to the expansion of the sheath area. The case of strongly positive probes is analogous, the roles of electrons and positive ions being interchanged. In the latter case, of course, the currents are very much larger. At intermediate potentials, $V_p \simeq V_e$, the sheath may disappear entirely and the quasi-neutral region may extend right up to the probe surface. However, strongly negative probes are of primary interest, because they permit the measurement of charge concentration and electron temperature without incurring the serious disturbance that occurs when the probe is drawing large electron currents.

As mentioned above, no unified probe theory has been produced so far that covers the wide range of experimental conditions, from the extreme at which there are no collisions in the perturbed regions to the other at which both the sheath and extra-sheath regions are collision dominated, although some progress along these lines has been made by Wasserstrom, Su, and Probstein.[15] Instead, specific theories have been developed to cover particular conditions. Three cases are of special interest.

I. Collisionless sheath and extra-sheath:

$$\lambda > r_s - r_p; \qquad \lambda > r_E - r_s.$$

II. Collisionless sheath and collision dominated extra-sheath:

$$\lambda > r_s - r_p; \qquad \lambda \ll r_E - r_s.$$

III. Collision dominated sheath and extra-sheath:

$$\lambda \ll r_s - r_p; \qquad \lambda \ll r_E - r_s.$$

The difficult case of just a few collisions has not yet been satisfactorily treated. However, Schultz and Brown[16] have provided a useful approximate treatment that has been amplified by Travers and Williams.[17]

In order to decide which theory to use in any particular case, it is necessary to have some measure of the importance of collisions. The Langmuir–Blodgett[18] equations of the vacuum diode valve can be used for this purpose. This gives the current potential characteristic for the passage of unipolar charge from one electrode to another, as a function of geometry, in the absence of collisions. In the case under discussion, for strongly negative probes, i.e. no electrons in the sheath, the outer edge of the sheath is taken as the emitter electrode and the probe as the collector. Then,

$$i = \frac{1}{9\pi} \left(\frac{2e}{M_+} \right)^{\frac{1}{2}} \frac{V_p^{\frac{3}{2}}}{z},$$ (5.11)

where all quantities are in CGS and here

planar; z = (sheath thickness)^2, i = current density,   (5.12a,b)

cylindrical; $z = r_p \beta^2$, i = current per unit length,   (5.13a,b)

spherical; $z = \alpha^2$, i = total current,   (5.14a,b)

the quantities α^2 and β^2 being functions of r_s/r_p as shown in Table 5.2. By substituting the values of i, r_p, and V_p the sheath thickness can be estimated. If it is much smaller than a mean free path, either I or II is appropriate, the distinction being made in terms of the consistency of the deduced concentration with the requirements of the theory used. Alternatively, if the calculated sheath thickness is much larger than a mean free path, III is appropriate.

(i) *Collisionless theory.* Collisionless theories are usually valid at pressures less than about 1 mmHg at room temperature, the maximum pressure being approximately proportional to the absolute temperature (c. 8 mmHg at final flame temperatures). Table (5.3), taken from Chen[19] presents the range of applicability of various collisionless theories for the measurement of positive ion concentration using strongly negative probes. The parameters used in the table are defined as follows:

$$\varepsilon = \frac{T_+}{T_e}, \qquad y_p = \frac{|V_p|}{V_e}, \qquad \lambda_D = \left(\frac{V_e}{4\pi n_0 e} \right)^{\frac{1}{2}}.$$ (5.15a, b, c)

λ_D is known as the Debye shielding length, and is in cm, if the quantities within the square root are in e.s.u. For combustion systems at the low pressures at which a collisionless theory is likely to be applicable (< 1 cmHg) λ_D lies approximately in the range 10^{-2} to 10^{-1} cm. It is worth noting here that quenching distances are inversely proportional to pressure and may be several centimetres at pressures of the order of 1 cmHg. Therefore if it is desired to make measurements on an unperturbed flame the burner may have to be very large.[20] Any interference of the probe with the flame reaction would also at low pressure have its range extended.

TABLE 5.2

The functions α^2 and β^2 for spherical and cylindrical diodes

r_s/r_p	α^2	r_s/r_p	β^2
1·0	0·0000	1·00	0·00000
1·05	0·0024	1·01	0·00010
1·1	0·0096	1·02	0·00040
1·15	0·0213	1·04	0·00159
1·2	0·0372	1·06	0·00356
1·25	0·0571	1·08	0·00630
1·3	0·0809	1·10	0·00980
1·35	0·1084	1·15	0·02186
1·4	0·1396	1·2	0·03849
1·45	0·1740	1·3	0·08504
1·5	0·2118	1·4	0·14856
1·6	0·2968	1·5	0·2282
1·7	0·394	1·6	0·3233
1·8	0·502	1·7	0·4332
1·9	0·621	1·8	0·5572
2·0	0·750	1·9	0·6947
2·1	0·888	2·0	0·8454
2·2	1·036	2·1	1·0086
2·3	1·193	2·2	1·1840
2·4	1·358	2·3	1·3712
2·5	1·531	2·4	1·5697
2·6	1·712	2·5	1·7792
2·7	1·901	2·6	1·9995
2·8	2·098	2·7	2·2301
2·9	2·302	2·8	2·4708
3·0	2·512	2·9	2·7214
3·2	2·954	3·0	2·9814
3·4	3·421	3·2	3·5293
3·6	3·913	3·4	4·1126
3·8	4·429	3·6	4·7298
4·0	4·968	3·8	5·3795
4·2	5·528	4·0	6·0601
4·4	6·109	4·2	6·7705
4·6	6·712	4·4	7·5096
4·8	7·334	4·6	8·2763
5·0	7·976	4·8	9·0696
5·2	8·636	5·0	9·8887
5·4	9·315	5·2	10·733
5·6	10·01	5·4	11·601
5·8	10·73	5·6	12·493
6·0	11·46	5·8	13·407

TABLE 5.3

Collisionless Collisionless probe theories and their ranges of validity

ε	Geometry	$r_p \ll \lambda_D$ all y_p	$r_p \approx \lambda_D$ $y_p \gg 1$	$r_p \gg \lambda_D$ $y_p \gg 1$	$1 < y_p < 5$
$\ll 1$	Spherical	22	19	19	21
	Cylindrical	22	19	19	21
$\simeq 1$	Spherical	22	23	$\begin{cases} 19 \\ 23 \end{cases}$	21
$\simeq 1$	Cylindrical	22	23	$\begin{cases} 19 \\ 23 \end{cases}$	21

The numbers refer to the appropriate references.

The method of determining the electron temperature by plotting a graph of $\ln j_e$ vs. V_p, which was described earlier (see eqn (5.10)), is still valid.[21] Thus ε can be determined for use in the measurement of positive ion concentration. In the case of electron concentration measurements, i.e. when using positive probes, the Langmuir orbital theory can be employed[19] but care must be taken to ensure that the current drain is not excessive.

In general it is preferable to use spherical rather than cylindrical probes because the theories are more reliable and easier to use.

(ii) *Collision dominated extra-sheath and collisionless sheath.* As a rough guide, this regime is likely to occur in the pressure range 1 mmHg to 1 cmHg at room temperature, these limits being approximately proportional to absolute temperature. The most rigorous theory available at the present time is based upon the simultaneous solution of the Boltzmann and Poisson equations for ions and electrons.[15] However, more easily usable, approximate theories have also been devised. These are based upon matching the current in the diffusion limited extra-sheath with that within the free fall sheath, the edge of which is assumed to be well defined.

The matching procedure outlined here is due to Waymouth[24] and has been chosen as an illustration because of its simplicity. The theory is for spherical probes and is valid when all the attracted charge carriers entering the sheath reach the probe, i.e. when the sheath thickness is small compared with the probe radius ($\{r_s - r_p)/r_p\} < 0.1$).

For an attractive potential, the particle flux reaching the probe is $\frac{1}{2}n_s\bar{c}$, where n_s is the concentration at the sheath edge and $\bar{c}$ the mean thermal velocity. It will be noticed that the numerical factor is $\frac{1}{2}$, rather than $\frac{1}{4}$ as in the expression for the random flux of particles across any plane drawn in the gas. The reason is that at an absorbing surface, since there are no returning particles, only that half of the distribution which has a velocity component directed towards the wall is present. Particles that are repelled, however, retain both halves of the distribution function and for them, therefore, the factor $\frac{1}{4}$ is appropriate.

In Waymouth's treatment, all the extra-sheath region is considered to be quasi-neutral ($n_+ \simeq n_e$). The currents at either side of the sheath edge are equal. Therefore, if G is the flux at the sheath edge, using the relation $K/D = e/kT = V^{-1}$ (Chapter 4) it follows that

$$G_e = -n_s \left(\frac{eV_e}{2\pi M_e}\right)^{\frac{1}{2}} \theta_e = -n_s K_e E - K_e V_e \left(\frac{dn}{dr}\right)_{r_s}, \qquad (5.16)$$

$$G_+ = -n_s \left(\frac{eV_+}{2\pi M_+}\right)^{\frac{1}{2}} \theta_+ = -n_s K_+ E - K_+ V_+ \left(\frac{dn}{dr}\right)_{r_s}, \qquad (5.17)$$

where

$$\theta_e = \exp\left(-\frac{|V_s|}{V_e}\right) \quad \text{and} \quad \theta_+ = 2 \text{ for } V_s < 0 \qquad (5.18a \text{ \& } b)$$

and

$$\theta_+ = \exp\left(-\frac{|V_s|}{V_+}\right) \quad \text{and} \quad \theta_e = 2 \text{ for } V_s > 0, \qquad (5.19a \text{ \& } b)$$

V_s being the potential drop across the sheath. The values of 2 for θ differ from those of Waymouth, who did not take account of the effect of the absorbing wall on the velocity distribution of the attracted species.

Quantities Q_e and Q_+ are defined as

$$Q_e = \frac{r_p}{K_e(V_e + V_+)}\left(\frac{eV_e}{2\pi M_e}\right)^{\frac{1}{2}} = \frac{3r_p}{4\{1 + (V_+/V_e)\}\lambda_e}, \qquad (5.20)$$

$$Q_+ = \frac{r_p}{K_+(V_e + V_+)}\left(\frac{eV_e}{2\pi M_e}\right)^{\frac{1}{2}} = \frac{3r_p}{4\{1 + (V_e/V_+)\}\lambda_+}. \qquad (5.21)$$

The relations on the far right-hand side arise because $K_+/D_+ = V_+^{-1}$ and $D_+ = \lambda_+ \bar{c}_+/3$, and similarly for the electrons. The current-potential characteristic, i.e. i_e or i_+ against V_p, is given by

$$\frac{i_+}{4\pi r_p^2} = \frac{n_0 e}{1 + Q_e\theta_e + Q_+\theta_+}\left(\frac{eV_+}{2\pi M_+}\right)^{\frac{1}{2}} \theta_+, \qquad (5.22)$$

n_0 being the ion concentration in the undisturbed plasma. The expression for i_e is the same, except that subscript e is written instead of +. V_s can be found from the following equation:

$$V_s = V_p - \left\{V_e - \frac{Q_e(V_e + V_+)\theta_e}{Q_e\theta_e + Q_+\theta_+}\right\} \ln(1 + Q_e\theta_e + Q_+\theta_+). \qquad (5.23)$$

V_s appears on both sides of eqn (5.23), as can be seen from eqns (5.18 and 5.19), and therefore can be found as a function of V_p and used in eqn (5.22) for the construction of the current potential characteristic. For a strongly negative potential,

$$i_+ = i_p = \frac{2n_0 e}{1+2Q_+} \left(\frac{eV_+}{2\pi M_+} \right)^{\frac{1}{2}}.$$ (5.24)

From this n_0, the undisturbed ion concentration, can be found.

In the quasi-neutral extra-sheath at radius r,

$$n = n_0 \left(1 - \frac{Q}{1+Q} \cdot \frac{r_p}{r} \right),$$ (5.25)

$$Q = Q_e \theta_e + Q_+ \theta_+.$$ (5.26)

From eqn (5.25) the extent of the disturbance caused to the plasma by the probe can be calculated.

It is easily shown from eqn (5.20) that the electron temperature can be found from the gradient of $\ln i_e$ against V_p at negative probe potentials.

The shortcomings of this approach are that the possibility of an appreciable region of space charge between the free fall sheath and the quasi-neutral extra-sheath is not taken into account and that the particle energy at the sheath edge is assumed to be thermal, i.e. that penetrating fields do not perturb the velocity distribution functions. These matters have been discussed by Waymouth, who has also mentioned other work based upon a similar matching technique.

(iii) *Collision dominated sheath and extra-sheath.* The treatments[25,26] outlined here apply to systems in which the motion of the charged species can be described in terms of the diffusion coefficient and the mobility right up to the probe surface (i.e. to pressures greater than about 1 cm at room temperature). Unlike in the previous case, here the probe characteristic is no longer an analytical expression, but it can be presented in graphical form.

When a spherical probe is at plasma potential, the ion current reaching it is due entirely to random motion and is given by

$$(i_+)_r = 4\pi r_p D_+ n_0 e.$$ (5.27)

A similar expression applies for electrons. For intermediate potentials, $0 \le |y_p| \le \ln(r_p/\lambda_D)$, Cohen[26] has presented results for $\varepsilon = 1.00, 0.10, 0.01$, and 0.00. His graph for $= 1.0$ is reproduced in Fig. 5.9. In this $J_+ = i_+/(i_+)_r$, $\rho_p = r_p/\lambda_D$ and $y_p = |V_p|/V_e$.

Su and Lam[25] have considered the case of more negative potentials. Their results are shown in Fig. 5.10:

$$\bar{\rho}_p = (1+\varepsilon)^{\frac{1}{2}} r_p/\lambda_D \quad \text{and} \quad \mu = \frac{i_e}{i_+} \frac{K_+}{K_e}.$$

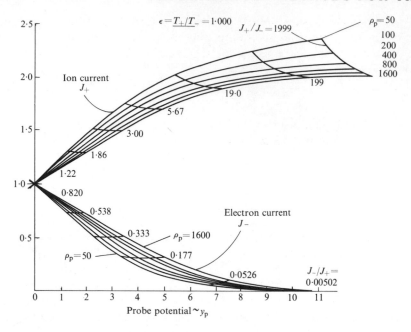

FIG. 5.9. Current/potential characteristic of high-pressure spherical probe. From Cohen.[26]

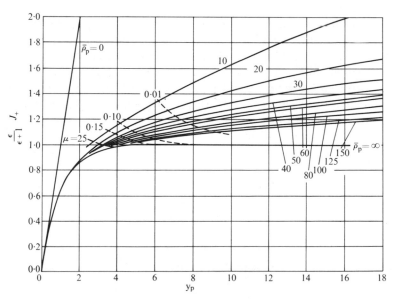

FIG. 5.10. Current/potential characteristic of high-pressure spherical probe. From Su and Lam.[25]

For large, very negative probes $\left[\text{i.e. } \ln\left\{\dfrac{\varepsilon}{1+\varepsilon}\cdot\dfrac{i_+}{(i_+)_r}\right\} \ll \ln\{\sqrt{(1+\varepsilon)\rho_p}\} \text{ and}\right.$

$\left.y_p \gg \dfrac{2}{3}\ln\dfrac{i_+\rho_p\varepsilon}{(i_+)_r\sqrt{(1+\varepsilon)}}\right\} > 1\right]$ the following expression for the current/

potential relation can be used:

$$y_p = \frac{r_p\sqrt{(1+\varepsilon)}}{\lambda_D}\, G\!\left(\frac{\varepsilon(1+\varepsilon)i_+}{(i_+)_r}\right) + \frac{2}{3}\ln\left\{\frac{r_p\sqrt{(1+\varepsilon)}}{\lambda_D}\right\} + 3. \qquad (5.28a)$$

The function G is given in Fig. 5.11.

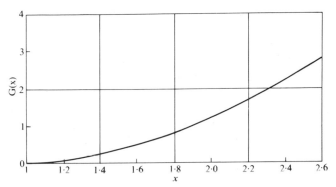

FIG. 5.11. The function G in the theory of Su and Lam. From Su and Lam.[25]

As an illustration of the use of these treatments, consider the construction of a probe characteristic from Fig. 5.10 for a given probe radius and unperturbed ion concentration, the values of ε and V_e having been determined by the method to be described in the next section. $\bar{\rho}_p$ is known from the data; therefore values of μ and $\dfrac{\varepsilon}{1+\varepsilon}\cdot\dfrac{i_+}{(i_+)_r}$ can be read from the graph as functions of y_p.

The positive ion current to the probe is deduced using eqn (5.27) to calculate $(i_+)_r$. The electron current follows directly, being equal to $(K_e/K_+)\mu i_+$. The probe current, i_p, is given by $i_+ - i_e$, and, since $V_p = y_p V_e$, the graph of i_p against V_p can be plotted for the given ion density. For the important case of the very negative probe, defined by the relation

$$y_p \gg \frac{2}{3}\ln\left\{\frac{i_+}{(i_+)_r}\cdot\frac{\rho_p\varepsilon}{\sqrt{(1+\varepsilon)}}\right\} > 1, \qquad (5.28b)$$

the disturbance to the quasi-neutrality of the plasma caused by the probe extends to a distance of the order of $r_p\left\{\dfrac{\varepsilon}{1+\varepsilon}\cdot\dfrac{i_+}{(i_+)_r}-1\right\}$ from the surface.

There is no rigorous theory† for the current/potential characteristic obtained with cylindrical probes. However, Calcote[27] has extended the treatment of Burhop, Bohm, and Massey[28] for a spherical probe at plasma potential to the case of finite and infinite cylinders. His expression for the positive ion current density to a probe of length l at plasma potential is

$$j_+ = n_0 e \left(\frac{kT_+}{2\pi M_+} \right)^{\frac{1}{2}} \left\{ 1 + \frac{1\cdot 5 l r_p}{4\lambda_+ B} \ln \left(\frac{X+B}{X-B} \right) \right\}^{-1}, \qquad (5.29)$$

where

$$X = l + 2\lambda_+ \quad \text{and} \quad B = \sqrt{\{X^2 - 4(r_p + \lambda_+)^2\}}.$$

Once j_+ is measured, the ion concentration can be found from eqn (5.29) from a knowledge of λ_+, T_+, and M_+. There is, however, some uncertainty in determining j_+ at plasma potential. In his work on flames, Calcote[27,29,30] has estimated this quantity by extrapolation of the positive ion saturation region back to plasma potential. The advantages of the cylindrical probe are that it can be made with a very small surface area and that end corrections can be made by taking measurements with probes of different lengths.

Double probes

So far the discussion has been limited to the current/potential characteristic of single probes. However, there must always be a second surface immersed in the plasma to complete the circuit. In practice, the second electrode can be ignored only if it is so large that positive ion saturation at its surface does not limit the electron current to the probe. Moreover, there is a very great advantage to be gained by using two probes of comparable area when measuring electron temperature,[31] for then the electron current to either of them cannot exceed the sum of the positive ion currents to both. This follows from charge conservation

$$(i_+)_1 + (i_+)_2 = (i_e)_1 + (i_e)_2, \qquad (5.30)$$

where the subscripts 1 and 2 denote the probes. In this way excessive drain of electrons from the plasma can be avoided.

The collisionless case will be considered first. In regions of equal plasma potential and in the absence of an applied field both probes assume the floating potential and no net current flows. If probe 2 is set at a positive potential with respect to 1, a current flows from 2 to 1, and the probes assume negative potentials V_1 and V_2 with respect to the plasma. Thus the electron currents vary exponentially with V_1 and V_2 whilst the positive ion currents

---

† *Note added in proof.* See, however, Su and Keil's (*J. appl. Phys.* **37**, 4907 (1966)) recent attempt at a rigorous treatment of high-pressure cylindrical probes.

are approximately constant (this last point is discussed later). From eqn
(5.30)

$$\sum i_+ = (j_e)_{r1} A_1 \exp\left(-\frac{|V_1|}{V_e}\right) + (j_e)_{r2} A_2 \exp\left(-\frac{|V_2|}{V_e}\right), \quad (5.31)$$

A_1 and A_2 being the probe areas.

Now,

$$V_p = V_1 - V_2, \quad (5.32)$$

therefore, from (5.31) and (5.32)

$$\ln\left(\frac{\sum i_+}{(i_e)_2} - 1\right) = -\frac{V_p}{V_e} + \ln\left\{\frac{(j_e)_{r_1} A_1}{(j_e)_{r_2} A_2}\right\}. \quad (5.33)$$

Thus the electron temperature, V_e, can be found from the gradient of the
graph of $\ln\left[\{\sum i_+/(i_e)_2\} - 1\right]$ against V_p. If $(i_+)_2$ and $(i_+)_1$ are constant they
can be found from the extreme positive saturation limbs of the characteristic
and $(i_e)_2$ follows from

$$i_p = (i_e)_2 - (i_+)_2 \quad (5.34)$$

as illustrated in Fig. 5.12.

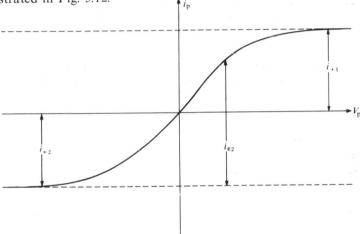

FIG. 5.12. Double probe current-potential characteristic.

Alternatively, the electron temperature can be found from the gradient
at the origin:

$$V_e = \frac{\sum i_+}{4}\left(\frac{dV_p}{di_p}\right)_{V_p=0}. \quad (5.35)$$

In practice there is some variation of $\sum i_+$ with applied potential; however,
this does not change the fundamentals of the argument. Details of the corrections to make when this variation is significant are given by Johnson and
Malter[31] and by Burrows.[32]

Recent work by Cozens and von Engel[33] and Bradley and Matthews[34] has shown that this technique can also be applied at high pressures using exactly the same expressions, the best results being obtained from eqn (5.33). Double probes have recently been used to measure electron temperatures in flames by several workers.[17,35,36]

A useful review of probe theories has been given by Chen[14] and details of probe construction can be found in refs. (14), (17), (27), and (28).

The theories that have been discussed above and in the references given are strictly valid only for stationary plasmas. It is only very recently that serious attempts have been made to extend the theory to include motion of the gas.[37,38] This work, however, is still developing rapidly and will not be detailed here. The validity of the stationary plasma idealization for a particular measurement can be checked by comparing the magnitudes of the field induced charge fluxes with those caused by convection. In high speed applications, for example detonation and shock tube studies, the stationary plasma theories are unlikely to be adequate.

Two experimental points deserve special emphasis. First, it is essential to avoid thermionic emission from the probe by keeping it sufficiently cool. This is especially important when substances of low work function, for example BaO, are present since these may collect on the probe surface. Second, it is essential that the current taken by the probe should be small compared to the rate of ion generation in the plasma under study. In the case of flames, additional difficulties may arise due to chemical (including catalytic), thermal, and aerodynamic disturbance of the reaction processes by the probes. Rates of free charge generation can be measured using the saturation current techniques discussed later in this chapter.

Measurements using resonant circuits and propagating waves

The techniques described in this section involve the measurement, by one means or another, of the a.c. conductivity of the plasma. Accordingly a brief discussion of a.c. circuits and a.c. conductivity of plasmas will be given before particular methods are considered.

In dealing with electric fields which vary harmonically with time, it is convenient to write $E_0 \cos \omega t$ as $E_0 e^{j\omega t}$, on the understanding that it is always the real part of the complex number that is referred to. This notation is valid so long as only linear differential equations are involved, for, as is easily verified,

$$\text{Real}\left(\frac{d^n e^{j\theta}}{d\theta^n}\right) = \frac{d^n}{d\theta^n}\{\text{Real}(e^{j\theta})\} \qquad (5.36)$$

but it breaks down if products are involved, since

$$\text{Real}(e^{j\theta_1} \times e^{j\theta_2}) = \cos(\theta_1 + \theta_2)$$
$$\neq \text{Real}(e^{j\theta_1}) \times \text{Real}(e^{j\theta_2}). \tag{5.37}$$

Thus when calculating power as the product of current and potential both must be written in the sine or cosine form.

A.c. conductivity

In the presence of a steadily propagating electromagnetic wave or an oscillating field between condenser plates, the field at any point can be written as

$$E = E_0 e^{j\omega t}, \tag{5.38}$$

where ω is the angular frequency in rad/s. The frequency, f, in c/s, is equal to $\omega/2\pi$. Charged carriers are accelerated by the field but they lose momentum by collision with neutral molecules. If v is the drift velocity imparted by the applied field and it is assumed that directed motion is randomized on collision, then from a momentum balance

$$Ee = eE_0 e^{j\omega t} = M\frac{dv}{dt} + Mvv, \tag{5.39}$$

where M is the mass of the charge carrier and v its collision frequency; the term Mvv is equal to the rate of loss of momentum as a result of randomizing collisions. Assuming v has the form

$$v = v_0 e^{j\omega t} \tag{5.40}$$

it is easily shown by substitution of this expression into eqn (5.39) that

$$v = \frac{E_0 e}{M}\left(\frac{v - j\omega}{v^2 + \omega^2}\right)e^{j\omega t} = \frac{E_0 e}{M}\frac{e^{j(\omega t - \alpha)}}{\sqrt{(v^2 + \omega^2)}}, \tag{5.41}$$

where

$$\alpha = \tan^{-1}\left(\frac{\omega}{v}\right). \tag{5.42}$$

The complex part of the middle expression in eqn (5.41) represents a phase difference between the applied field and the motion of the charge. The current density is given in phase and magnitude by

$$j = nev. \tag{5.43}$$

The conductivity, σ_k, for species k, is defined as being equal to the ratio j_k/E; therefore

$$\sigma_k = \frac{n_k e^2 (v_k - j\omega)}{M_k(v_k^2 + \omega^2)}. \tag{5.44}$$

The expression for conductivity applies to ions and electrons alike, differences arising only in the values of mass and collision frequency used. Clearly, the total conductivity is equal to the sum of the conductivities of all the species present. It is convenient to write the conductivity as

$$\sigma = \sigma_x - \mathbf{j}\sigma_y, \tag{5.45}$$

where, by inspection,

$$\sigma_x = \sum_k \frac{n_k e^2 v}{M_k(\omega^2 + v_k^2)}, \qquad \sigma_y = \sum \frac{n_k e^2 \omega}{M_k(\omega^2 + v_k^2)}. \tag{5.46a \& b}$$

The variation of σ_x and σ_y with ω and v for a single charged species is shown schematically in Fig. (5.13). The quantities in brackets refer to the case when σ_y is measured along the vertical axis. The locus of maxima ($\sigma_{x,\max} = \sigma_{y,\max} = ne^2/2Mv$) lies along the line $v = \omega$.

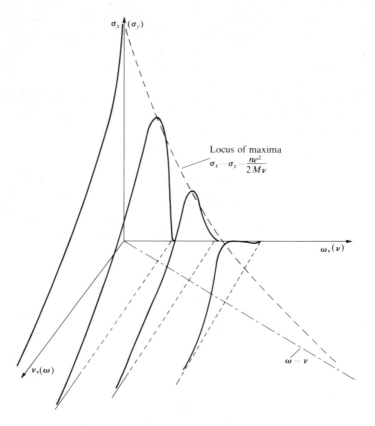

FIG. 5.13. Variation of σ_x and σ_y with v and ω.

The expressions for σ_x and σ_y are approximate inasmuch as they are based upon a simplified treatment of momentum transfer on collision. A more exact treatment has been given by Margenau[39] in which the Boltzmann transport equation is solved. Margenau's expression is

$$\sigma_x - j\sigma_y = \frac{4\pi}{3} \frac{ne^2}{M} \int_0^\infty \frac{\partial}{\partial c}\left\{\frac{c^3}{v(c)+j\omega}\right\} f(c)\, dc, \qquad (5.47)$$

where $f(c)$ is the normalized velocity distribution, the form it takes for a Maxwellian distribution being given by eqn (4.48). For a constant collision frequency, the values of σ_x and σ_y are identical with those in eqns (5.46a and b). Data on the variation of electron collision frequency is given by Frost.[46] However, for the present purpose it will be assumed that σ_x and σ_y are given by eqns (5.46a and b) on the understanding that this is strictly true only if the collision frequency is constant. It is worth noting that if $\omega \gg v$, σ_y tends to the value $ne^2/M\omega$ whatever the nature of the variation of v with velocity.

The real part of the complex conductivity, σ_x, gives rise to motion in phase with the field and is therefore responsible for ohmic losses. The significance of the imaginary part, σ_y, can be seen in terms of the dielectric constant of the plasma. The current density at any point in the plasma is equal to the sum of the current density due to electrons and that due to the changing electric field. In one dimension,

$$j = E\sigma + \frac{1}{4\pi}\frac{\partial E}{\partial t} \qquad (5.48)$$

but $\partial E/\partial t = j\omega E$. Therefore setting $\sigma = \sigma_x - j\sigma_y$ in eqn (5.48)

$$j = E\sigma_x + \left(\frac{1}{4\pi} - \frac{\sigma_y}{\omega}\right)\frac{\partial E}{\partial t}. \qquad (5.49)$$

Now the current density in a leaky dielectric of conductivity Σ and dielectric constant D, is given by

$$j = E\Sigma + \frac{D}{4\pi}\frac{\partial E}{\partial t}. \qquad (5.50)$$

From a comparison of eqns (5.49) and (5.50) it follows that, when in an a.c. field, the plasma behaves as a leaky dielectric with conductance σ_x and dielectric constant $1 - (4\pi\sigma_y/\omega)$.

Ions and electrons respond differently because of their different masses and collision frequencies. Consider the real part of the conductivity first:

$$\frac{\sigma_{i,x}}{\sigma_{e,x}} = \frac{n_i}{n_e}\frac{M_e}{M_i}\frac{\left(v_e^2 + \omega^2\right)}{\left(v_i^2 + \omega^2\right)}\frac{v_e}{v_i}. \qquad (5.51)$$

If the ions and electrons are at comparable temperatures, $M_i v_i^2 \simeq M_e v_e^2$, to a first approximation, and $v_e \gg v_i$ (at 1 atm $v_i \simeq 10^8$, $v_e \simeq 10^{11}$). Therefore, unless the ions outnumber electrons by at least two orders of magnitude, $\sigma_{e,x} \gg \sigma_{i,x}$ and the measurement of the real part of the conductivity provides information about electrons only. From eqn (5.46), if $\sigma_{e,x} \gg \sigma_{i,x}$,

$$\frac{e^2}{M_e \sigma_x} = \frac{v_e}{n_e} + \frac{\omega^2}{n_e v_e}. \tag{5.52}$$

By measuring σ_x as a function of angular frequency, ω, both v_e/n_e and $1/n_e v_e$ can be found from the intercept and gradient, respectively, of the graph of $e^2 M_e \sigma_x$ against ω^2. Consequently v_e and n_e can be deduced.

Consider now the complex part of the conductivity. At angular frequencies equal to, and greater than, the electron collision frequency the contribution of the ions is negligible, unless $n_i > (M_i/M_e)n_e \simeq 10^4 \, n_e$, since

$$\frac{\sigma_{e,y}}{\sigma_{i,y}} = \frac{n_e}{n_i} \frac{M_i}{M_e} \frac{v_e^2 + \omega^2}{v_i^2 + \omega^2} \simeq \frac{n_e}{n_i} \frac{M_i}{M_e} \frac{v_e^2 + \omega^2}{\omega^2}. \tag{5.53}$$

Clearly, v_e and n_e can be determined by measuring σ_y over a range of applied frequencies of the same order as v_e—at atmospheric pressure this corresponds to the microwave range. On the other hand, at angular frequencies of the order of the ion collision frequency, $\sigma_{e,y}$ and $\sigma_{i,y}$ are comparable for equal concentrations. This can be shown from eqn (5.47) by setting $M_e v_e^2 \simeq M_i v_i^2$ and $\omega \simeq v_i \ll v_e$. In this region $\sigma_{e,y}$ is approximately constant, whilst $\sigma_{i,y}$ is strongly frequency dependent, permitting the determination of v_i and n_i from the variation of $\sigma_{i,y}$ with ω. At atmospheric pressure the frequencies required are in the 100 Mc/s range.

Various techniques that have been developed for the measurement of σ_x and σ_y will be discussed next. They differ principally according to whether resonating circuits or propagating electromagnetic waves are used.

Resonance of a.c. circuits

If a potential $V_0 \, e^{j\omega t}$ is applied to a resistance R, a capacitance C, and an inductance L joined in series, the current is given by

$$i = \frac{V_0 \, e^{j(\omega t - \alpha)}}{\sqrt{[R^2 + \{\omega L - (1/C\omega)\}^2]}} = \frac{V_0}{Z} \cdot e^{j(\omega t - \alpha)}, \tag{5.54}$$

where

$$\alpha = \tan^{-1}\left(\frac{\omega L - 1/\omega C}{R}\right). \tag{5.55}$$

The current reaches a maximum when $Z = R$, i.e. at an angular frequency ω_0, given by

$$\omega_0 = 1/\sqrt{(LC)}. \tag{5.56}$$

Defining parameters

$$Q_0 = \frac{\omega_0 L}{R} \text{ and } \Omega = \frac{\omega}{\omega_0}, \tag{5.57a \& b}$$

$$\frac{R}{Z} = \left\{1 + Q_0^2 \Omega^2 \left(1 - \frac{1}{\Omega^2}\right)^2\right\}^{-\frac{1}{2}} \propto \frac{V_0}{Z} \propto i \tag{5.58}$$

for given values of R, L, and C. The variation of i with frequency is shown schematically in Fig. 5.14. The larger the value of Q_0 the sharper the peak, i.e. the more selective the system. In particular for $Q_0 \gg 1$,

$$Q_0 = \frac{1}{\Delta\Omega}, \tag{5.59}$$

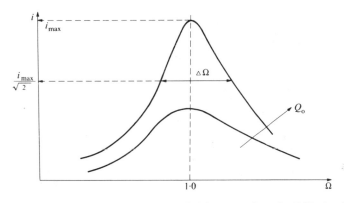

FIG. 5.14. Variation of current with applied frequency in series RCL circuit.

where $\Delta\Omega$ is the width of the resonance curve at a current equal to the peak current divided by $\sqrt{2}$. Given the inductance, therefore, C can be found from the resonant frequency (eqn (5.56)) and R can be deduced from Q_0 (eqn (5.57a)), which is determined by the width of the resonance curve (eqn (5.59)). In practice the circuits are more complex than that just described. However there is no basic difference in the determination of C and R, from which σ_y and σ_x follow directly; the former from the dielectric constant, the latter from the resistance.

As was shown, it is possible to determine ion collision frequencies and concentration by using values of $\omega \approx \nu_i$. A radio-frequency system for making this measurement is shown in Fig. 5.15(a). The plasma is made part of the circuit by placing it within a self-resonant coil. Numerous investigations at radio frequencies have been carried out. In some a parallel plate condenser was used,[42,43] Fig. 5.15(b), in others the plasma was made part of the circuit by placing it within a coil.[41,44] The main disadvantage of the technique is its poor spatial resolution.

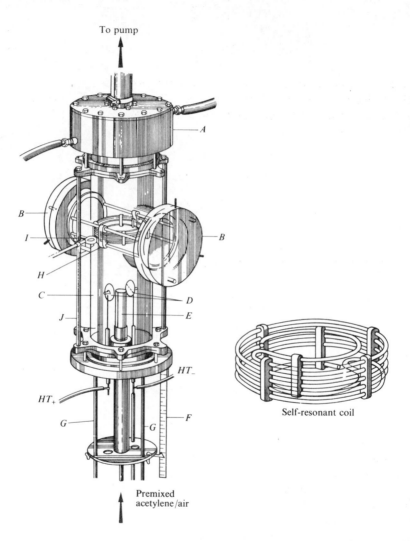

To pump

Premixed
acetylene/air

FIG. 5.15(a). Self-resonant cell for the measurement of flame ionization at radio frequencies. A, water-cooled head; B, Perspex window; C, Pyrex low-pressure vessel; D, ignition discharge electrodes; E, burner tube assembly; F, burner position scale; G, burner guide rods; H, self-resonant coil; I, generator coil; J, coil supports (polystyrene). From Williams.[41]

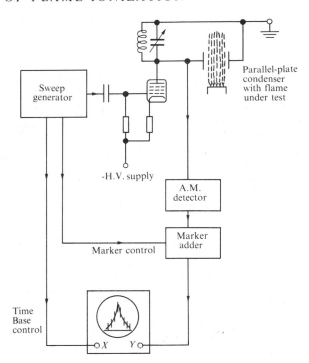

FIG. 5.15(b). Parallel-plate condenser circuit for the measurement of flame ionization at radio frequencies. From Borgers.[42]

At angular frequencies of the order of the electron collision frequency, i.e. in the microwave range at atmospheric pressure, only electrons contribute significantly to σ_x and σ_y from which, in principle, both v_e and n_e can be found. At these frequencies the plasma is inserted within a resonant cavity, the change in the Q of which is used to determine σ_x. In early experiments a long cylindrical cavity operating in the TE_{011} mode was used,[45] but this gave poor spatial resolution, of the order of 3 cm. More recently, Padley and Sugden[46] have obtained a resolution of about 3 mm using a cavity operating in the TM_{010} mode at about 3000 Mc/s (Fig. 5.16). It is worth noting that these workers were able to deduce n_e from the measurement of σ_x without evaluating v_e explicitly, by calibrating the cavity with flames seeded with known concentrations of caesium, which was completely ionized at the prevailing temperature.

Methods based upon the resonance of a.c. circuits are useful down to electron densities of the order of 10^8 cm^{-3} at atmospheric pressure. Below this concentration σ_x is too small to be measured. However in view of the increase in conductivity that accompanies a decrease in collision frequency, (see Fig. 5.13), it should be possible to obtain greater sensitivity at lower pressures by applying lower frequencies.

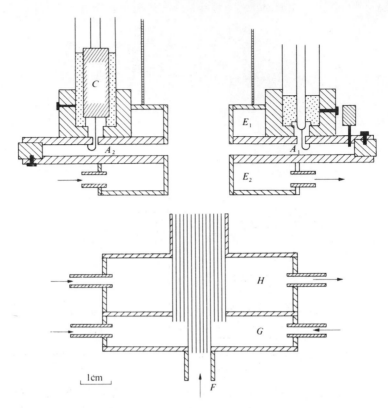

FIG. 5.16. The 10-cm cavity and burner. A_1, A_2, input and output loops; C, crystal, E_1, E_2, and H cooling water F, inner flame + additives; G, shield flame. From Padley and Sugden.[46]

Propagation of electromagnetic waves through plasmas

When a plane electromagnetic wave propagates through medium or complex conductivity, σ, the electric field strength

$$E = E_0\, e^{\mathbf{j}\omega(t - z/v)}, \tag{5.60}$$

where t is the time, z the distance measured in the direction of propagation and v is given by

$$\frac{1}{v^2} = \frac{\{1 - (\mathbf{j}4\pi\sigma/\omega)\}}{c^2}, \tag{5.61}$$

c being the velocity of light *in vacuo*. The reflection coefficient, R, for a sharply bounded plasma is given by

$$R = \frac{\text{reflected power}}{\text{incident power}} = \text{mod}\left|\left(\frac{1 - v/c}{1 + v/c}\right)^2\right|. \tag{5.62}$$

From eqn (5.61), constants c_p and β can be found such that

$$v^{-1} = \frac{1}{c_p} - j\beta. \tag{5.63}$$

On evaluation, using eqns (5.46a and b) and (5.61), it is found that

$$\frac{1}{c_p}(+) \text{ or } \beta(-) = \frac{1}{c\sqrt{2}}\left[\left(1 - \frac{\omega_p^2}{\omega^2 + v_e^2}\right)^2 \right.$$

$$\left. + \frac{v_e^2}{\omega^2}\frac{\omega_p^4}{(\omega^2 + v_e^2)^2}\right\}^{\frac{1}{2}} \pm \left(1 - \frac{\omega_p^2}{\omega^2 + v_e^2}\right)\right]^{\frac{1}{2}}, \tag{5.64}$$

where, by definition,

$$\omega_p = \sqrt{\left(\frac{4\pi e^2 n_e}{M_e}\right)} = 5\!\cdot\!64 \times 10^4 \sqrt{n_e} \ \text{s}^{-1} \tag{5.65}$$

and the bracketed plus and minus signs on the left-hand side of eqn (5.64) refer to those within the square root. The quantity ω_p is known as the plasma frequency.

From eqns (5.60) and (5.63)

$$E = E_0\, e^{-\beta\omega z} . e^{j\omega(t - z/c_p)}. \tag{5.66}$$

The first exponential term leads to attenuation. Since the power is proportional to the square of the electric field

$$\frac{W}{W_0} = e^{-2\beta\omega z} = e^{-\chi z}, \tag{5.67}$$

where W is the power at z, W_0 the incident power, and χ is known as the attenuation coefficient. It is usual to quote attenuation in dB/cm, β'. By definition,

$$\beta' = -10\log_{10}\frac{W_1}{W_0} = 4\!\cdot\!35\chi \ \text{dB/cm}, \tag{5.68}$$

where W_1 is the energy flux remaining after a path of 1 cm. (1 dB is equivalent to an attenuation of about 20 per cent.)

By inspection of eqn (5.66) it is clear that c_p is the wave velocity in the plasma. Therefore the refractive index is given by

$$\mu = c/c_p. \tag{5.69}$$

The value of c/c_p can be obtained from eqn (5.64). It is useful to note that, when $\omega^2 \gg \omega_p^2$ and $\gg v_e^2$, the expressions for μ and χ become

$$\mu = 1 - \frac{1}{2}\frac{\omega_p^2}{\omega^2}, \tag{5.70}$$

$$\chi = \frac{v_e\omega_p^2}{c\omega^2}. \tag{5.71}$$

The value of χ is approximate in that, being a function of v_e, it is dependent on the manner in which v_e varies with electron velocity. The expression is only strictly correct for a constant collision frequency. Equation (5.70) for μ is accurate, being independent of v_e.

Balwanz[47] has calculated attenuation and wave velocity from eqn (5.64) over a range of plasma properties. His results for attenuation are shown in Fig. (5.17) in which β' is in dB/m, f is the wave frequency ($\omega/2\pi$), and the electron concentration is in m^{-3}. The units are the same for Fig. (5.18), which gives the wave velocity in m/s as a function of v_e, f, and n_e.

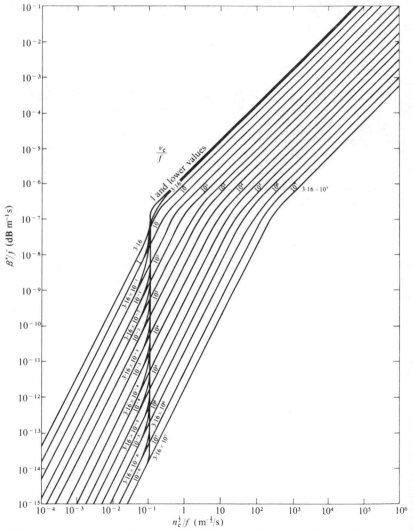

FIG. 5.17. Attenuation as a function of collision frequency, electromagnetic wave frequency, and electron concentration. From Balwanz.[47]

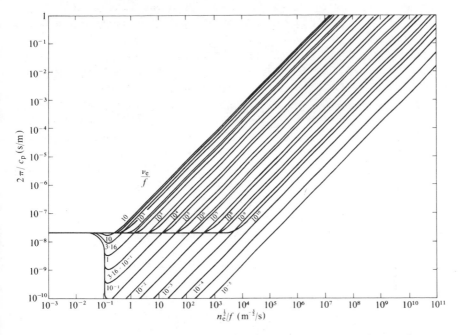

FIG. 5.18. Wave velocity as a function of collision frequency, electromagnetic wave frequency, and electron concentration. From Balwanz.[47]

It is obvious from eqn (5.70) that the velocity of the wave in the plasma can exceed that *in vacuo*. In Fig. (5.18) the horizontal line represents the vacuum condition; below it $c_p > c$. This is not a contradiction of special relativity, however. Relativity limits the velocity of a group of waves to that of light *in vacuo* but sets no limitation upon the speed of a continuous monochromatic wave train since this can carry no information.

It is possible to study plasmas by reflection, attenuation, or refractive index change. These will now be discussed in turn with particular reference to combustion systems.

Reflection. If a plasma has a sharp boundary and $\omega_p \gg \nu_e$, the reflection coefficient is very small for $\omega > \omega_p$ and rises steeply in the region $\omega \simeq \omega_p$. Thus by measuring the reflection coefficient as a function of frequency, ω_p can be estimated and the electron concentration deduced (see eqn (5.65)). This technique has been used by Gardner[48] to study a low pressure, 95 per cent ionized helium plasma, the electron concentration being of the order of 10^{13} cm^{-3}. However, although useful for order of magnitude estimates of electron concentrations, reflection methods suffer from uncertainties because of the diffuse nature of the plasma boundary. This may be a serious limitation in combustion systems, especially at low pressure, but may be less serious in shock and detonation tubes, for, in these, the boundary is relatively sharp initially. As a general guide, the thickness of the diffuse layer should be much less than $1/\chi$ (the skin depth), i.e. the distance over which the power is reduced to $1/e$ of its original value.

Attenuation. Attenuation has been widely used for the study of ionization, both within and outside the field of combustion. In order to avoid spurious effects arising from reflection at the plasma boundary, it is necessary to ensure that $\omega > \omega_p$. There will also be strong diffraction if the frequency is so low that the wavelength of the radiation is comparable to the plasma dimensions. On the other hand, if the frequency is too high, attenuation is negligible—see eqn (5.71). Although ideally, the frequency employed should provide a wavelength short enough to achieve the resolution required by the phenomenon under study, in practice the attainment of a measurable level of attenuation sets the limit to the spatial resolution.

The greatest amount of information can be obtained when ω and v_e are comparable. Under these circumstances, when $\omega_p/\omega < 0.3$,

$$\beta' \simeq 0.46 \frac{n_e v_e}{v_e^2 + \omega^2} \text{ dB/cm.} \tag{5.72}$$

Thus by measuring β' as a function of ω, both n_e and v_e can be determined. Belcher and Sugden[49] measured n_e and v_e in coal gas–air flames at atmospheric pressure in this manner. Because $v_e \simeq 10^{11} \text{ s}^{-1}$ in atmospheric pressure flames, they used microwave frequencies. A sketch of the apparatus is shown in Fig. 5.19.

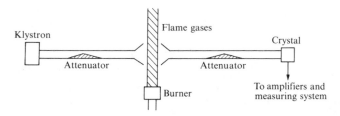

FIG. 5.19. Direct attenuation measurement. From Belcher and Sugden.[49]

Because microwave frequencies are the highest that can be used in order to obtain appreciable attenuation, at the electron concentrations found naturally in flames (c. 10^{12} cm^{-3}), they have been employed by numerous other workers in the field of combustion; for example, Smith and Sugden,[43] Shuler and Weber,[50] Balwanz, Headrick, and Ahern,[51] and Belcher and Sugden.[52] Those that have not made use of the variation with frequency to obtain electron collision frequency have had to assume, or to calculate, a value of v_e in order to deduce the electron concentration.

For results to be obtainable, β' must be greater than about 0.05 (1.25 per cent attenuation/cm) for plasmas of the order of a centimetre thick—

Belcher and Sugden[49] were able to measure to ± 0.01 dB. Therefore, in the region $\omega \simeq v_e$, n_e must be greater than $0.2\ v_e$. Even if $\omega \ll v_e$, i.e. when β' is largest, n_e must still be greater than $0.1\ v_e$. At atmospheric pressures this requirement limits the use of the method to electron concentrations in excess of 10^{10} cm^{-3}.

As an alternative to varying ω, it is possible to measure attenuation in a variable magnetic field. In this method the electric field vector of the wave is aligned at right angles to an applied magnetic field, H. If $\omega^2 \gg \omega_p^2$ and $\gg v_e^2$ the attenuation for a path length of z cm is given by

$$\beta' = 0.23\ n_e v_e \left\{ \frac{1}{v_e^2 + (\omega + \omega_b)^2} + \frac{1}{v_e^2 + (\omega - \omega_b)^2} \right\} z \text{ dB}, \qquad (5.73)$$

where ω_b, the electron cyclotron frequency, is $eH/M_e c$ (see Chapter 4). The second term on the right shows a strong resonance at $\omega = \omega_b$. If attenuation is measured at constant ω as a function of H, n_e and v_e can be determined from the attenuation maximum and the width of the resonance curve of β' against H:

$$\Delta H/H_0 = 2 v_e/\omega \qquad (5.74\text{a})$$

and

$$n_e = 3.8 \times 10^7 \beta_0' \Delta H/z \text{ cm}^{-3}, \qquad (5.74\text{b})$$

where β_0' is the attenuation in dB when $\omega = \omega_b$, i.e. when $H = H_0$, and ΔH is the width of the curve between points where $\beta' = \beta_0'/2$. In using these expressions it is essential that the inequality $\omega^2 \gg \omega_p^2$ be strictly fulfilled—a discussion of this point has been given by Gray.[53] Employing the cyclotron resonance technique Bulewicz and Padley[54] have investigated a variety of low pressure hydrocarbon and cyanogen flames (1–10 cmHg) and Schneider and Hofmann[55] have studied acetylene-air flames in the pressure range 4.6–10.6 cmHg.

It is interesting to compare direct and cyclotron resonance attenuation. In the presence of a magnetic field, if $\omega \gg v_e$, maximum attenuation is independent of ω, being equal to $0.23\ n_e z/v_e$ dB (eqn 5.73). It is therefore theoretically possible to use high-frequency waves to achieve good spatial resolution. However, ω is limited by the maximum attainable cyclotron frequency. Taking 5×10^4 G as an upper limit for the magnetic field, the cyclotron frequency, and therefore ω, is limited to a maximum of about 8×10^{11} s^{-1} ($\lambda = 0.3$ cm). This not only limits the spatial resolution but also renders the method unsuitable for combustion studies at pressures of the order of 1 atm or above, because already at 1 atm $v_e \simeq 10^{11}$ s^{-1}.

On the other hand, in the absence of a field ω cannot greatly exceed v_e if there is to be appreciable attenuation (see eqn (5.72)). For this reason direct attenuation is unsuitable for low pressure studies, the wavelength becoming so long that it invalidates the measurement. This is in contrast with cyclotron resonance measurements that become more sensitive as the pressure is reduced.

Refractive index. The optical path difference between a ray travelling z cm in medium 1 and the same distance in medium 2 is given by

$$p\lambda = z(\mu_2 - \mu_1),\qquad(5.75)$$

where λ is the wavelength and p the number of wavelengths by which the paths differ (the 'fringe shift' in an interferogram). If one medium is the test space and the other a reference path, for example a flame in air, the fringe pattern formed when the beams are made to interfere, provides information about the refractive index field in the test space.

When using interferometry to explore the refractive index fields in plasmas it is necessary to avoid excessive reflection losses, i.e. $\omega \gg \omega_p$. In order to illustrate the sensitivity of the method the reference path will be assumed to be in vacuum, i.e. $\mu_2 = 1$. Now if $\omega > 3\omega_p$ the relation given in eqn (5.70) can be used, the approximation improving as the ratio ω_p/ω diminishes. The fringe shift due to the electrons in the test space then becomes

$$p\lambda = \frac{z}{2}\frac{\omega_p^2}{\omega^2}.\qquad(5.76)$$

Substituting for ω_p^2 and putting $\omega = 2\pi c/\lambda$

$$\frac{n_e}{\omega} = -118\cdot5\,\frac{p}{z}.\qquad(5.77)$$

The sensitivity of an interferometric method can be expressed as the minimum fringe shift detectable, say p_{min}. Thus for measurements to be possible, ignoring the sign of p,

$$\frac{n_e}{\omega} > 118\cdot5\,\frac{p_{min}}{z}.\qquad(5.78)$$

For example, if $p_{min} = 0\cdot1$, $z = 1$ cm and $n_e = 10^x$ cm^{-3}, then ω must be less than $8 \times 10^{(x-2)}$ s^{-1}, i.e. λ must be greater than $2 \times 10^{(12-x)}$ cm. Thus if $n_e = 10^{12}$ cm^{-3}, $\lambda > 2$ cm. The value of p_{min} chosen is somewhat arbitrary but of the right order of magnitude in the absence of special techniques. It is worth noting, however, that in the case of optical interferometry detection of differences of the order of 10^{-2} of a fringe has been claimed.[56]

The dispersive nature of the electron component to the overall refractive index makes it possible to distinguish the electrons from other species present. Whereas the optical path for neutral species is almost independent of frequency (except near an absorption band), the electron contribution varies as ω^{-2}. Thus by making measurements at two frequencies, the concentrations of both electrons and neutrals can be determined.

In most flame systems the spatial resolution attainable is very poor because of the long wavelengths required for electrons to produce a measurable contribution to the phase shift. Optical methods to measure refractive index changes caused by density variations have, of course, been used extensively in combustion (see Weinberg[57]). The large temperature and composition-induced changes in refractive index tend to contribute to the difficulty of 'electrical' measurements. For electron densities found naturally in flames (10^{12} cm^{-3}) the minimum wavelength is 2 cm for a 1-cm path in terms of the above sensitivity criterion and decreases in inverse proportion to z (see eqn (5.78)). The prospects for interferometry of augmented and high temperature seeded flames and detonations, to determine electron concentrations in excess of 10^{13} cm^{-3}, are very much better. For example, Brown and Bekefi[58] have used a far infra-red interferometer ($\lambda = 0.03$ cm) and measured electron densities in excess of about 5×10^{13} cm^{-3} over path lengths of the order of 1 cm. Interferometry is particularly convenient at optical frequencies, i.e. for electron densities in excess of about 10^{16} cm^{-3}, but such dense plasmas are rare in combustion. It is worth noting though, that Friedman and Macek[59] calculated an equilibrium electron density of 10^{16} cm^{-3} in the combustion products of a solid propellant composed of $C_2(CN)_4$, $C_2(NO)_6$, and CsN_3 (see Chapter 3). Plasma interferometry at optical frequencies has been surveyed in detail by Alpher and White.[56]

Apart from interferometry, deflection mapping, schlieren, and shadow methods are also available for the study of refractive index fields.[57] The application of schlieren methods to plasmas has been discussed by Lovberg.[60]

'EXTERNAL' METHODS

The model envisaged for those 'external' methods, in which ions are transported by a field, is that of a flame between plane electrodes, across which a large potential difference is applied and from which the flame is well separated. In each of the two electrode spaces, charges are all of one kind (because field strengths are high enough to induce ion drift velocities much higher than those due to convection and other effects) and they travel to the electrode in gases that often differ from those in the flame. It might appear more favourable, at first sight, to immerse the electrodes into the flame gases and mass spectrometric measurements, for instance, generally do attempt to approach this ideal. Since, however, this produces large quenching regions of indeterminate composition and temperature variations, the region of non-equilibrium ionization itself being so thin, for some measurements it is considered preferable to recognize the essential heterogeneity of the electrode regions and allow the flame to be acted on by fields alone.

It will be convenient to consider first the measurement of amount of charge, followed by the identification of its carrier and discuss the discrimination of position of origin with each method in turn.

Current densities

It is in principle possible to measure flame-ion concentrations from currents withdrawn, in terms of conductivity. An example of such a method is provided by the work of Van Tiggelen[9,10] mentioned in the preceding section. The distinction between electrodes internal and external to the flame becomes blurred when this function is fulfilled by the burner edges; although the electrode spaces were cold in this work, their effect was elegantly subtracted out of the system by using two lengths of flame path. However, it will be shown in Chapter 8 that the theory expressing conductivity in terms of charge concentration applies strictly only for vanishingly small fields, in the presence of electrons. Furthermore, since such concentrations are the result of competition between a rate of generation and a rate of recombination, this is perhaps not the most direct use of current measurement, particularly since both these rates vary rapidly with position in the flame and all the possible methods average over appreciable distances. The alternative aim has been[61,62]† to determine the rate of ion generation per unit area of flame front in a unidimensional system—the flame-front being perpendicular to the single dimension. This rate, like other flame parameters, is a unique property of the composition and state of an inflammable reactant mixture and, in the usual case where equilibrium product ionization does not contribute appreciably, the main zone of ion generation is narrowly confined towards the end of the region of temperature rise; see Fig. 6.1.

The principle of the method is to determine j_s, the saturation current per unit area of flame. The essential property of j_s is that, unlike smaller and larger values of current (the latter occurring at breakdown) it does not rise with increasing applied potential. The method must then be devised so as to produce a clearly discernible plateau in the current-voltage relationship and, in order to consider how this can be achieved, some of the theory discussed in more detail, in Chapter 8, will be anticipated here.

The system to be considered consists of a flame between two electrodes, all three surfaces being plane and parallel to one another. Suitable practical approximations to this ideal can be obtained by the use of Botha–Spalding[63] or Egerton–Powling[64] burners for premixed reactants and the flat counter-flow diffusion flame[65] for initially separate reactants. These were discussed in Chapter 3. Edge effects can be further minimized by use of the guard-ring principle and this is perhaps the main asset of these types of burner. Thus

† *Footnote added in proof.* See also papers by Boothman, Lawton, Melinek, and Weinberg, and by Peeters and Van Tiggelen, 12th International Symposium on Combustion, 1968.

other flame configurations—such as those of Fristrom's spherical flame or of Wolfhard and Parker's flat diffusion flame (see Chapter 3)—would seem to offer equally suitable geometries, except that parts of the burner into which lines of force could crowd are situated near the periphery of the flame.

When a field is applied, the generation—recombination equilibrium in the flame is altered by the addition of a charge withdrawal term. As the field is increased, the latter competes against recombination with increasing success and the current rises with applied potential. Eventually, ions are removed so rapidly that they have no time to recombine. The current then ceases to increase with applied potential, reaching its saturation value, which is the integral of the generation rate across the flame thickness and is virtually independent of electrode separation, except for flames of high product ionization. This state persists until the field is high enough for secondary ionization to occur somewhere in the system. Thereafter the current rises with applied potential once more. The main consideration in the design of the experiment[62] is to defer secondary ionization to high enough potentials so that j_s can be measured. Since, at these field intensities, the ion cloud in each electrode space may be assumed to be unipolar, we have

$$j = neKE \qquad (5.79)$$

on either side, if ion generation is negligible outside the flame. Substituting into Gauss's Law (eqn (4.3a)) gives the distribution of field strength as

$$\frac{dE}{dx} = 4\pi ne = \frac{4\pi j}{KE}, \qquad (5.80)$$

i.e.

$$E^2 = E_0^2 + \frac{8\pi jx}{K}, \qquad (5.81)$$

where E_0 is the field at the flame ($x = 0$). Thus, because of space charge, E rises continuously with distance from the flame, its rate of increase proportional to (j/K). Since j is constant, the highest field, leading to breakdown, occurs at the more distant electrode, if the mobilities are the same. If they are not (e.g. because the anode space is in hot products in which electrons have a longer life), the relevant quantity is (a/K), where a is the electrode separation. As E_0 is close to zero up to saturation (Chapter 8), the strongest ion source for which j_s can be measured is given by

$$j_s = \frac{E_B^2}{8\pi a}, \qquad (5.82)$$

where E_B is the breakdown field and the equation is limiting in that electrode space in which ($E_B^2/8\pi a$) is less. For $E_B = 30\,\text{kV/cm}$, $a = 5\,\text{mm}$, and K for H_3O^+ at s.t.p., the above gives

$$j_s = 0.25\,\text{mA/cm}^2.$$

The values for stoichiometric hydrocarbon-air flames[62] are of the order of one-fifth of this.

It follows that, although there is no difficulty about measuring saturation currents of weak ion sources[61] (see Fig. 8.3), for strong ion sources it is essential to minimize electrode separation, particularly on the side of the ion of lower mobility. For this purpose the Botha–Spalding burner, with the porous plate serving as cathode, is ideal, since it automatically minimizes the electrode distance on the cold, low-mobility side to a value considerably below the 5 mm used in the estimate above. It has the added advantage that the flame temperature can be varied, at constant partial pressures of the reactants, by varying total flow rate and hence the heat loss to the burner. It is by this method that most of the rates of generation and effective activation energies discussed in the next chapter were obtained.[62]

The chief advantages of the method are that it yields generation rates unaffected by recombination and that it is not impaired by charge transfer since, no matter how much the ion identities change on the way to the electrodes, charge flux is conserved. This is most valuable in studies of mechanism and kinetics. As regards practical applications, it provides data required for all those which involve drawing a current, in the form of the maximum charge that is available from unit area of flame-front in unit time. For processes that do not involve drawing a current, for example the attenuation of electromagnetic waves by rocket exhausts, direct measurement of electron concentrations by one of the methods discussed in the preceding section is more relevant.

An interesting case should arise if flame ionization were due chiefly to collisions with excited electrons, as has been suggested by Cozens and von Engel[35]; see Chapter 6. Since electron temperature must increase with the applied field, the rate of ionization would be expected to follow and this would provide a new additional reason for current to increase with field strength. Under such conditions, the current that can be drawn from flames for practical purposes could not be deduced at all from measurements of ion concentration (normally it is simply $j_s = \int r_e \, dx = \int \alpha n^2 \, dx$). In fact, no saturation current would be predicted on this model—a smooth transition to secondary ionization at the electrode would be expected to take its place. The existence of well-defined saturation plateaux even at low pressures has been put forward as an argument against the predominance of such a mechanism of ionization in flames studied by this method hitherto.

The next question is to what extent the distribution of current drawn can be used to deduce profiles of ion generation, or of their identity. The problems here are quite different from those that arise in the use of probes which, although they can easily be moved with respect to the flame, are liable to interfere in a distinct way with each zone they sample. If the only agency interacting with the flame is a field, the question is, how does the distribution of current density (or ion species) arriving at the electrode differ from that leaving the flame? The main causes of displacement are diffusion and convection—which cause deviations from the lines of force—and, chiefly, distortion of the lines of force themselves, as a result of space charge. In the case of ion identity there is, in addition, charge transfer, clustering, etc. As regards the two-dimensional distribution of current arriving at an electrode from a flame, a qualitative, or at best semi-quantitative, method for portraying the distribution of j over a very large number of stations has been developed.[66,67] This is, essentially, an attempt to 'photograph' flames by the ions they emit in an electric field.

If a flame, or other ion source under study, is placed between two plane and parallel electrodes that are reasonably distant and cold (Fig. 5.20(a)), and

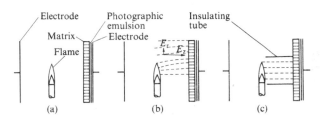

Fig. 5.20. Recording ions on photographic emulsions. From Weinberg.[67]

the aim is to display the distribution of current density over the electrode face as a measure of the ionization distribution in the flame, it follows from the above remarks that three specific problems arise. The first concerns the method of displaying current density, the second, how closely ions can be made to follow lines of force, and the third, how distortion of the lines of force themselves can be minimized.

Some method of visualizing the distribution of current density is desirable, to obviate the need of sectioning the electrode and measuring current to each of an enormous number of elements. This could be done by a television-camera type of device using high speed scanning by an electron beam, but much simpler and less expensive methods offer themselves. These are based on visible changes caused by small amounts of electrolysis of a suitable substance. (At normal pressures, the additional kinetic energy of ions at the largest fields short of breakdown are too small to be utilized in this way.) An obvious example would be to cover one electrode with paper impregnated with potassium iodide-starch solution which allows detection, by the blue colouration produced, of $2 \cdot 5$ μg of iodine—equivalent to the passage of about $1 \cdot 8 \times 10^{-3}$C of charge. It transpires, however, that photographic emulsions, subsequently developed as usual are very much more sensitive. Their sensitivity to small amounts of electrolysis, moreover, appears to lie in the same sequence for different emulsions as does their light sensitivity index.[67] All flame light must, of course, be rigorously excluded.

A suitable screen, opaque to light but conducting electricity in the direction perpendicular to the electrode, consists of a slab of insulating material, transfixed perpendicularly by a matrix of individual wires that are ground flat with both sides. It can be made, for instance, by embedding a uniform wire brush in plastic and grinding the sides flat, after setting. The photographic paper is then sandwiched between the slab and the electrode. To ensure good contact, a layer of sponge rubber covered with metal foil may be used to back the matrix. The extent of blackening and spread of each spot increases with the current carried by the particular wire. A record of current density distribution is thus produced in the manner of a half-tone illustration. A considerable advantage is the increase in sensitivity due to confining all current to the cross-sectional area of the wires. Since marking depends on charge transfer per unit area, the smallest detectable current density is magnified by the factor (opaque area per conducting filament) $\div$ (cross-sectional area of each filament), which can be made of the order 10^3 without much loss in resolution. The absolute limit to the useful minimum cross-sectional area of the wires is set by the grain of the photographic emulsion.

The deviation of ion trajectories from lines of force due to gas flow and to diffusion is minimized by using small electrode separations and large (sub-breakdown) applied potentials—conditions that also ensure maximum currents and hence maximum intensity of marking—i.e. minimum exposure times. Under these conditions, both effects can be rendered negligible. Thus it has been shown[66] that because of high ion velocities, deviations due to flow and diffusion can be kept to 1 mm and 10^{-1} mm respectively, for voltages of 3×10^4 V at electrode separations of 5 cm and mean gas flow velocities up to 60 cm/s. Other possible errors and spurious effects due to bruising of the emulsion, electrolytic cell action, light transmission past, and heat conduction along the wires have been investigated and eliminated.

In this manner an adequate record of current distribution at the electrode can be produced and the method has obvious potentialities for direct 'ion photography', mobility, and other measurements. However, because of space charge, lines of force must always diverge somewhat between flame and electrode (Fig. 5.20(b)) so that the current density distributions at the two stations are never quite the same. Figure 5.21 is an 'ion photograph' of a diffusion flame of hydrogen with a small pre-mixed inner cone and its similarity to the flame indicates that for such weak ion sources this effect is not serious. On the other hand, Fig. 5.22(a) shows the record for a relatively strong source—a 'bat's wing' flame of ethylene. The plane of the flame was parallel to the electrodes: the exposed area, however, is the entire disc of the wire matrix. If the current distribution leaving the flame, rather than that

FIG. 5.21. Partly premixed hydrogen/air flame. From Weinberg.[67]

reaching the electrode, is to be recorded it thus becomes necessary to maintain the lines of force parallel to each other, i.e. the 'ion camera' requires a 'collimator' in addition to an ion-sensitive 'plate'.

A relatively simple method of achieving this will now be briefly discussed because the principle is of more general applicability when ions are to be guided in a field by electrostatic lenses.

An insulating surface placed in the ion stream will become charged to an equilibrium value such that in the steady state no further ions will attach to it. Each part of it will then be at such a potential that the adjacent lines of field will become parallel to it locally. Thus, if the ion stream in the above method of recording is surrounded by an insulating tube (Fig. 5.20(c)), the outer lines of force become collimated. Figure 5.22(b) shows the record obtained

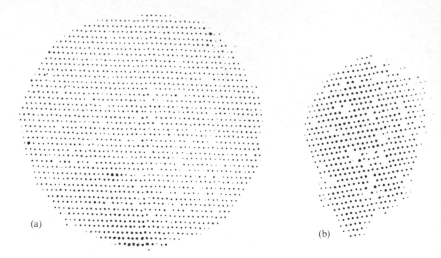

FIG. 5.22. 'Bat's wing' flame of ethylene, (a) without collimator and (b) with collimator. From Weinberg.[67]

in the experiment previously described, when a cylinder made of mica sheet is placed between flame and electrode with its axis perpendicular to both. The outline of the 'bat's wing' flame is now correctly portrayed. This principle is applicable irrespective of the tube shape and ion streams can be 'ducted' in this way to diverge or converge, the insulating tube acting as a former for a self-generated electrostatic lens. Thus magnification or diminution should be achievable. If large variations in current density occur across the field of 'view' it might be necessary to collimate individual tubes of force, for example by using a honeycomb array. It is thought that such a method could be used for guiding and confining charged droplets or particles (for example, of fuel or carbon—see Chapter 7).

Identification of ions: mass spectra and mobility determination

Outstandingly the most accurate device for identifying small ions is the mass spectrometer, which resolves ionic species according to their atomic mass numbers. The chief difficulty of applying the method to combustion systems, aside from expense, is that it identifies species at the detector, and these may be very different from those present in the flame. Two generic types of mass spectrometer are available—that in which resolution by mass is achieved by different lateral displacements (as in an optical spectrometer) and that in which it is based on different times of flight. In general, the latter has the advantage over the former of a potentially more extended sampling zone. It would, however, be difficult to turn this to full account in the case of flame studies, because both types operate at pressures very much lower than those at which flames can be burnt, so that a small entrance aperture is always necessary. This adds interaction of the ion beam with the entrance slit and its boundary layer to the possibility of charge transfer and other possible changes of ion identity between the reaction zone and the mass spectrometer.

Mass spectrometry was used[68] as early as 1947 to analyse samples of gases withdrawn by probes from flames and subsequently ionized. Such analytical uses have been reviewed by Fristrom and Westenberg.[69] In recent years, however, a large number of studies with the purpose of identifying species present as ions already in the flame, have been undertaken by this technique.[29,30,70–79] The majority of these were carried out using conventional dispersing mass-spectrometers, in which a combination of magnetic and electric fields constrains ions to curved paths such that the point of termination of the trajectory at the detector depends only on the charge to mass ratio, i.e. on the mass, for singly charged ions (see Figs. 5.23 and 5.24). For details of such instruments, the reader is referred to appropriate texts. The relevant differences between the several schools of research relate chiefly to the flame and the sampling system, i.e. the method of inducing flame ions to enter the detecting apparatus that operates at 10^{-6} or, at most, 10^{-5} mmHg. This pressure is, of course, many orders of magnitude less than those at which anything resembling a true flame can be burnt and an essential part of the method lies in pumping away the neutral gas while confining as many of the flame ions as possible to the beam that enters the mass spectrometer. Figure 5.24 illustrates the system used by Van Tiggelen's school and Fig. 5.23 that of Sugden. They differ as regards the principle adopted for selectively concentrating the ions by a series of chambers of progressively decreasing pressure, the ion beam being encouraged to pass through the confining slits by fields applied between auxiliary electrodes and the slits themselves. In the first system illustrated, the flame burns at atmospheric pressure, the other three chambers being at approximately 10^{-3}, 3×10^{-5}, and 2×10^{-6} mmHg pressure. Very high rates of pumping are required, particularly to maintain the pressure in the first chamber. A rate of 250 1/s was employed with an entrance aperture (a hole in metal foil) of 0·005-cm diameter. In the work of van Tiggelen and his school, the flame burned at 40 mmHg pressure, and the first aperture opened only periodically for a short time (1/1300 s), which was effected by using a rotating disk containing an opening sliding over a stationary aperture (Fig. 5.24).

Calcote and his colleagues reduced the pressure at which their flames burned even further (to 1–10 mmHg) in order to obtain a thicker flame and better spatial resolution (further discussed below). As an aid to this, the entrace aperture to their radio-frequency mass spectrometer[79] was situated

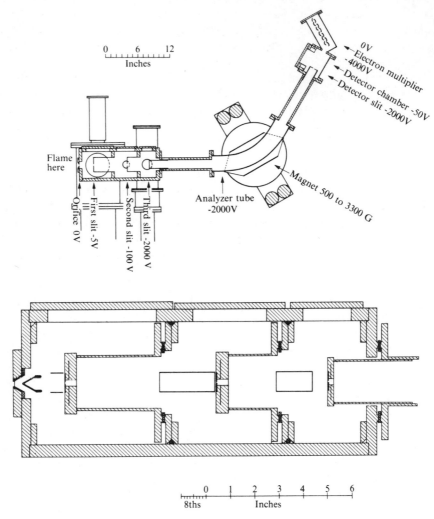

FIG. 5.23. Mass spectrometer and sampling system—Sugden's school.[72-75] From Knewstubb and Sugden.[72] Detail below shows successive slits and focusing cylinders.

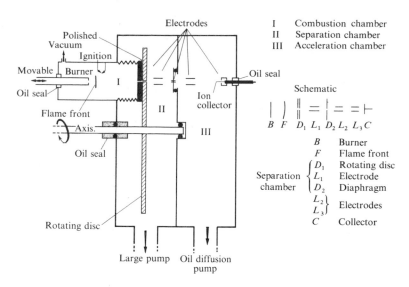

FIG. 5.24. Mass spectrometer sampling system—van Tiggelen's school.[76-78] From Deckers and van Tiggelen.[76]

in the mouth of a conical probe. The pressure there was about 10^{-4} mmHg and dropped to 10^{-5} mmHg in the analyser.

In general, little difficulty has been encountered in identifying most of the peaks obtained. Calibrations have been carried out introducing alkali metals (Li, K, Na) into the flame and using the additional peaks their ions produced. Identification has further been aided by displacements due to isotopes, either occurring in natural proportions or deliberately introduced (e.g. by the use of deuterated acetylene[76]). The main problem however, is to relate ions so identified to those generated within the flame, as distinct from species that may have been produced elsewhere in the apparatus. This will be discussed in some detail, paying attention to the work of Sugden and his co-workers, who have considered this problem in several studies.

First, most of the ions never reach the detector. When the 'electrode lenses' in Fig. 5.23 are made inactive, the ion current falls to about 1/100. Even in the presence of these fields, however, nine-tenths of the ions are lost across the three slits. This loss is unlikely to be a random one, since it must depend on the properties of the ions. Some confirmation of this selectivity derives from a study of ions, primarily in nitrogen after glows,[80] in which the extraction potential (between the entrance slit and a grid in the plasma) was varied. The relative abundance of ions was found to vary with this potential. Long-term (of the order of 1 min) drifts in applied potentials occur in practice and are apparently caused by the build-up of surface charge on the electrodes, once some fouling has occurred. Spokes and Evans[80] have demonstrated long-term variations in individual ion collection rates, using an aluminium system (in which di-electric surface layers would be expected to occur without fouling). Sugden and his co-workers observed such drifts even in a gold-plated sampling system, and frequent cleaning, together with speeding up each run, proved necessary to obtain meaningful results.

Apart from effects of applied fields on relative abundance, the ion can change identity during interaction with surfaces, with the gas, and particularly with that part of the gas that may itself be modified thermally or chemically by the vicinity of surfaces—the boundary layer. The interaction with surfaces and boundary layers is likely to be the most serious within the apertures and such channels should therefore be as short and wide as possible. Eltenton,[68] who in 1947 pioneered the application of mass spectrometry to flames (though with somewhat different aims), gives the average number of collisions of a molecule with the wall as $(4 \times \text{length})/(\text{diameter})$. It is for such reasons that apertures have been made by pricking holes in thin foil. However, the maximum size of hole is determined by the need to maintain the large pressure drop required, for acceptable flow rates, whilst the minimum foil thickness is limited by its mechanical strength against such a pressure drop and by its ability to conduct heat to a supporting cooling coil, if flame gases are allowed to impinge against it. With both the hole diameter and the foil thickness at 0.005 cm,[72] each molecule still collides with the wall about four times during its passage into the first chamber. In spite of this, Knewstubb and Sugden observed no appreciable change in composition of the ion beam when the length of the channel was varied. This question, however, is tied up with the effect of the boundary layer discussed below.

Possible effects in the gas phase have been[72] considered under two headings: events in the first chamber, where the pressure is still quite high, and subsequent changes. The former includes changes in identity due to charge transfer, as well as clustering and dissociation. The attitude towards such events has been that, since they are likely to occur in the flame gases anyway, the first chamber is merely a trivial (because of its low pressure) extension of the former path. This argument presupposes that all the relevant reactions are affected equally by changes in pressure. Subsequent events *en route* in the spectrometer, which could include most of the above, plus secondary ionization (including any due to electrons knocked out of the metal parts) were shown[72] not to be important because all peaks were sharp, corresponded to integral mass numbers and the composition proved independent of the accelerating potential.

The boundary layers associated with surfaces in contact with hot combustion gases are regions not only of different velocity and pressure (as in the case of 'aerodynamic boundary layers') but also of quite different gas temperature and composition. It is this that makes their contribution so important. The ions in the flame are themselves mostly not the 'parent' ions, but they are in some sort of equilibrium with the flame gases, as regards charge exchange, clustering, etc. Now, since the apertures in mass-spectrometric sampling are so small, it is possible for the boundary layer to 'seal the aperture' so that all the ions are filtered through a zone of gas totally unlike flame gases. Table 5.4[74] shows ratios of ion counts, obtained with two

TABLE 5.4

A comparison of maximum ion-counts of various ions with large $(0.7 \text{ cm}^3\text{s}^{-1})$ *and small* $(0.07 \text{ cm}^3\text{s}^{-1})$ *sampling holes*. From Bascombe, Green, and Sugden[74]

Ion	Count (large)/count (small)
18	1·4
19	60
36	1·0
37	1·7
55	3·0
29	20
31	50
33	3·0
39	No result because of potassium impurity difficulties
43	2·0
49	20
51	0·8
53	2·0
59	1·0

different aperture sizes, for various mass numbers between 18 and 59. The conclusion that emerges is that, since the larger aperture must have very considerably favoured flame ions over those created in the boundary layer, many of the bewildering array of ions that have been found in fact originate in the sampling process. Only four of the thirteen ions listed show a sufficient increase to be unambiguously identified as flame ions.

The results of mass spectrometric sampling and the conclusions drawn regarding mechanisms of flame ionization are discussed in another chapter. However, it is just worth mentioning that H_3O^+ is the most abundant species found, because the properties of this ion are used in calculations in other sections of this monograph.

Turning now to the distance resolution of mass spectrometry, this is determined by the sampling aperture. The pressure drop there is always greater than that required to induce sonic flow. Some of the aspects of this problem are common to sampling from flames in general (i.e. sampling of neutral species followed by general chemical analysis) which has been

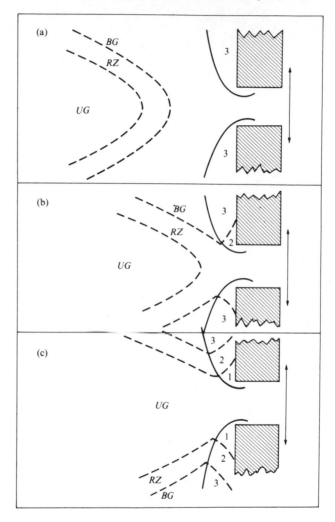

FIG. 5.25. Diagramatic representation of flame, boundary layer, and sampling hole when (a), direct collection of burnt gas (*BG*) is being made; (b), direct collection from the reaction zone is being made; and (c), direct collection of unburnt gas is being made. The arrows to the right represent the diameter over which effective sampling occurs. Regions 1, 2, and 3 represent parts of the boundary layer in which the composition will be approximately that of *UG*, *RZ*, and *BG* respectively, with modification due to cooling and catalysis. From Bascombe, Green and Sugden.[74]

reviewed elsewhere.[69] The volume flow for a single gas inducted is proportional to the orifice area, to the pressure ratio, and to $T^{\frac{1}{2}}$; this defines the size of the sampling zone. This theory applies for a sampling aperture large in comparison with the mean free path, which represents the conditions of Sugden and his school, even at the highest temperatures. It is worth noting, however, that in going to pressures of a few millimetres of mercury, in order to obtain a thick reaction zone, one can run into conditions where mean free paths can locally exceed the size of sampling apertures even as large as several tens of microns. If this happens, the sample will be biased towards constituents of lower molecular weight (proportionally to the inverse square root of molecular weight) because of their greater molecular velocities.

Sampling followed by analysis, as used in flame structure studies, however, is carried out with fine uncooled quartz micro-probes (quenching follows upon expansion) and is a much more delicate method, of higher distance resolution, than that which can be achieved in the presence of a relatively massive water-cooled entrance aperture in the mass-spectrometer system. This is particularly to be borne in mind in conjunction with the role played by cool boundary layers, as discussed above. It is well illustrated by Fig. 5.25[74] from the work of Bascombe, Green, and Sugden in which various parts along the axis of the small conical flame were permitted to impinge upon the sampling aperture. One complete traverse represents a flame displacement of 0·35 mm. It will be seen that, in addition to the complexities already discussed, the gases from which the boundary layer is initially formed vary from products via reacting species, to reactants in a rather indeterminate manner, in going from (a) to (c). In view of the high rate-constants of ion-molecule reactions and the profound and specific influence of each boundary layer composition and state on the ion constitution discussed above, these authors conclude that, as regards ion profiles within such primary reaction zones, '... detailed interpretation of results like those shown ... becomes prohibitively difficult unless further information can be obtained'. It seems possible that near limit flames that attain thick flame zones even at normal pressures are more amenable to this kind of analysis of the distribution of ionic species. Profiles in the recombination region downstream of the primary reaction zone are, of course, much more gradual and, at reduced pressures, the entire structure expands—see Chap. 3.

Mobilities outside flame gases

The physical principles of the 'direct methods' of measuring ion mobilities resemble those of mass spectrometry, except in that a flow of cold, inert gas replaces the magnetic field as the agency displacing ion trajectories from the lines of electric field. The consequent advantages are that the apparatus is not at low pressure (thus avoiding sampling difficulties) and is very considerably cheaper, but there are disadvantages in that the long paths in gas at normal pressures afford every opportunity for attachment, clustering, and charge transfer. This is somewhat mitigated by the freedom to choose the constitution, temperature, and pressure of the carrier gas.

The literature of the measurement of mobilities of ions from sources other than flames is much too extensive to be surveyed here—a review is given, for instance, by Loeb[81]—and again only a summary of the two more relevant principles analogous to the two types of mass spectrometer will be given here. The use of a uniform flow of carrier gas is exemplified by the Erikson air-blast method[82]—Fig. 5.26. A and B are the rectangular plate electrodes

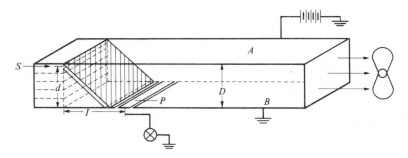

FIG. 5.26. Measurement of mobilities by the air-blast method. After Loeb.[81]

and form two sides of a box that is completed by two plates of a non-con-taminating, insulating material, such as plate glass. The control field is applied between A and B but only that current passing to the thin insulated strip, P, is recorded. With ions entering from a source at S, as shown, a fast but non-turbulent flow of gas is passed along the rectangular tube and may be recirculated through purifiers; air was used in Erikson's method, but another gas may often be more suitable. If the point of arrival of one species of ions, under the competing effects of field and flow, occurs at a distance l (see Fig. 5.26) downstream, their K can be deduced.

The time-of-flight mass spectrometer has a counterpart in the time-of-flight methods for measuring ion mobilities, for example that due to Tyndall, Starr, and Powell[83]—Fig. 5.27. Here the gas is stationary and resolution by mobility is achieved by pulsing the ion supply. The method has been the subject of many refinements, but essentially consists of four parallel gauzes, B, C, D, E in that sequence. The main d.c. field under which the ion drift velocities are measured is established over an appreciable and variable distance between C and D. The ion source, A, which may be extensive, is to the left of B, while the ion current detector, F, is to the right of E. The two combinations of closely-spaced gauzes B and C, D and E are made to act as periodic gates by applying an alternating potential between the partners in each pair so that ions can pass only during pulses of short duration. The

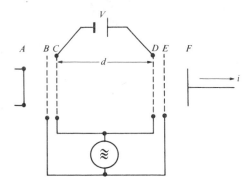

FIG. 5.27. Electric-shutter method for measuring on mobilities. From von Engel.[84]

frequency of both 'gates' is the same. Thus charges reach the receptor beyond
E only if a burst of ions leaving C and drifting to D arrives there at a period of
transmission by the second gate. The accuracy of the method is greatly
enhanced by repetition of such coincidences at fixed incremental distances,
d, as the assembly of D, E, and detector is gradually displaced from C.

In addition to such direct methods, there are several indirect ones that
are consequent upon mobility being the principal variable in quantities such
as the ionic wind and field distribution for a given current density—see
Chapter 4. These are less discriminating as diagnostic methods because they
yield some mean mobility, if several ions are involved.

The 'diagnostic' measurement of ion mobility in a standard gas outside
flames has so far received only scanty attention, for reasons discussed below.
Effective mean mobilities were deduced, almost incidentally, in an investigation[61] correlating current densities with ionic winds from flames. The
method has been discussed in Chapter 4.

A more serious attempt[85,86] has been based on an adaptation of the air
flow method—see Fig. 5.28. A small flame burning around an axial wire
was surrounded by two co-axial cylinders. The effective ion source was a
circular slot cut in the inner cylinder, the field between the inner cylinder
and axial wire merely serving to draw the flame ions to it. The main control
field was arranged between the co-axial cylinders and the carrier gas was
induced to flow in this annular space. The points of ion arrival at the outer
tube were ascertained by recording the current to the insulated collector
ring, which could be moved vertically. It can be shown that, for a given
volume flow rate of carrier gas, the point of arrival of ions of a particular
mobility is independent of the gas velocity distribution (although the shape
of the trajectory does depend on it). In spite of considerable aerodynamic
difficulties occasioned by stabilizing a small flame in the vicinity of a rapid
air flow and attempting to extract a representative sample of ions through the
slit, two to three distinct ion peaks were usually obtained. This was very

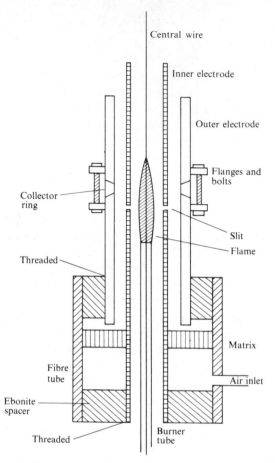

FIG. 5.28. Measurement of mobility of ions drawn from flames.[85,86]

much a preliminary attempt aimed at establishing the possibility of the method and no obstacle was encountered that could not be overcome with some development.

The more intractable difficulties are theoretical, as concerns certain groups of diagnostic application of the method. The theory of mobility of small ions is discussed in Chapter 4; Figure 4.1 shows the variation of mobility with mass for small ions in nitrogen at s.t.p. It will be seen that for large ionic masses (still within the range of small ions) the mobility varies too gradually for ion identification by mobility to be reliable. The smallest ions, however, are those most likely to change during the inevitably long trajectory in gas at normal temperatures and pressures, which mobility measurement entails. This is because their strong field divergence is most conducive to clustering. We therefore conclude that the most profitable applications of the mobility method in combustion research are likely to arise in the case of much larger charged particles, or droplets, for example particles of carbon in flames (see Chapter 7). For these the method would be most sensitive, changes by charge transfer in a carrier gas, or by clustering, are unlikely; and mass spectrometry would become a much less superior competitor than it is in the case of small ions.

As regards the distance resolution of mobility methods, no sampling into a region of lower pressure is involved and the limitations to resolving power are therefore set primarily by field configuration, as discussed under 'ion photography'.

REFERENCES

1. CALCOTE, H. F. *Combust. Flame* 1, 385 (1957).
2. WORTBERG, G. *10th Int. Symp. Combust.*, p. 651. The Combustion Institute, Pittsburgh (1965).
3. CALCOTE, H. F. and KING, I. R. *5th Int. Symp. Combust.*, p. 423. Reinhold, New York (1955).
4. WILSON, H. A. *Rev. Mod. Phys.* 3, 156 (1931).
5. THOMSON, J. J. and THOMSON, G. P. *Conduction of electricity through gases*, 3rd edn (1928).
6. KINBARA, T. and IKEGAMI, H. *Combust. Flame* 1, 199 (1957).
7. DIBELIUS, N. R. *et al. Engineering aspects of magnetohydrodynamics* (edited by C. Mannal and N. W. Mather, p. 307. University of Columbia Press (1962)).
8. PAIN, H. G. and SMY, P. R. *J. Fluid Mech.* 10, 51 (1961).
9. PONCELET, J., BERENDSEN, R., and VAN TIGGELEN, A. *7th Int. Symp. Combust.*, p. 256. Butterworths, London (1959).
10. VAN TIGGELEN, A. *Ionization in high temperature gases* (edited by K. E. Shuler and J. B. Fenn, p. 165. Academic Press, New York and London (1963)).
11. SUITS, C. G. *Phys. Rev.* 55, 561 (1939).
12. VAN WONTERGEM, J. and VAN TIGGELEN, A. *Bull. Soc. chim. Belg.* 63, 235 (1954).
13. MOTT-SMITH, H. M. and LANGMUIR, I. *Phys. Rev.* 28, 727 (1926).
14. CHEN, F. F. *Plasma diagnostic techniques* (edited by R. H. Huddlestone, and S. L. Leonard, Chap. 4. Academic Press, New York (1965)).
15. WASSERSTROM, E., SU, C. H., and PROBSTEIN, R. F. *Physics Fluids* 8, 56 (1965).
16. SCHULTZ, G. J. and BROWN, S. C. *Phys. Rev.* 98, 1642 (1955).
17. TRAVERS, B. E. L. and WILLIAMS, H. *10th Int. Symp. Combust.*, p. 657. The Combustion Institute, Pittsburgh (1965).
18. LANGMUIR, I. *Collected works* (edited by G. Suits, Vol. 3, pp. 115 and 125. Pergamon Press, Oxford (1961)).
19. CHEN, F. F. *J. Nucl. Energy*, Part C, 7, 47 (1965).
20. Friedman, R., *4th Int. Symp. Combust.*, p. 259, Williams and Wilkins, Baltimore, 1953.
21. LAM, S. H. *Physics Fluids* 8, 73 and 1002 (1965).
22. LANGMUIR, I. *Collected works* (edited by G. Suits, Vol. 4, p. 99. Pergamon Press, Oxford (1961)).

23. HALL, L. S. *University of California Radiation Laboratory Report*, *UCRL*-7660-*T* (1964).
24. WAYMOUTH, J. F. *Physics Fluids* **7**, 1843 (1964).
25. SU, C. H. and LAM, S. H. *Physics Fluids* **6**, 1479 (1963).
26. COHEN, I. *Physics Fluids* **6**, 1492 (1963).
27. CALCOTE, H. F. *8th Int. Symp. Combust.*, p. 184. Williams and Wilkins, Baltimore (1962).
28. BURHOP, E. H. S., BOHM, D., and MASSEY, H. S. W. *Characteristics of electrical discharges in magnetic fields* (edited by A. Guthrie and R. K. Wakerling), Chap. 2. McGraw-Hill (1949).
29. CALCOTE, H. F. *9th Int. Symp. Combust.*, p. 622. Academic Press, New York (1963).
30. CALCOTE, H. F., CURZIUS, F. C., and MILLER, W. J. *10th Int. Symp. Combust.*, p. 605. The Combustion Institute, Pittsburgh (1965).
31. JOHNSON, E. O. and MALTER, L. *Phys. Rev.* **80**, 58 (1950).
32. BURROWS, K. M. *Aust. J. Phys.* **15**, 162 (1962).
33. VON ENGEL, A. and COZENS, J. R. *Proc. phys. Soc.* **82**, 85 (1963).
34. BRADLEY, D. and MATTHEWS, K. J. *Physics Fluids* **10**(6), 1336 (1967).
35. COZENS, J. R. and VON ENGEL, A. *Int. J. Electronics* **19**, 61 (1965).
36. BRADLEY, D. and MATTHEWS, K. J. *11th Int. Symp. Combust.*, p. 359. The Combustion Institute, Pittsburgh (1967).
37. BRUNDIN, C. L. *Institute of Engineering Research, University of California, Report* *AS*-64-9 (1964).
38. CHUNG, P. M. *Physics Fluids* **7**, 110 (1964).
39. MARGENAU, H. *Phys. Rev.* **69**, 508 (1946).
40. FROST, L. S. *J. appl. Phys.* **32**, 2029 (1961).
41. WILLIAMS, H. *7th Int. Symp. Combust.*, p. 269. Butterworths, London (1959).
42. BORGERS, A. *10th Int. Symp. Combust.*, p. 268. The Combustion Institute, Pittsburgh, (1965).
43. SMITH, H and SUGDEN, T. M. *Proc. R. Soc.* **A211**, 31 (1952).
44. WILLIAMS, H. *8th Int. Symp. Combust.*, p. 179. Williams and Wilkins, Baltimore (1962).
45. SUGDEN, T. M. and THRUSH, B. A. *Nature, Lond.* **168**, 703 (1951).
46. PADLEY, P. J. and SUGDEN, T. M. *8th Int. Symp. Combust.* p. 164. Williams and Wilkins, Baltimore (1962).
47. BALWANZ, W. W. *AGARD conference proceedings*, No. 8, p. 699 (1965).
48. GARDNER, A. L. *Engineering aspects of magnetohydrodynamics* (edited by C. Mannal and N. W. Mather), p. 438. University of Columbia Press (1962).
49. BELCHER, H. and SUGDEN, T. M. *Proc. R. Soc.* **A201**, 480 (1950).
50. SHULER, K. E. and WEBER, J. *J. chem. Phys.* **22**, 491 (1954).
51. BALWANZ, W. W., HEADRICK, J. M., and AHERN, J. A. *U.S. Naval Research Lab. Report No. AC.SIL/57/529* (1956).
52. BELCHER, H. E. and SUGDEN, T. M. *Proc. R. Soc.* **A202**, 17 (1950).
53. GRAY, E. P. *9th Int. Symp. Combust.*, p. 654. Academic Press, New York and London, (1963).
54. BULEWICZ, E. M. and PADLEY, P. J. *9th Int. Symp. Combust.*, pp. 638 and 647. Academic Press, New York and London (1963).
55. SCHNEIDER, J. and HOFMANN, F. W. *Phys. Rev.* **116**, 244 (1959).

56. ALPHER, R. A. and WHITE, D. R. *Plasma diagnostic techniques* (edited by R. H. Huddlestone, and S. L. Leonard), p. 431. Academic Press, New York and London (1965).
57. WEINBERG, F. J. *Optics of flames.* Butterworths, London (1963).
58. BROWN, S. C. and BEKEFI, G. *J. opt. Soc. Am.* **53**, 448 (1963).
59. FRIEDMAN, R. and MACEK, A. *10th Int. Symp. Combust.,* p. 731. The Combustion Institute, Pittsburgh (1965).
60. LOVBERG, R. H. *I.E.E.E. Trans. nucl. Sci.* **11**, 187 (1964).
61. PAYNE, K. G. and WEINBERG, F. J. *8th Int. Symp. Combust.,* p. 207. Williams and Wilkins, Baltimore (1962).
62. LAWTON, J. and WEINBERG, F. J. *Proc. R. Soc.* **A277**, 468 (1964).
63. BOTHA, J. P. and SPALDING, D. B. *Proc. R. Soc.* **A225**, 71 (1954).
64. POWLING, J. *Fuel, Lond.* **28**, 25 (1949).
65. PANDYA, T. P. and WEINBERG, F. J. *Proc. R. Soc.* **A279**, 544 (1964).
66. WARD, F. J. and WEINBERG, F. J. *8th Int. Symp. Combust.,* p. 217. Williams and Wilkins, Baltimore (1962).
67. WEINBERG, F. J. *Combust. Flame* **10**, 267 (1966).
68. ELTENTON, G. C. *J. chem. Phys.* **15**, 455 (1947).
69. FRISTROM, R. M. and WESTENBERG, A. A. *Flame structure.* McGraw-Hill, New York (1965).
70. KNEWSTUBB, P. F. and SUGDEN, T. M. *Nature, Lond.* **181**, 474 (1958).
71. KNEWSTUBB, P. F. and SUGDEN, T. M. *7th Int. Symp. Combust.,* p. 247. Butterworths, London, (1959).
72. KNEWSTUBB, P. F. and SUGDEN, T. M. *Proc. R. Soc.* **A255**, 520 (1960).
73. SUGDEN, T. M. and GREEN, J. N. *9th Int. Symp. Combust.,* p. 607. Academic Press, New York, (1963).
74. BASCOMBE, K. N., GREEN, J. N., and SUGDEN, T. M. *Adv. Mass Spectrom.* **2**, 66 (1962).
75. SCHOFIELD, K. and SUGDEN, T. M. *10th Int. Symp. Combust.,* p. 589. The Combustion Institute, Pittsburgh (1965).
76. DECKERS, J. and VAN TIGGELEN, A. *Combust. Flame* **1**, 281 (1957).
77. DE JAEGERE, S., DECKERS, J., and VAN TIGGELEN, A. *8th Int. Symp. Combust.,* p. 155. Williams and Wilkins, Baltimore (1962).
78. FEUGIER, A. and VAN TIGGELEN, A. *10th Int. Symp. Combust.,* p. 621. The Combustion Institute, Pittsburgh (1965).
79. CALCOTE, H. F. and REUTER, J. C. *J. chem. Phys.* **38**, 310 (1963).
80. SPOKES, G. N. and EVANS, B. E. *10th Int. Symp. Combust.,* p. 639. The Combustion Institute, Pittsburgh (1965).
81. LOEB, L. B. *Basic processes of gaseous electronics.* University of California Press, Berkeley (1961).
82. ERIKSON, H. A. *Phys. Rev.* **17**, 400 (1921); **18**, 100 (1921); **19**, 275 (1922); **23**, 110 (1924); **24**, 502 (1924).
83. TYNDALL, A. M., STARR, L. H., and POWELL, C. F. *Proc. R. Soc.* **A121**, 172 (1928).
84. VON ENGEL, A. *Ionized gases,* p. 133. Clarendon Press, Oxford (1965).
85. PAYNE, K. G. and WEINBERG, F. J. *Proc. R. Soc.* **A250**, 316 (1959).
86. PAYNE, K. G. Ph.D. Thesis, University of London (1958).

6. Ionization in Flames

THE discussion that follows is divided into two parts. The first deals with the ionization found naturally in flames, the second with that arising from additives introduced into the reactants or burnt products. A considerable amount of reliable data has been amassed on the identity, concentration, and rates of generation and recombination of ions. Yet, although various chemical and physical processes have been put forward to account for their presence, there is still uncertainty over a number of fundamental matters, for example the origin of non-equilibrium ionization in premixed and diffusion flames and the mechanism of ionization of alkali metal additives. These and other questions with regard to flame ionization will be discussed in detail. The uncertainties are mentioned here just to indicate that the subject is in a state of rapid growth and that, in consequence, conclusions drawn with regard to mechanism must still be treated as tentative.

NATURAL FLAME IONIZATION

Occurrence, concentration, and identity of flame ions

Ion concentrations found in the reaction zones of unseeded premixed hydrocarbon/air and hydrocarbon/oxygen flames at pressures between 2 and 760 mmHg lie in the range 10^9–10^{12} ions cm^{-3},[1-4] concentrations being highest when acetylene is the fuel.[5] Not all flames, however, are observed to contain such large numbers of ions. Carbon monoxide, hydrogen sulphide, hydrogen, and carbon disulphide, for example, produce only minute concentrations of ions when burned in air or oxygen; what little there is probably arises from impurities.[5-7]

Figures 6.1 and 6.2 show how the ion concentration profile, as measured by electrical probes, fits into the structure of a flat flame. Similarly shaped profiles have also been found by Bradley and Matthews,[10] who, in addition, deduced local rates of ion generation from their results. The concentrations and rates of generation of ions reach maxima close to the final temperature; the emission intensities of CH, OH, and C_2 and the rate of heat release also reach maxima in this region.

Figure 6.3 shows the variation of maximum ion concentration (expressed as a mole fraction) with equivalence ratio for a number of low-pressure

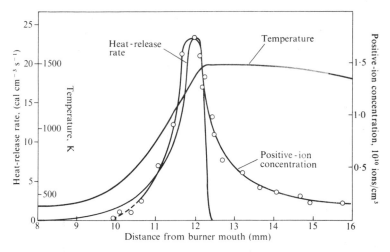

FIG. 6.1. Positive ion concentration, heat release rate and temperature versus distance from burner mouth. Dashed line indicates the uncertain region in ion-concentration measurements. From Wortberg.[8]

flames. Other experiments on the effect of pressure on ion concentration at fixed equivalence ratio have been carried out on propane/air[11] and acetylene/oxygen mixtures[5] using electrical probes and cyclotron resonance, respectively. All show the mole fraction of ions present to be virtually independent of pressure, at a fixed equivalence ratio, ϕ.

Information on the identities of flame ions has been obtained by sampling flame gases directly into a mass spectrometer (see Chapter 5). A very great variety of species has been detected. Table 6.1 shows the positive ions reported by Green and Sugden,[6] Calcote,[9] van Tiggelen,[12] and Cooper.[13] Although the list cannot claim to be exhaustive, it is thought to contain most of the positive ions between mass numbers 2 and 67 present in appreciable quantities. Green and Sugden were able to divide the positive ions they found into three categories: true flame ions, ions formed in the mass spectrometer, and those of uncertain origin (see Chapter 5). Unfortunately other workers have not always made these distinctions. It is noteworthy that only four mass numbers were found to be present in *all* the flames examined, viz., 15, 19, 29, and 39, which most probably correspond to CH_3^+, H_3O^+, CHO^+, and $C_3H_3^+$, respectively; other assignments are possible in some cases but are less likely.[6]

Profiles for some of the positive ions present in a low-pressure acetylene/oxygen flame are shown in Fig. 6.4, which also serves to illustrate the general features found in most hydrocarbon flames. The dominant ion is H_3O^+, which reaches a peak concentration downstream of the others. There are appreciable amounts of $C_3H_3^+$, which is found to peak ahead of the other

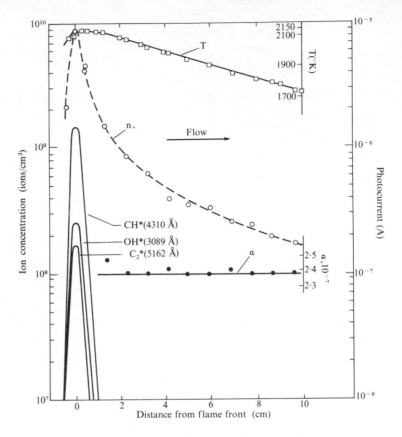

FIG. 6.2. Centre-line profiles for propane–air flame. Equivalence ratio $= 0.93$; pressure $= 40$ torr; total flow (s.t.p.) $= 170\ cm^3/s$; probe Pt—40% Rh; diam. $= 0.025\ cm$; length $= 0.5\ cm$. From Calcote, Kurzius, and Miller.[9]

ions at low pressures. Finally, there is a small peak corresponding to CHO^+. However, various exceptions to this pattern occur, an interesting one being the case of methane burning in oxygen where concentrations of CH_3^+ and CHO^+ equal to about one-tenth that of H_3O^+ have been found.[14]

Results obtained from studies of negative ions in flames are contradictory, and no clear overall picture has yet emerged. Calcote,[9] using electrical probes, deduced that about 98 per cent of the negative charge is present as negative ions. This conclusion is supported by the work of Feugier and van Tiggelen[15] who used a mass spectrometer. However, other measurements carried out with a mass spectrometer by Green,[16] Green and Sugden,[6] and Calcote[9] indicate that there are only very small concentrations of negative ions present, up to 99 per cent of the charge being carried by

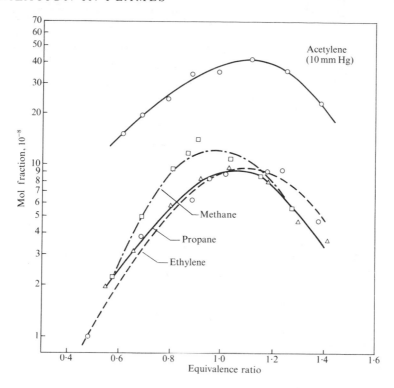

FIG. 6.3. Maximum positive ion concentration, fuel in air at 33 mmHg. From Calcote.[1]

free electrons. Studies by Knewstubb[17] show appreciable negative ion formation only in the pre-heat zone, their concentration falling rapidly in the reaction zone, attaining an extremely low value in the products. There is also considerable divergence between Calcote[9] and Feugier and van Tiggelen[15] with regard to the identities and relative abundances of the negative ions found. Whether these differences represent real discrepancies in the flame systems studied is not known. A possible explanation is suggested by the work of Knewstubb and Sugden[18] who found no negative ions in alkali seeded $H_2/O_2/N_2$ flames except in the cold outer regions of the product gases. Thus it is possible that the detection of large concentrations of negative ions depends upon the flame gases being pre-cooled by the sampling system before entering the mass spectrometer, thereby providing more opportunity for attachment. This is the explanation given by Green[16] for the relatively large amounts of CO_2H^- detected by his mass spectrometer system. He found that increasing the size of the sampling orifice increased the count of CO_2H^- by a factor only one-third of that found for other negative ions, among which C_2H^- and O_2^- predominated. Although there

TABLE 6.1

Mass numbers and identities of positive ions in hydrocarbon flames

Ref.	Mass No.	Identity	Ref.	Mass No.	Identity
2	1	H^+	7,8	38	$C_3H_2^+$
2	2	H_2^+	1,2,5,6,7,8	39	$C_3H_3^+$
2	3	H_3^+	3	40	?
2,6,7,8	12	C^+	5	41	$C_3H_5^+$, C_2HO^+
2,6,7,8	13	CH^+	4	42	$C_2H_2O^+$, $C_3H_6^+$
2,6,7,8	14	CH_2^+	1,2,4,6,7,8	43	$C_2H_3O^+$, $C_3H_7^+$
1,2,5,6,7,8	15	CH_3^+	2,4	45	$C_2H_5O^+$, CHO_2^+
2	17	OH^+	6,8	46	$CH_2O_2^+$
2,4,7	18	H_2O^+, NH_4^+	3	47	$C_2H_7O^+$, $CH_3O_2^+$
1,2,3,6,7,8	19	H_3O^+	6	48	$CH_4O_2^+$
7,8	24	C_2^+	3,7,8	49	$CH_5O_2^+$, C_4H^+
7,8	25	C_2H^+	4	51	$C_4H_3{}^+$
3,7,8	26	$C_2H_2^+$, CN^+	1,4,7	53	$C_4H_5^+$, C_3HO^+
8	27	$C_2H_3^+$	4	54	$C_4H_6^+$, $C_3H_2O^+$
4,7	28	$C_2H_4^+$, CO^+	4	55	$H_3O^+ . 2H_2O$
1,2,3,6,7,8	29	CHO^+	4	56	?
2,5	30	NO^+	4	58	$C_2H_2O_2^+$, $C_3H_6O^+$
1,2,3,6	31	CH_3O^+	4	59	$C_2H_3O_2^+$
4	32	CH_4O^+	3	61	$C_2H_5O_2^+$
1,4,6	33	CH_5O^+, HO_2^+	5	63	$C_5H_3^+$, $C_2H_7O_2^+$
4	36	$NH_4^+ . H_2O$	5	65	$C_5H_5^+$
1,2,4,6,7,8	37	$H_5O_2^+$, C_3H^+	5	67	$C_5H_7^+$, $C_4H_3O^+$

1—Calcote[9] stoichiometric C_2H_2/O_2, at 2 torr.

2—Cooper[13] H_2/air with admixtures of CH_4, C_2H_4, C_2H_2 at 1 atm.

3, 4, 5—Sugden[6], respectively, flame ions, input system and ions of uncertain origin in a
   $H_2/O_2/N_2$ with admixtures of C_2H_2 at 1 atm.

6, 7, 8—van Tiggelen[12] respectively, $CH_4/O_2/N_2$, $(CH_3)_4C/O_2/N_2$ and $C_2H_2/O_2/N_2$ at 1 atm.

are almost certainly negative ions arising naturally in flames, present evidence indicates that their formation is contingent upon the primary production of electrons.[16]

Rates of ion generation and the influence of pressure, temperature, and composition

We have measured[19,20] rates of ion generation per unit area of flame front for a variety of premixed flames at atmospheric pressure stabilized on a porous disk burner (see Chapter 3). Using this burner it is possible to vary

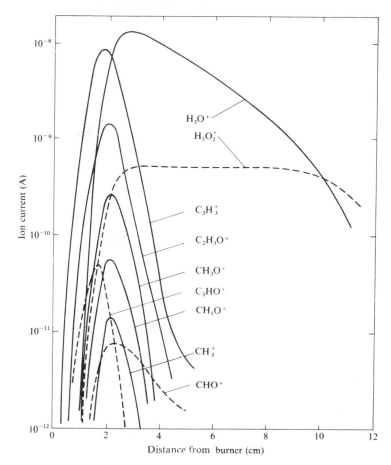

FIG. 6.4. Positive-ion profiles for acetylene–oxygen flames. Equivalence ratio = 1; pressure = 2 torr; total flow (s.t.p.) = 70 cm³/s. From Calcote, Kurzius, and Miller.[9]

the flame temperature at constant composition by varying the total flow rate, the rates of ion generation being measured by the saturation current method. A discussion of this technique can be found in Chapter 5. Measurements carried out in the range 1450–1900°K for methane/air, ethylene/air, and propane/air flames and for flames of H_2/air with admixtures of up to 1 per cent of these hydrocarbons showed that an increase of 100°K caused increases in the rate of ion generation ranging from two to four times. It was also found that a change in equivalence ratio from 0·6 to 0·9 for each of the hydrocarbon/air mixtures caused the same increase in saturation current as a temperature increase of between 50 and 100°K. By plotting the logarithm of the saturation current against the reciprocal of the final flame temperature straight lines were obtained (see Fig. 6.5), from which effective activation

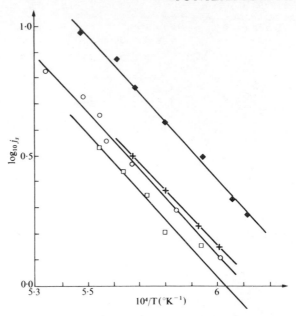

FIG. 6.5. $Log_{10} (j_s)$ plotted against $10^4/T(°K^{-1})$ for propane/air mixtures. ϕ: ◆, 1·06; ○, 0·92; □, 0·81; +, 0·64. From Lawton and Weinberg.[19]

energies for the overall process were found (Table 6.2). The final flame temperature was taken to be appropriate for this calculation because it is there that the rate of ion generation reaches a sharp maximum. Poncelet, Berendsen, and van Tiggelen[3] have obtained activation energies in acetylene

TABLE 6.2

Activation energies, E (kcal/mol)

	hydrogen/hydrocarbon/air									
hydrocarbon	propane					methane		ethylene		
% hydro-carbon	0·18	0·29	0·40	0·50	0·63	0·55	0·99	0·29	0·45	0·60	0·82
E (kcal/mol)	58 ±4	59 ±2	58 ±1	57 ±2	56 ±2	60 ±4	55 ±3	60 ±3	55 ±3	53 ±1	54 ±4
mean E	58					57·5		56		

	hydrocarbon/air									
hydrocarbon	propane				methane				ethylene		
% hydro-carbon	2·7	3·4	3·9	4·3	6·7	8·5	9·1	10·0	4·1	5·3	5·9
ϕ	0·64	0·81	0·92	1·06	0·71	0·89	0·96	1·05	0·63	0·81	0·91
E (kcal/mol)	48 ±0	50 ±6	50·5 ±2	52 ±2	67 ±4	51 ±7	52 ±2	46 ±0	46 ±2	37·5 ±2	40·5 ±2
mean E	50				54				42		

flames by measuring the electrical resistance of the reaction zones of flames at various temperatures; their technique is described in Chapter 5. However, these were not based upon the final flame temperature but upon a specially weighted mean equal to $[0.74 + 0.26\ T_u/T_b]T_b$, where T_u and T_b are unburnt and burnt temperatures, respectively†. The present authors have deduced rates of ion generation in free burning flames by extrapolation of their saturation current results to the adiabatic flame‧ temperature—Fig. 6.6.

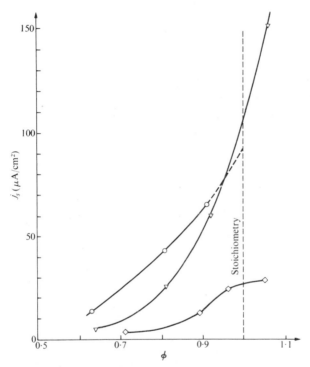

FIG. 6.6. Saturation current plotted against composition for hydrocarbon/air mixtures. ◇, methane; ▽, propane; ○, ethylene. From Lawton and Weinberg.[19]

Values range from 3×10^{13} for the leanest flames to 6×10^{14} ions s^{-1} cm^{-2} for near stoichiometric mixtures. At stoichiometry, one ion pair is generated for every 2×10^5 to 2×10^6 carbon atoms, for the fuels used. The figures agree well with those found by other workers. Hand and Kistiakowsky[21] found one ion pair was produced per 2×10^6 carbon atoms in C_2H_2/O_2 reactions in shock tubes, and Sternberg, Galloway, and Jones[22] found one ion pair per 2×10^6 carbon atoms for a H_2/air flame to which small amounts of

† *Footnote added in proof.* Recent developments in the use of the saturation current technique were presented by Boothman, Lawton, Melinek, and Weinberg, and by Peeters and van Tiggelen at the 12th International Symposium on Combustion, Poitiers, July, 1968.

propane were added, this value being constant to within a few per cent over a very wide range of propane concentrations.

Calcote[1] interpreted ion concentration profiles to yield approximate values for the maximum rates of ion generation for a propane/air flame over a range of pressures from 33 to 760 mmHg. Charge conservation in a one-dimensional system requires that

$$\frac{\mathrm{d}}{\mathrm{d}x}\left(D_{\mathrm{a}}\frac{\mathrm{d}n}{\mathrm{d}x}\right) - \frac{\mathrm{d}(nv)}{\mathrm{d}x} - \alpha n^2 + r_{\mathrm{c}} = 0, \qquad (6.1)$$

where D_{a}, v, α and r_{c} are, respectively, the ambipolar diffusion coefficient, the local gas velocity, the recombination coefficient and the local rate of ion generation. The concentration reaches a maximum very near the final flame temperature, and there $\mathrm{d}n/\mathrm{d}x$ and $\mathrm{d}v/\mathrm{d}x$ are zero, i.e. $\mathrm{d}(nv)/\mathrm{d}x$ is zero. Furthermore, at pressures above about 30 mmHg, ambipolar diffusion does not play an important role. Therefore at the higher pressures, at the point where $n = n_{\max}$, $r_{\mathrm{c}} \simeq r_{\mathrm{c,max}}$ and

$$r_{\mathrm{c,max}} \simeq \alpha n_{\max}^2, \qquad (6.2)$$

the value of α being known from other experiments, to be discussed in the next section. Table 6.3 shows that $r_{\mathrm{c,max}}/P^2$ is approximately constant over a

TABLE 6.3

Rate of ion formation for propane-air flames at $\phi = 0.875$

Pressure	α	q	q/P^2
			ions/(cm^3 s)
	cm^3/s	ions/(cm^3 s)	(mm Hg)^2
33	1.6×10^{-7}	5.6×10^{12}	5.0×10^9
66	2.4×10^{-7}	2.8×10^{13}	6.4×10^9
520*	1.6×10^{-7}	2.1×10^{15}	7.8×10^9
760†	$\sim 2 \times 10^{-7}$	$\sim 2 \times 10^{15}$	$\sim 3.5 \times 10^9$

twentyfold range of pressure. This indicates that ionization is a second order process (the temperature increase, as the pressure is raised from 33 to 760 mmHg, is not itself large enough to increase greatly the rates of ion generation and so does not affect this general conclusion). This inference is also supported by the observations that the mole fraction of ions present[5,11] and the recombination coefficient[9] are independent of pressure.

Recombination and ambipolar diffusion

The falling off in ion concentration in the products is due to the simultaneous effects of recombination and ambipolar diffusion, the former predominating at pressures in excess of about 30 mmHg.[1] By measuring the decay of charge concentration in the burnt gases, it is possible to deduce the coefficients of charge recombination and ambipolar diffusion. Extending the unidimensional charge conservation equation (eqn (6.1)) to three dimensions and restricting it to the usual case of the flat flame in which the velocity is only a function of distance, x, in the products where there is a negligible rate of generation and D_a is constant, one may write

$$\frac{d(1/n)}{dt} - \frac{1}{n}\frac{dv}{dx} = \alpha - \frac{D_a}{n^2}\nabla^2 n, \tag{6.3}$$

where $v \equiv dx/dt$. This latter substitution can be used to transform the distance travelled by an element of product gas into a time scale.

In the burnt gases v is very nearly constant, so that above about 0·1 atm, when ambipolar diffusion is unimportant, the equation can be integrated to give

$$\frac{\alpha}{v}(x - x_0) = \left(\frac{1}{n} - \frac{1}{n_0}\right) = \alpha(t - t_0). \tag{6.4}$$

Thus, when ambipolar diffusion and the variation of velocity are negligible, α can be deduced from the slope of $1/n$ against t, where $t = (x - x_0)/v$. However, in the general case, the recombination coefficient and ambipolar diffusion coefficient can be found from the intercept and slope, respectively, of a graph of $\{d(1/n)/dt - (1/n)(dv/dx)\}$ plotted against $(1/n^2)\nabla^2 n$, as in Fig. 6.7. (The factor $T^{\frac{3}{2}}$ included by Calcote to take account of the variation of

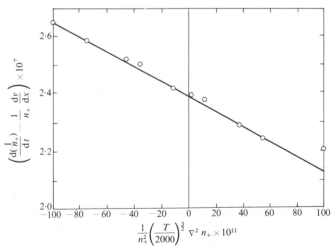

FIG. 6.7. Determination of α and D_a. Equivalence ratio = 0·93; pressure = 40 torr; total flow (s.t.p.) = 170 cm³/s; probe, Pt—40% Rh; diam. = 0·025 cm; length = 0·5 cm. From Calcote, Kurzius, and Miller.[9]

ambipolar diffusion coefficient with temperature, is only approximate (see Chapter 4), but is unlikely to lead to serious error in the determination of D_a at flame temperatures.) Reducing his values to 1 mmHg and 298°K, using an inverse dependence on pressure and a three halves power temperature dependence, Calcote found D_a in propane/air flames to be 41, 60, and 90 cm^2/s for equivalence ratios of 0·87, 0·93, and 1·0, respectively.

The recombination coefficient has been measured by a number of workers and found to lie in the range $(2 \pm 1) \times 10^{-7}$ cm^3/s, being sensibly independent of pressure, temperature, and gas composition,[1,2,6,9,23] although a marked variation with pressure has been reported by King.[24]

Mechanism of ion formation and recombination

The discussion so far has been restricted to actual measurements made on flame ions. However, there has also been a great deal of speculation concerning the detailed chemical mechanisms by which the processes of ion formation and recombination occur. Although these matters are by no means resolved, some clear patterns are beginning to emerge from the experimental work. The investigation of ion formation divides into two parts. First, there is the question of the primary ionization step that relates to identifying the primary ion or ions and establishing the mechanism of their formation. Second, there is the question of how the wide variety of ions found in flames derives from these parent ions.

The subsequent discussion is divided into two parts, the first dealing with ion formation, the second with ion recombination.

Ion formation

Very many suggestions have been advanced to account for the high levels of ionization in the reaction zones of flames. Calcote in 1957[7] published a comprehensive survey of the mechanisms put forward up to that date and, with the information then available, was able to eliminate those that were unlikely to make a significant contribution. Amongst these were mechanisms based upon thermal ionization. Clearly thermal ionization of impurities such as the alkali metals and of species of low ionization potential stable at flame temperatures, such as NO, cannot be the cause of the ionization peak in the flame, in view of the rapid decay of free charge downstream of the reaction zone. Although thermal ionization of transient species such as the radicals C_2, CH, and OH would account for the rapid decay, it was found, after an extensive survey of species known to exist, and likely to exist, in flames, that they would need to be present in impossibly large concentrations to account for the observed ion concentrations. It had also been suggested that small carbon particles, of a work function close to that of graphite,

4.35 eV, could be the chief source of ions. This was deemed unlikely on three counts. First, there is a great deal of ionization in the leanest of flames. Second, in diffusion flames the maximum charge concentration is found in the relatively cool, premixed base, not in the hot sooty tip of the flame. Lastly, Calcote calculated that, even if all the carbon in the fuel were agglomerated into 100-Å particles, the resulting ion concentration would still be two orders of magnitude lower than that observed.

Other types of mechanism considered by Calcote depend upon either translational or electronic excitation. Saenger, Goercke, and Bredt[25] suggested that the fast-moving radicals emerging from highly exothermic reactions cause ionization on collision with other neutral species. This seems unlikely, however. Momentum considerations limit the conversion of kinetic to internal energy to about one-half, for bodies of comparable mass (see Chapter 2). There is also a high probability that these energetic fragments will lose their energy in collision with a stable species before colliding with one of the relatively rare ones that have a sufficiently low ionization potential. Even when an energetically suitable collision does occur, the probability of the molecule–molecule collision resulting in ionization is very low—see Chapter 2.

The case for ionization as a result of the exchange of electronic excitation is very much better. For example, the radiative lifetimes of $CH(A^2\Delta)$ and $CH(B^2\Sigma^-)$ are about 10^{-6} s,[26] enabling them to undergo about 5000 collisions before returning to the ground state by a radiative transition. Momentum considerations do not limit energy exchange here, as they do in the conversion of kinetic to internal energy. Under suitable conditions, if the energy of excitation of a molecule exceeds the ionization potential of the species with which it collides, ionization will proceed with high efficiency, as was discussed in Chapter 2. An example of the process is

$$Ne^* + Ar \rightarrow Ne + Ar^+ + e^-.$$

Finally, Calcote considered chemi-ionization. In processes of this kind, species undergo a chemical rearrangement that releases sufficient energy to ionize one of the products (see Chapter 2). Thus in the case of a flame it would be expected to occur as a side reaction between species involved in the main flame propagation reaction. In view of the relatively large number of such reactions that are energetically feasible and involve either two ground states, or a ground state and an excited species, chemi-ionization either with or without the exchange of excitation is thought to be the most likely source of flame ionization.

Since this survey, a considerable amount of experimental work has been carried out, which has done much to confirm the importance of chemi-ionization. However, before considering subsequent work, a suggestion, put forward since the publication of Calcote's survey, by von Engel and Cozens[27] will be discussed, in order to complete the account of the types of mechanism proposed. Electrons readily acquire energy by collision with vibrationally excited species. It has been calculated[27] that average electron energies of between 0·2 and 1·0 eV (2320 and 11 600°K) could be expected in flames, as a result of the balance between energy gained from excited species and energy lost in elastic collisions. The existence of elevated electron temperatures in some flames has been suggested by a number of electrostatic probe measurements.[2,10,27] For example, in the recent work of Bradley and Matthews[10] in which double probes were used at low pressures, electron temperatures of up to 30 000°K were reported. Von Engel and Cozens[27] have suggested that, since electrons with energies slightly in excess of the ionization potential are very efficient at ionizing atoms and molecules, they are likely to be the source of the ionization found in those flames with elevated electron temperatures. Indeed, there seems little doubt that electrons at temperatures of the order of 30 000°K would lead to very rapid rates of ionization. Recent work[28] on ionization in carbon monoxide/oxygen flames containing hydrocarbons suggests that in addition to chemi-ionization considerable amounts of O_2^+ are produced which may result from the presence of electrons at elevated temperatures. The latter is thought[28] to derive from interaction with excited CO_2 molecules produced by recombination of carbon monoxide molecules with atomic oxygen. However, elevation of electron temperature has not been found in all flames exhibiting excess ionization.[10,29] Moreover, we were able to obtain[19,20] flat saturation current plateaux at atmospheric pressure† over a wide range of applied potentials in measurements of rates of ion generation, in spite of the fact that the fields present within the flame zone itself were of the order of kV/cm and thus able to raise the electron temperature significantly. This suggests that more than one important ionizing mechanism may be active in different flames; the elucidation of the role of energetic electrons, as one of the possible sources, is clearly going to be an important field of research in the future.

The process most favoured at the present time by experimental evidence is chemi-ionization, only reactions that are either exothermic or slightly endothermic being considered as possibilities. Two specific mechanisms have been advanced which have a favourable thermochemistry,

$$CH + O \rightarrow CHO^+ + e^- \quad \Delta H \simeq 20 \, kcal^{[6]} \qquad\qquad I$$

and

$$CH(A^2\Delta) + C_2H_2 \rightarrow C_3H_3^+ + e^-. \quad \Delta H \simeq -7 \, kcal.^{[30]} \qquad\qquad II$$

One of the most important reasons for supposing that one or other of these mechanisms may be responsible is that, as can be seen from Table 6.1, the

---

† *Footnote added in proof.* And very recently at reduced pressures also; Ref. Boothman, D., Lawton, J., Melinek, S., and Weinberg, F. J., Paper 96, 12th International Symposium on Combustion, Poitiers, July, 1968.

only ions common to every flame investigated are CH_3^+, CHO^+, $C_3H_3^+$, and H_3O^+. Of these H_3O^+ is formed later than the others and is therefore unlikely to be a primary ion. The only ion remaining unaccounted for is CH_3^+. So far no thermochemically feasible primary ionization process has been advanced to account for its formation. Moreover, it is found to be present in concentrations less than one-thousandth of CHO^+ and $C_3H_3^+$ in atomic flame studies[31] and so has generally been discounted. Left with only two possible primary ions, the cases for mechanisms I and II will now be considered.

CHO^+:   The formation of CHO^+ by mechanism (I) is only of the order of 20 kcal[32] endothermic and involves only ground state species that are known to exist in flames. It is also noticeable that CHO^+ peaks early on in the flame structure. If CHO^+ has a precursor in an excited state, the addition of an excitation quencher such as CO_2 should greatly reduce its concentration. However, it was found by one group of investigators[9] that addition of CO_2 to acetylene/oxygen flames had only a very small effect on the concentration of CHO^+ and by another,[31] that the addition of CO_2 to an atomic flame of oxygen and ethylene reduced the excitation whilst the ion concentration was found to increase slightly. The first result indicates that CHO^+ has a ground state precursor and the second that the overall ionization also originates from ground state species in the C_2H_4 flame. These conclusions, taken together with the observation that CHO^+ is one of the two possible primary ions, strongly favours reaction (I). Green and Sugden[6] have found that mechanism (I) also agrees well quantitatively with their measurements of ion concentrations in $H_2/O_2/N_2$ flames at atmospheric pressure, with admixtures of up to 1 per cent of acetylene in the hydrogen. On the assumption that H_3O^+ is formed by a charge transfer reaction involving the primary ion and water, they were able to show that it was necessary for the square of the concentration of H_3O^+ to be directly proportional to the concentration of the ion from which it derives. Only in the case of CHO^+ was this criterion fulfilled. Moreover, using a reaction rate constant for mechanism (I), assumed to be equal to that for the isoelectronic reaction $N + O \rightarrow NO^+ + e$, they were able to show that the estimated concentrations of CH and O in their flames were capable of producing the required rates of ionization. Other evidence in favour of this reaction has been summarized by, for example, Calcote.[32] Assuming that CHO^+ is the primary ion, various charge transfer schemes have been devised to account for the presence of the other ions found, e.g.

$$CHO^+ + H_2O \rightarrow H_3O^+ + CO \qquad \Delta H = -34 \text{ kcal,}[6]$$

$$H_3O^+ + CH_2O \rightarrow CH_3O^+ + H_2O \qquad \Delta H = +5 \text{ kcal,}[6]$$

$$CHO^+ + CH_2O \rightarrow CH_3O^+ + CO \qquad \Delta H = -42 \text{ kcal,}[9]$$

$$CH_3^+O + C_2H_2 \rightarrow C_3H_3^+ + H_2O \qquad \Delta H = -12 \text{ kcal.}[6]$$

Many other possible mechanisms are to be found in the references quoted.

$C_3H_3^+$: There is also some evidence pointing to $C_3H_3^+$ as one of the parent ions and as the dominant primary ion in certain circumstances. Experiments carried out by Kistiakowsky and Michael[30] on the oxidation of methane in shock tubes have shown that there are considerable quantities of $C_3H_3^+$ formed, but no detectable amounts of CHO^+. In addition, in premixed flames, $C_3H_3^+$ peaks early on, sometimes before CHO^+, as can be seen from Fig. 6.4, and even reaches concentrations comparable to that of H_3O^+.

Fontijn, Miller, and Hogan,[31] working with atomic flames, found no $C_3H_3^+$ in C_2H_4/O mixtures but large amounts of CHO^+, whilst in C_2H_2/O flames approximately equal concentrations of the two ions were found. The addition of CO_2 to the acetylene flame, to quench excitation, reduces CH radiation and the overall ionization in the same proportion, while additions of CO_2 to the ethylene flame, although quenching CH radiation, cause a rise in the overall ion concentration. Experiments[9] on the addition of CO_2 to low-pressure acetylene/oxygen flames have shown that whereas the concentration of $C_3H_3^+$ is greatly reduced, that of CHO^+ is barely affected.

These studies seem to indicate that the ionization processes in ethylene and acetylene flames are different, i.e. two mechanisms rather than just one are required to explain the observed ionization, and that CHO^+ and $C_3H_3^+$ are ions of independent origin, the one with a precursor in the ground state, the other with a precursor in an excited state.

Accepting that $C_3H_3^+$ is a primary ion, the evidence in favour of mechanism (II) includes the following: the formation of this ion has been shown to be inhibited by the addition of CO_2, the other discernible effect of which was to reduce the concentration of CH^*[9]; $C_3H_3^+$ is formed in especially large quantities in acetylene flames[31]; $C_3H_3^+$ is the only ion to have a maximum concentration on the rich side of stoichiometry in C_2H_2/O_2 flames[9]; finally, reaction (II) is nearly thermoneutral.

Various reaction schemes have been put forward to account for the formation of secondary ions from $C_3H_3^+$,[9] e.g.

$$C_3H_3^+ + O_2 \rightarrow C_2H_3O^+ + CO$$

$$C_2H_3O^+ + O \rightarrow CH_3O^+ + CO,$$

$$C_2H_3O^+ + OH \rightarrow CH_3O^+ + CHO,$$

$$C_3H_3^+ + O \rightarrow CHO^+ + C_2H_2.$$

If some CHO^+ is indeed formed by the last reaction, then all those charge-transfer reactions involving CHO^+ that were discussed above, must also be considered.

Summarizing, it is necessary at present to consider on the one hand the possible existence of two primary chemi-ions, namely CHO^+ and $C_3H_3^+$, the relative importance of which may vary from flame to flame and from which the great variety of ions formed in flames probably derive as a result of charge transfer reactions and, on the other, the possibility of ionization by electron collision.†

Charge recombination

The charge recombination coefficient, α (c. 2×10^{-7} cm^{3}s^{-1}), has been found by a number of workers to be independent of both pressure and temperature, as was discussed earlier in this chapter. The lack of pressure dependence is significant because it points to a dissociative electron–ion recombination process (Chapter 2). Further support for this type of mechanism comes from mass spectrometric work. Thus most studies have shown that all but a minute amount of the negative charge is carried by electrons and that the most abundant ion is H_3O^+, which can undergo a highly exothermic dissociative-recombination with an electron:

$$H_3O^+ + e^- \rightarrow H + H_2O \qquad \Delta H = -145 \text{ kcal.}^{[6]}$$

This reaction is currently believed to be the principal charge recombination mechanism operative in flames.

IONIZATION IN THE PRESENCE OF ADDITIVES

It is well known that the addition of small amounts of substances of low ionization potential, such as alkali or alkaline earth metals, can greatly enhance ionization in the product gases of flames. As regards practical considerations—see Chapter 7—this can be either a nuisance or an advantage depending upon the circumstances. For example, in rocket exhausts it can cause interference with radio communications, whilst deliberate seeding with alkali metals has been used to raise the electrical conductivity of combustion products to the level required to operate MHD generators.

Non-equilibrium ionization of additives

In this section the kinetic, as distinct from equilibrium, aspects of the influence of various additives on flame ionization, and the enhancement of ionization of seeded flames by applied electric fields, is considered. The additives that have been studied in some detail are lead, alkali, and alkaline earth metals and alkali metals in presence of chlorine.

Lead. The addition of lead to a premixed $C_2H_2/O_2/N_2$ flame greatly reduces the rate at which charge recombination occurs in the burnt gases.[33]

---

† *Footnote added in proof.* The roles of electrons and excited species have been reviewed in a recent article; von Engel, A. *Br. J. appl. Phys.* **18**, 1661 (1967).

This has the effect of increasing the local ion concentration at all points downstream of the reaction zone. The decay is still found to follow a second order law

$$\frac{1}{n_e} - \frac{1}{(n_e)_0} = \alpha t \qquad (6.5)$$

but now α is of the order of 10^{-9} cm^3/s compared to 2×10^{-7} cm^3/s without lead. Padley and Sugden[33] have suggested that the great reduction in recombination rate occurs because of charge transfer to the lead atoms, to form Pb^+, which prevents the fast dissociative recombination between H_3O^+ and electrons from taking place. The lead ions then recombine with either electrons or negative ions. In either case the rate of recombination would be greatly reduced. This is because, on the one hand, electron–atomic ion recombination requires a third body to remove the excess energy, dissociation being impossible and, on the other hand, negative ions are present in much lower concentrations than electrons.

The charge transfer process is probably

$$Pb + H_3O^+ \rightarrow Pb^+ + H_2O + H \qquad \Delta H = 26 \text{ kcal}$$

followed by one of

$$Pb^+ + e^- + M \rightarrow Pb + M, \qquad\qquad\qquad A$$

$$Pb^+ + OH^- + M \rightarrow Pb + OH + M, \qquad\quad B$$

$$Pb^+ + OH^- \rightarrow Pb + OH^*. \qquad\qquad\qquad C$$

Reaction A is likely to be the most important of the three, for several reasons. It has a rate constant of the right magnitude, which increases with increasing concentrations of triatomic species (H_2O and CO_2) as would be expected in view of their relatively high efficiency as third bodies. Then, even if it were found that B and C have suitable rate constants, the low concentrations of negative ions makes these reactions only minor contributors to the overall recombination process. There is no reason to suppose that lead is unique in its behaviour. The energy released when H_3O^+ is neutralized is 6·3 eV, which means that electron transfer is energetically feasible for any metal atom with an ionization potential up to c. 7·5 eV, an endothermicity of about 1 eV being no great impediment at flame temperatures. This is discussed further in the next section in relation to the alkali metals.

Alkali metals. The mechanism and rate of ionization of alkali metals in flames is greatly affected by the presence of hydrocarbons. The case of alkali additions to $H_2/O_2/N_2$ and $CO/O_2/N_2$ free of hydrocarbon is considered first. It is found that the rate of ionization in these flames is a steeply decreasing function of ionization potential, Li < Na < K < Rb < Cs, as would be expected. Thus, from the work of Jensen and Padley[34] it can be deduced that at $2250°K$ in a $H_2/O_2/N_2$ flame, for concentrations at which sodium would be fully ionized in equilibrium, it takes about 1.4×10^{-1} s to reach 63 per cent of the equilibrium value; the corresponding figure for caesium is 2.5×10^{-4} s. Earlier experiments[33] indicated that the activation energy of the ionization process in $H_2/O_2/N_2$ flames is only half the ionization potential of the metal atom. This conclusion provoked the suggestion that a chemi-ionization mechanism is responsible, e.g.

$$Na + H_2O \rightarrow NaH_2O^+ + e^-.$$

However more recent work shows that the activation energy for both CO and $H_2/O_2/N_2$ flames is equal to the ionization potential within experimental error[34,35] (see Table 6.4) and that the reaction is first order, depending only on the alkali atom concentration. This indicates that ionization occurs via a thermal rather than a chemical route, e.g.

$$Na + X \rightarrow Na^+ + X + e^-,$$

where X is some major constituent of the product gases. If the process is completely described by this reaction then

$$\frac{d[Na]}{dt} = k_i[Na][X].$$

From eqn (2.9) the expression for the rate constant is

$$k_i = \sqrt{\left\{\frac{8kT}{\pi M_1 M_2}(M_1 + M_2)\right\}Q}\, e^{-eV_i/kT}.$$

Inserting an average molecular weight, M_1, for the combustion products, the atomic weight of the metal, M_2, and a value for the concentration of X, the collision cross-sections, Q, have been deduced from experiment[34,35] (see Table 6.4). The values quoted differ from those in the original papers in that the partial pressure of X is taken as 1 atm (Jensen and Padley arbitrarily assumed $\frac{1}{3}$ atm), and the collision cross-section itself is quoted (Hollander, Kalff, and Alkemade quote their results as Q/π). This normalization is necessary in order to bring out the good general agreement between the two different series of experiments. The reason for the discrepancy in the case of caesium has not been established. These cross-sections are greater by two to three orders of magnitude than the gas kinetic values, and indicate therefore that, although the reaction is first order with respect to the metal, it cannot be a simple single-step process like that given above. Nevertheless, for practical applications, for example MHD generation, eqn (2.9) with the appropriate values of Q and activation energy taken from Table 6.4, can still be used to estimate the rate at which alkali metals ionize. Hollander, Kalff, and Alkemade[35] have attempted to explain these anomalous results by suggesting that ionization of the metal atoms occurs from electronically excited levels rather than from the ground state. This explanation, however, has not yet been put on a quantitative footing.

A number of workers have measured coefficients of charge recombination in flames containing alkali metals which, because of the preponderance of electrons, is thought to occur by the process

$$A^+ + e^- + M \rightarrow A + M.$$

Values are listed in Table 6.5 and are similar to those found in the case of lead addition to hydrocarbon flames.

TABLE 6.4

Activation energies and collision cross-sections for ionization of alkali metals in H_2 and CO flames

Metal	Ionization potential (kcal/mol)	Activation energy (kcal/mol)		Collision cross-section ($Å^2$)	
		$H_2/O_2/N_2{}^{(34)}$	$CO/O_2/N_2{}^{(35)}$	$H_2/O_2/N_2{}^{(34)}$	$CO/O_2/N_2{}^{(35)}$
Li	124	—	—	$2\cdot3 \times 10^4$	—
Na	117·5	$1,15 \pm 5$	117 ± 3	$3\cdot0 \times 10^4$	$4\cdot3 \times 10^4$
K	100	99 ± 4	100 ± 3	$2\cdot7 \times 10^4$	$1\cdot8 \times 10^4$
Rb	96·5	97 ± 4	—	$2\cdot3 \times 10^4$	—
Cs	89·5	92 ± 6	90 ± 3	$3\cdot7 \times 10^4$	$4\cdot4 \times 10^3$

TABLE 6.5

Charge recombination coefficients in flames seeded with alkali metals

Metal	Flame[36]	α ($cm^3s^{-1} \times 10^9$)	Flame[37]	α ($cm^3s^{-1} \times 10^9$)	Flame[35]	α ($cm^3s^{-1} \times 10^9$)
Li		7·8–9				
Na		1·9–4·0				8·7
K	C_3H_8/air at 1 atm	0·18–0·7	$H_2/O_2/N_2$ at 1 atm	3·3	$CO/O_2/N_2$ at 1 atm	3·2
Rb	1970°K	0·19–0·4				
Cs		0·17–0·53				0·68

Padley and Sugden[33] observed that the addition of small amounts of hydrocarbon (for example, 1 per cent methane) to a hydrogen/air flame greatly increase the rate at which alkali metals ionize. They suggest that the mechanism is the same as for lead,

$$Na + H_3O^+ \rightarrow Na^+ + H + H_2O.$$

This is in accord with the fact that if electron transfer is possible for lead, it is energetically feasible for the alkali metals which have lower ionization potentials.

Halogens. The addition of small amounts of halogen also has a profound effect upon the ionization of alkali metals in $H_2/O_2/N_2$ flames. It might seem at first that, since chlorine is strongly electronegative, its addition would reduce the electron concentration because of electron attachment. Quite the contrary is found,[38] in fact. The addition of up to 1 per cent of Cl_2 to a sodium laden $H_2/O_2/N_2$ flame causes the local electron concentration to increase twofold in the product gases, reaching a maximum a short distance downstream of the reaction zone. Increasing the Cl_2 addition beyond 1 per cent causes a gradual reduction in electron concentration, presumably owing to electron attachment.

The effect of chlorine can be explained in terms of a series of fast, balanced, bimolecular reactions:

$$Na + Cl \rightleftharpoons Na^+ + Cl^-$$

$$Cl^- + H \rightleftharpoons HCl + e^-$$

$$\frac{HCl + H \rightleftharpoons H_2 + Cl}{Na + H + H \rightleftharpoons Na^+ + e^- + H_2}$$

The totalled reaction at the bottom is a three-body recombination of hydrogen atoms in which sodium atoms, acting as third bodies, become ionized. The function of chlorine, which does not appear explicitly in the overall reaction, is to catalyse this otherwise slow recombination. The eventual decrease in electron concentration downstream of the maximum is due to the drop in hydrogen atom concentration.

Alkaline earths. It has been known for some time that the alkaline earth metals—calcium, strontium, and barium—ionize very rapidly in flames.[39] Thus strontium with an ionization potential of 5·65 eV ionizes more rapidly than potassium, whose ionization potential is 4·32 eV.[37] The possibility that it is SrOH that ionizes has been considered, but is rendered improbable by a comparison between actual kinetics and those implied by the direct ionization of this species.[37] However, in view of the predominance of Sr^+ and $SrOH^+$ ions, shown by mass spectrometric studies,[37,40] any acceptable mechanism must be capable of accounting for both of them. One scheme that does this is[37]

$$Sr + OH \rightarrow SrOH^+ + e^- \qquad \Delta H = 16 \pm 10 \text{ kcal/mol}$$

and/or

$$SrO + H \rightarrow SrOH^+ + e^- \qquad \Delta H = 37 \pm 10\,kcal/mol$$

followed by

$$SrOH^+ + H \rightarrow Sr^+ + H_2O \qquad \Delta H = -4 \pm 10\,kcal/mol.$$

The chemi-ionization step suggested here is in sharp contrast to the purely thermal process believed to prevail in the case of the alkali metals.

It has also been found[37] that mixing strontium with sodium greatly increases the rate at which sodium ionizes. In this case a charge transfer process is probably operative, such as

$$SrOH^+ + Na \rightarrow Na^+ + SrOH.$$

The effect of additives on flame ionization is very complex. Although a large number of phenomena have been described in the literature, the explanations for them must in most cases still be regarded as tentative. However, there is little doubt of the eventual value of this branch of ionization studies to the practical systems mentioned at the outset, in which it is desirable either to enhance or diminish electron concentrations. Moreover, it is probable that a deeper understanding of the processes of ionization in the absence of additives will also emerge from investigations of this kind.

Ionization augmented by applied electric fields. The effect of electric fields upon ionized gases was discussed in Chapter 4. It was shown there that electrons, because of their high charge to mass ratio, are raised to high temperatures by applied fields too small to influence the thermal energy of the ions appreciably. Moreover, electrons with an energy in excess of the ionization potential are very efficient at causing ionization of the atoms or molecules with which they collide. Thus applied fields might provide a means of increasing flame ionization without the need to raise the temperature of all the species present. Although possibly effective in the reaction zone itself which contains only species of rather high ionization potential, fields are likely to be much more effective in the products of flames seeded with easily ionized materials, for example alkali and alkaline earth metals. Thus, we were able[19] to draw saturation currents from unseeded flames† by applying high electric fields, clearly indicating the absence of field augmented ionization in the reaction zone. Methods for calculating electron concentrations in the presence of an electric field will now be discussed.

Electron energies can be calculated by the techniques referred to in Chapter 4. However, only in the cases where electron–electron collisions dominate over those between electrons and other species is it correct to speak of an electron temperature; for it is only then that electrons have a Maxwellian energy distribution. Otherwise it is necessary to calculate the electron energy

† And recently also from lightly seeded ones (Boothman, Lawton, Melinek, and Weinberg at the 12th International Symposium on Combustion, Poitiers, July, 1968.

distribution from the Boltzmann equation, balancing energy losses and energy gains in detail.

If the electron energy distribution is Maxwellian, and the effects of molecule–molecule collisions and excited species on the ionization rate can be neglected, electron concentrations can be calculated by using the electron temperature in the Saha equation (Chapter 4). This has been done by Kerrebrock[41] for potassium-seeded argon. He finds good qualitative agreement between theory and experiment, an exact comparison being hampered by the presence of impurities in the argon he used.

Freck[42] has carried out calculations for potassium-seeded combustion systems using non-Maxwellian electron energy distributions deduced for gas mixtures of similar composition to that of the flame products. He balanced the rate of ionization by electron collision (eqn (2.7)) against the rate of recombination, assuming a three-body electron-atomic ion recombination process. The recombination coefficient was evaluated from Thomson's theory; see Chapter 2. Once again, quite good agreement between theory and experiment was found.

In such work it is common practice to ignore molecule–molecule collisions and the effect of excited species. However, in view of the unexpectedly large overall cross-sections for ionization probably resulting from molecule–molecule collision (one molecule may be excited) found in flame systems by Jensen and Padley[34] and Hollander et al.[35], this omission needs to be reviewed, especially at lower electron concentrations. In addition, account must be taken, when making theoretical predictions, of the possibility of an arc discharge forming in the gases. The effect of this is to set a ceiling to the field strength that can be applied.

Equilibrium ionization of additives

Homogeneous gas phase

The equilibrium concentrations of ions and electrons can be calculated from the Saha equation, as detailed in Chapter 2, given the relevant ionization potentials, electron affinities, and statistical weights. For example, at $2500°K$ and 1 atm, the presence of only 2 p.p.m. of caesium metal results in an electron concentration of about $3 \times 10^{12} \, cm^{-3}$ in the absence of electron attachment; in this case 50 per cent of the metal is ionized.

In general, when calculating electron and ion concentrations and electrical conductivity (Chapter 4) it is necessary to take account of equilibrated reactions that result in electron attachment (see Chapter 2) and in removal of the free metal by the formation of salts. The importance of these side reactions can be seen from calculations carried out by Dibelius, Luebke, and Mullaney[43]—see Fig. 6.8—for CsCl sprayed into atmospheric pressure propane/oxygen flames at a partial pressure of $6.25 \times 10^{-3} \, atm$. (The

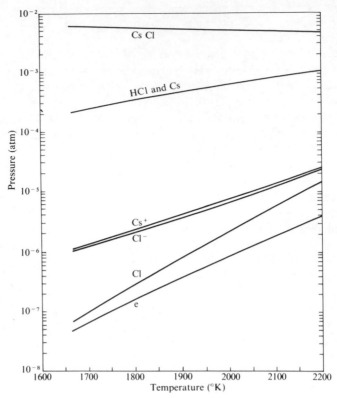

FIG. 6.8. Calculated pressures of the gaseous species due to CsCl additive to flame. From Dibelius, Luebke, and Mullaney.[43]

absolute values of electron concentration agree with those deduced from conductivity measurements to within a factor of 2.) Even at 2200°K only 16 per cent of the CsCl is dissociated into metal atoms, and the electrons are outnumbered by negative ions six to one. A number of other calculations of electron concentration and conductivity in seeded flames have been carried out, taking electron attachment and salt formation into account,[44–46] and these also show the importance of such factors in determining levels of equilibrium ionization.

It should be noted that there is still considerable uncertainty with regard to the electron affinities of OH and CO_2. Some measured values for OH are 1·78,[47] 2·1,[48] 2·65,[49] and 2·8 eV.[50] A difference of 1 eV as between the highest and the lowest values gives rise to considerable differences in the calculated concentration of OH^-, especially at the lower flame temperatures. Buchel'nikova[51] obtained a value of 3·8 eV for the electron affinity of CO_2. This is very large and is not supported by mass spectrometric measurements on negative ions, CO_2^- not being found in large quantities. However, if this value proved to be correct, calculations of equilibrium ionization levels in seeded flames carried out hitherto would be in serious error in their neglect of CO_2^-. For not only is the quoted electron affinity as large as that of Cl, but CO_2 is also a major component of the product gases of all but very rich, or very hot, flames.

Particle suspensions

The ionization produced by particles of low work function in combustion products has, on the one hand, been advanced as a practical method of raising electrical conductivity in MHD generators, and, on the other, is a potential source of interference to radio communications with solid propellant rockets. Thermal ionization of particles was discussed from a fundamental point of view in Chapter 4, where equations for the calculation of electron concentration and data on thermionic properties were given.

Two aspects that are of particular relevance to flames are discussed here. The first is the comparison of levels of ionization resulting from additives giving rise to gaseous as opposed to particulate products; the second is the ionization of solid carbon particles in combustion products.

The reason for supposing that dust suspensions can give much greater concentrations of ions than can gaseous additives is that the work functions of a number of substances are very much lower than the lowest ionization potentials, as discussed in Chapter 2. However, to be set against this is the fact that, as the particle loses charge to the gas phase by thermionic emission, its effective work function is increased by an amount $1.44 \times 10^{-7} N/r_p$ eV, where N is the number of residual positive charges and r_p the radius in centimetres.

If r_0 is the radius of the molecules comprising the particle, then the number of molecules in each particle is given, to an order of magnitude, by $(r_p/r_0)^3$. If the fraction of molecules that have lost an electron is f', then

$$N \simeq f'(r_p/r_0)^3. \qquad (6.6)$$

Thus the increase in work function due to loss of electrons is given by

$$V \simeq 1.44 \times 10^{-7} f' r_p^2/r_0^3 \text{ eV}. \qquad (6.7)$$

In order to make a comparison with a gaseous additive, it will be assumed that the gaseous seed is sodium (ionization potential 5·15 eV), since this has been suggested as one of the more economical seed materials for MHD generation; it is also found naturally in rocket exhausts. Typically, sodium atoms would be ionized to the extent of 0·1 to 1 per cent at normal flame temperatures. In order to make a comparison, it will be assumed that the particle has zero work function when uncharged, is 10^3 Å in diameter (i.e. very small particles) and its constituent molecules have a diameter of 10 Å. These values are chosen so as to over-estimate the ionization likely from particles, as can be seen by reference to eqn (6.7). Once the work function has increased beyond the ionization potential of sodium atoms, the particles are less efficient at releasing electrons. Thus, for the above dimensions and an increase in work function of 5 eV, the fraction of the particle molecules ionized is seen from eqn (6.7) to be of the order of 10^{-4}.

Even though the calculation is only rough it shows clearly that even very fine suspensions (c. 10^3 Å diam.) of substances of very low work function are less efficient sources of ionization than alkali metals on a weight for weight basis. This is an important conclusion with regard to the choice of seed materials to enhance conductivity, as for example in M.H.D. applications, and in assessing the origin of ionization in contaminated flame products, as in some rocket exhausts. If, however, very small particles (for example, 100 Å in diameter) can be introduced into the flame gases, the prospect for particulate seeds is rather better; soot particles formed in flames may come into this category†.

The presence of soot particles in the products of hot rich flames are known to give rise to appreciable concentrations of free charge.[52] If detailed knowledge is available with regard to the size of the particles and their thermionic properties (work function and pre-exponential factor) then electron concentrations can be calculated using the methods discussed in Chapter 2. In the case of soot particles it is usually assumed, in the absence of reliable data, that the pre-exponential factor and the work function‡ are

TABLE 6.6

Calculated radii and degrees of ionization of free carbon particles in rich oxyacetylene flames, and measured electron concentrations[52]

Fuel ratio C_2H_2/O_2	T_b (°K)	n_e (cm^{-3} × 10^{-10})	r_p (Å)	N per particle
1·5/1	3285	3·47	60	11
1·75/1	3250	3·32	56	10
2/1	3210	2·74	54	9
2·25/1	3195	1·42	74	13
3/1	3135	0·696	85	16
4/1	3090	0·644	81	14

† Pulverized fuel and seeding with emitting particles in MHD generators has recently been discussed by Fells and co-workers (Vol. II and Vol. IV, *Electricity from MHD*, International Atomic Energy Agency, Vienna, 1968).

‡ *Footnote added in proof.* In recent work on ionization in the products of sooting premixed flames a value of $4·6 \pm 0·4$ eV was established for the work function of soot particles; ref. Boothman, D. Lawton, J., Melinek, S., and Weinberg, F. J., Paper 96, 12th International Symposium on Combustion, Poitiers, July, 1968.

those for graphite, viz. 48 A/cm^2 and 4·35 eV, respectively. However, there is very little information available on the numbers and size distributions of soot particles, which makes fully predictive calculation of levels of ionization due to their presence very tentative.[52–54] At present the only reliable technique is still to measure electron concentrations. Table 6.6 shows measured values of electron concentration, estimated values of particle size, and the charge per particle in the product gases of sooty oxyacetylene flames. It is interesting to note that Place and Weinberg,[55] by applying an axial electric field to a counterflow diffusion flame (see Chapter 3), were able to collect particles on the anode even though the particles were formed in a region of positive space charge, thus showing that thermionic charging can prevail even over bombardment charging, under some circumstances. Moreover, since the carbon particles can be collected, weighed, their size determined and the current measured, it is in principle possible to deduce the charge to mass ratio and the charge per particle using this method†.

REFERENCES

1. CALCOTE, H. F. *8th Int. Symp. Combust.*, p. 184. Williams and Wilkins, Baltimore (1962).
2. CALCOTE, H. F. *9th Int. Symp. Combust.*, p. 622. Academic Press, London and New York (1963).
3. PONCELET, J., BERENDSEN, R., and VAN TIGGELEN, A. *7th Int. Symp. Combust.*, p. 256. Butterworths, London (1959).
4. KING, I. R. *J. chem. Phys.* **31**, 855 (1960).
5. BULEWICZ, E. M. and PADLEY, P. J. *9th Int. Symp. Combust.*, p. 638. Academic Press, London and New York, (1963).
6. GREEN, J. A. and SUGDEN, T. M. *9th Int. Symp. Combust.*, p. 607. Academic Press, London and New York (1963).
7. CALCOTE, H. F. *Combust. Flame* **1**, 385 (1957).
8. WORTBERG, G. *10th Int. Symp. Combust.*, p. 651. The Combustion Institute, Pittsburgh (1965).
9. CALCOTE, H. F., KURZIUS, S. C., and MILLER, J. *10th Int. Symp. Combust.*, p. 605. The Combustion Institute, Pittsburgh (1965).
10. BRADLEY, D. and MATTHEWS, K. J. *11th Int. Symp. Combust.*, p. 359. Combustion Institute, Pittsburgh (1967).
11. KING, I. R. *J. chem. Phys.* **29**, 681 (1958).
12. VAN TIGGELEN, A. and FEUGIER, A. *Revue Inst. fr. Pétrole* **20**, 1135 (1965).
13. COOPER, A., Ph.D. Thesis, University of Sheffield, 1965.
14. VAN TIGGELEN, A. *Ionization in high temperature gases* (edited by K. E. Shuler and J. B. Fenn), p. 165. Academic Press, London and New York (1963).

† *Footnote added in proof.* In our most recent investigations on carbon formation in counterflow diffusion flames, using applied electric fields, Dr. P. J. Mayo determined particle size and mobility, from which it has been deduced that there is a single charge per particle over the complete experimental range of particle diameters of 90 to 200 Å.

15. FEUGIER, A. and VAN TIGGELEN, A. *10th Int. Symp. Combust.*, p. 621. The Combustion Institute, Pittsburgh (1965).
16. GREEN, J. A. *AGARD conference proceedings No.* 8, p. 191 (1965).
17. KNEWSTUBB, P. F. *10th Int. Symp. Combust.* p. 623. The Combustion Institute, Pittsburgh (1965).
18. KNEWSTUBB, P. F. and SUGDEN, T. M. *Nature, Lond.* **196**, 1312 (1962).
19. LAWTON, J. and WEINBERG, F. J. *Proc. R. Soc.* **A277**, 468 (1964).
20. LAWTON, J., Thesis, University of London, 1963.
21. HAND, C. W. and KISTIAKOWSKY, G. B. *J. chem. Phys.* **37**, 1239 (1962).
22. STERNBERG, J. C., GALLOWAY, W. S., and JONES, D. T. L. *3rd Int. Symp. Gas Chromatography*, p. 231. Academic Press, London and New York (1962).
23. SEMENOV, E. S. and SOKOLIK, A. S. *Zh. tekh. Fiz.* **32**, 1074 (1962).
24. KING, I. R. *Ionization in high temperature gases* (edited by K. E. Shuler and J. B. Fenn) p. 197. Academic Press, London and New York (1963).
25. SAENGER, E., GOERCKE, P., and BREDT, I. *Z. phys. Chem.* **46**, 199 (1952).
26. BENNETT, R. R. and DALBY, F. W. *J. chem. Phys.* **32**, 1716 (1962).
27. VON ENGEL, A. and COZENS, J. R. *Proc. phys. Soc.* **82**, 85 (1963).
28. HURLE, I., NUTT, G., and SUGDEN, T. M. Private communication (1967).
29. TRAVERS, B. E. L. and WILLIAMS, H. *9th Int. Symp. Combust.*, p. 657. Academic Press, London and New York (1963).
30. KISTIAKOWSKY, G. B. and MICHAEL, J. V. *J. chem. Phys.* **40**, 1447 (1964).
31. FONTIJN, A., MILLER, W. J., and HOGAN, J. M. *9th Int. Symp. Combust.*, p. 545. Academic Press, London and New York (1963).
32. CALCOTE, H. F. *AGARD conference proceedings No.* 8, p. 1 (1965).
33. PADLEY, P. J. and SUGDEN, T. M. *8th Int. Symp. Combust.*, p. 164. Williams and Wilkins, Baltimore (1962).
34. JENSEN, D. E. and PADLEY, P. J. *Trans. Faraday Soc.* **62**, 2140 (1966).
35. HOLLANDER, Tj., KALFF, P. J., and ALKEMADE, V. T. J. *J. chem. Phys.* **39**, 2558 (1963).
36. KING, I. R. *J. chem. Phys.* **36**, 553 (1962).
37. SCHOFIELD, K. and SUGDEN, T. M. *10th Int. Symp. Combust.*, p. 589. The Combustion Institute, Pittsburgh (1965).
38. PADLEY, P. J., PAGE, F. M., and SUGDEN, T. M. *Trans. Faraday Soc.* **57**, 1552 (1961).
39. SUGDEN, T. M. and WHEELER, R. C. *Discuss. Faraday Soc.* **19**, 76 (1955).
40. KNEWSTUBB, P. F. and SUGDEN, T. M. *Nature, Lond.* **181**, 474 (1958).
41. KERREBROCK, J. L. *Engineering aspects of magnetohydrodynamics* (edited by C. Mannal and N. W. Mather), p. 327. Columbia Press, New York and London (1962).
42. FRECK, D. V. *Phil. Trans. R. Soc.* **A261**, 471 (1967).
43. DIBELIUS, N. R., LUEBKE, E. A., and MULLANEY, G. J. *Engineering aspects of magnetohydrodynamics* (edited by C. Mannal and N. W. Mather), p. 307. Columbia Press, New York and London (1962).
44. BINCER, H. *International symposium on magnetohydrodynamic electrical power generation*, p. 31. ENEA (1961).
45. ROSTAS, F. ibid., p. 91.
46. FROST, L. S. *J. appl. Phys.* **32**, 2029 (1961).
47. BRANSCOMB, L. *Atomic and molecular processes* (edited by D. R. Bates), p. 100. Academic Press, London and New York (1962).

48. JAMES, C. G. and SUGDEN, T. M. *Proc. R. Soc.* **A227**, 312 (1955).
49. PAGE, F. M. and SUGDEN, T. M. *Trans. Faraday Soc.* **53**, 1092 (1957).
50. GOUBEAU, J. and KLIMM, W. *Z. phys. Chem.* **1336**, 362 (1937).
51. BUCHEL'NIKOVA, J. *Zh. eksp. teor. Fiz.* **35**, 1119 (1958).
52. EINBINDER, H. *J. Chem. Phys.* **26**, 948 (1957).
53. SHULER, K. E. and WEBER, J. *J. chem. Phys.* **22**, 491 (1954).
54. SUGDEN, T. M. and THRUSH, B. A. *Nature, Lond.* **168**, 703 (1951).
55. PLACE, E. R. and WEINBERG, F. J. *Proc. R. Soc.* **A289**, 192 (1965).

7. Practical Consequences and Applications

THE aim of this chapter is to summarize the practical effects of applying fields, both direct and oscillating at frequencies up to those of electromagnetic waves, to flame plasma, irrespective of present usefulness. Some of these effects are, indeed, the opposite of useful (for example, the interaction between rocket exhaust gases and radio signals), some are methods of detection or measurement, others are at various stages along the scale from laboratory curiosities to established industrial practice. One reason for this approach is that the subject is developing so rapidly, on the time-scale of producing a book, that it seems quite possible that topics which currently attract most of the research interest may have exhausted their promise by the time this monograph becomes available, and vice versa. As before, it is convenient to subdivide the subject according to whether the movement of charge, or that of the gas as a whole, is involved.

MOVEMENT OF CHARGE CARRIERS

Electrolysis

At first sight the most obvious suggestion for the use of steady fields applied to flame plasma is to electrolyse out the ions, causing them either to deposit on electrodes or to traverse gas in which they can induce chemical effects. We are here discussing specifically the transport of matter—the associated flow of charge will be considered in connection with methods of direct generation and other applications involving the passage of current. Such electrolytic schemes are, however, limited by the build-up of space charge that leads to an increase of field strength towards the electrodes. This results in secondary ionization and breakdown, before useful mass fluxes can be attained, in the case of small ions. The theory is given in Chapter 8 and it is shown there that, as a rough guide, the maximum rate of deposition without secondary ionization is the molecular weight of the ion in mg/h, for 100 cm² of flame front. This result presupposes a unidimensional system with a breakdown field of 30 kV/cm and is confined to small ions for which the effect of mass on mobility (Chapter 4) is small. If this rate is exceeded, the flame plasma ceases to be the sole source of charge and its ion current meets, and is progressively neutralized by, the charge carriers from the discharge at the electrodes. This effect is a very real limitation when the ion species matters—much more so than in the case of, for example, transport of charge, where any additional ion current contributes.

'Electrolytic' applications are therefore confined to large charge carriers, unless the small ion is so active as to induce chemical effects in minute concentrations, or so desirable that such unfavourable methods of collection would be considered. These are somewhat academic possibilities that could arise, for example, if the ion were a free radical required to initiate a reaction chain or to be 'frozen' on a cooled electrode; in the latter case an increase in pressure could be used, in addition to the low electrode temperature, to increase the flux maximum appreciably—see Chapter 8. At the time of writing, evidence that the chemistry or rates of flame reactions can be considerably altered by moving ions (as distinct from larger charged particles) by means of fields is rather scanty. It appears[1,2] that ions manipulated by fields in a flame can affect nucleation of carbon formation in the pyrolysis zone. It has been suggested[3,4] that field-induced changes in the burning velocities of premixed flat flames and in the blow-off conditions of counter-flow diffusion flames (Chapter 3) are due to modified reaction rates.† However, in many instances where chemical effects have been postulated in order to account for the then inexplicable experimental observations, better understanding of the underlying aerodynamics has made such an explanation unnecessary.

The limitations regarding transport of charged masses by fields rapidly diminishes in importance as the mass/charge ratio of the carrier increases and, by the time we are concerned with particles of many thousands of atomic masses per electronic charge, the possible rates of deposition, for example, of soot or oxide products, can attain their rate of formation per unit area of flame.

Large charge-carriers

Electrostatic precipitators have been used industrially for many years; see for example Lodge,[5] Roseff and Wood,[6] and White.[7] Their ion source is usually a localized region of breakdown in the form of a corona discharge, rather than a flame, and they are therefore of only marginal interest here. Because of the similarity of the subject, however, their general theory is, in fact, covered by the various sections of this monograph. Conditions required for controlled breakdown can be deduced from Chapter 8, the subsequent attachment of the resulting ions to the particles and droplets is fully discussed in Chapter 2, and the mobilities and hence trajectories of the particles in the field are dealt with in Chapter 4.

† *Footnote added in proof.* For recent work see also ref. (117).

The question of whether the ion source is a region of breakdown or a flame is relatively unimportant; what is unnecessarily restrictive is the application to precipitation alone. This is of value in combustion only as a method of collecting fly-ash or a desirable product (for example, a metal oxide or carbon black). Another possibility is to precipitate the fuel as a safety measure, when this is an explosive dust, for example by associating precipitators with the ventilation system. The principle of controlling the movement of particles or droplets involved in combustion processes by means of fields, however, has potentially much wider applications. Research is currently in progress on a number of topics under this heading[†] and work on the management of the burning of suspensions and on the control of flame carbon will be briefly summarized.

The electrically controlled system for burning suspensions of fuel may begin conveniently with an electrical method of producing the dispersion. In the case of liquid fuels, atomization occurs[8–11] in a strong divergent electric field. According to Drozin[10] the existence of this phenomenon was discovered as early as 1745 by Bose. The simplest laboratory system is a hypodermic syringe needle from which a trickle of the liquid emerges, and which acts as one electrode. The other electrode may be a coarse gauze, a ring, or just an earthed surface in the proximity of the jet. An open electrode is most useful in the present context since one of the most valuable aspects of the system is that the droplets, of which only a small proportion is lost to the surface, appear to retain their charge on passage through the second electrode, so that their subsequent movement can be further influenced by fields.

The mechanism of the process is that a dipole tends to be induced by the field, and the surface charge, which then migrates to the liquid tip, overcomes by self-repulsion the confining forces due to surface tension. For a spherical droplet of radius r_p, the equilibrium condition is

$$\frac{2\gamma}{r_p} - P' - \frac{\sigma}{2D} = 0, \tag{7.1}$$

where γ, P', σ, and D are its surface tension, excess internal pressure, surface charge density and dielectric constant, respectively. Instability sets in when P' increases faster than the confining forces, for an elemental disturbance. Thus when σ exceeds a critical value, the effective surface tension is neutralized, the jet loses the property that is characteristic of a liquid—that of trying to minimize its surface area for a given volume—and rapidly elongates into a thin filament. At its tip, surface tension reasserts itself, probably because of an insufficient rate of charge supply due to reduced conduction and/or leakage of charge, and fine droplets are seen to form (Fig. 7.1). At large divergent fields, for suitable liquids, several short fine filaments are formed and

[†]*Footnote added in proof.* See also ref. (118).

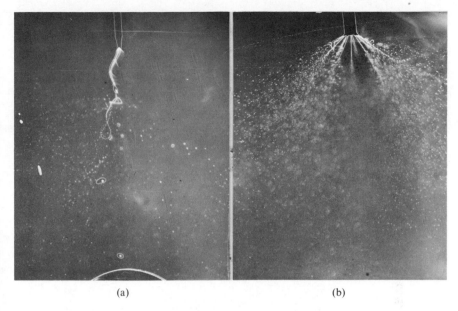

(a)   (b)

FIG. 7.1. Instantaneous schlieren micrographs of spraying in an electric field; (b) shows effect of strongly divergent field on acetone.

the spray appears more uniform. This is shown in Fig. 7.1(b); both (a) and (b) are for acetone and were obtained here by Dr. G. P. Tewari, using a spark-schlieren system at high magnification.

Although the mechanism can be described physically, at the time of writing there is no quantitative theory that correctly and fully accounts for the droplet size, the charge/mass ratio, and their distribution. The theory of the stability of jets and droplets and the influence on it of electric fields has been extensively investigated (see for example Raleigh,[12] Taylor,[13] Nayyar and Murty,[14] Peskin and Lawler,[15] Graf,[16] and Hendricks and Schneider[17]), but the setting-up of a unified analysis of the spraying process is hampered by a number of intractable factors. The mechanism of charge flow along the jet does not appear to be a case of conduction uniformly across the cross-section; the dielectric properties and hence surface stresses can become important in the case of some liquids and the space charge, with its dependence on geometry and influence on field distribution is rather involved. Furthermore, as can be seen from Fig. 7.1(a), any extended 'trunk' tends to whip around rapidly because of electric wind effects at its tip and associated space charge effects. This leads to the breaking away of large satellite drops at convolutions of large curvature. Lastly, transition from one spraying regime to another (for example Fig. 7.1(a) as compared with (b)) is discontinuous and influenced by incidental factors such as specks of impurities attaching to the electrodes.

It has been demonstrated here by Dr. K. Gugan that a similar effect can be induced in powered solid fuels. The method was to apply a high potential to a fluidized bed of the powder, whereupon a proportion of particles, which increases with the applied field, is carried away by the fluidizing gas. The particle size required for this effect to become appreciable decreases as the density of the solid increases.

Providing the second electrode offers a large open area, most of the droplets or particles pass through without giving up their charge, even in the absence of a superimposed gas flow. The reason must lie ultimately in the low particle mobility, which allows droplet trajectories to overshoot lines of force once their momentum is sufficiently high. The low mobility is responsible also for the large induced ionic winds that contribute to this effect. The consequence is most desirable, because it permits the manipulation of the droplets by fields in the region beyond the second electrode, i.e. it is not necessary to construct the entire combustion system between the two primary electrodes.

Such subsequent electrical control can be applied for various purposes. Thus, by applying a transversely oscillating field the droplets can be induced to proceed in sine waves, or zig-zags, in order to increase the length of their trajectories within a given distance downstream. This is shown in Fig. 7.2, taken from the work of Gugan. The increase in effective path lengths

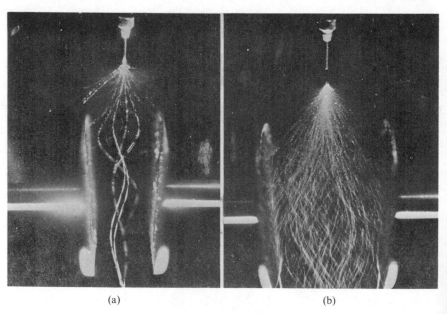

(a)   (b)

FIG. 7.2. Individual droplets (a) and a spray (b) oscillating in an a.c. field following electrical dispersion of liquid.

and of relative velocities with respect to the air increases vaporization and hence decreases the combustion length. In addition to the movements of the droplets themselves, secondary air can be inducted by fields along the axis of the apparatus. This effect is discussed below in the context of ionic winds. An insulating tube can act in the manner of the 'electrostatic collimator' discussed under direct ion recording (Chapter 4). Its walls acquire such a charge everywhere as to make lines of force parallel to them, so that further charges are repelled. In this way they help to maintain the dispersion uniform.

Although this section concentrates on electrical control of dispersion of reactants, intermediates, or products, it may sometimes be possible to introduce inert particles into an otherwise gaseous system, in order to achieve the appreciable transfer rates that space-charge limitations preclude in the case of ions. An example would be the use of a heat exchanger consisting of a cloud of particles made to oscillate between hot and cold gas by an alternating field. The usefulness of this, for example, to increase combustion intensity by arranging such an interchange between cold reactants and hot flame products, must be investigated and weighed against the additional radiation losses that such a scheme entails.

In many ways the most interesting applications arise when the product or an intermediate of a gaseous combustion reaction is a particulate solid. The most common case, which is of interest for hydrocarbon fuels, is the formation of carbon. (In accordance with previous usage, the term 'carbon' will be used, although the solid is carbon-black or soot, rather than elemental carbon.) This can be regarded either as a product or an intermediate, depending on stoichiometry and the conditions of burning. It is also a particle of low work-function that has, at various times, been associated with the process of charge generation in flames.

The observation that carbon particles are in some way associated with the electrical properties of flames that contain them, emerges incidentally from a wide variety of studies. Thus, in the work described later, of augmenting heat transfer from flames to surfaces by means of electric fields,[18] the increase observed for a given geometry rises systematically with percentage of hydrocarbon fuel. Figure 7.3 is a graph of maximum percentage increase for ethylene–air flames plotted against volume per cent of ethylene in the mixture. The rapid increase in slope occurs at the composition at which flame luminosity, characteristic of free carbon particles, is first observed to set in.

In experiments on flame deflection induced by ionic winds, carbon collects, usually on the negative electrode, if a deflected carbon-bearing flame (see Plate I(ii)) is allowed to touch or approach the electrode surface. The deposit is strikingly different from that collecting on a cold surface in contact with the flame in the absence of a field. The rate of deposition seems greatly increased—but it will be shown below that this is true only of volume, not

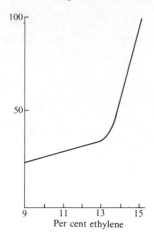

FIG. 7.3. Field-induced improvement in heat transfer (as percentage at maximum) plotted against percentage ethylene in reactant mixture. From Payne and Weinberg.[18]

of mass, and is occasioned by a change in density—with filaments of carbon growing visibly, quite unlike the continuous layer obtained without a field.

Again, a procedure widely used in the carbon-black industry is the addition of small amounts of alkali metals—usually in the form of their salts—to the process in which asphaltic oils are burnt with an inadequate air supply in a long tunnel. These additives alter the 'structure' of the black in minute concentrations and their effectiveness appears to vary in the order of the ionization potential of the metal.

Early quantitative work on the effect of fields on carbon deposition was carried out[18] on simple ethylene diffusion flames burning in a long steam jacket that acted as the cathode. Steam rather than water cooling is employed in all these experiments to prevent water condensing on the deposits. The anode was a metal rod along the axis. Although the ratio $\dfrac{\text{deposition on tube}}{\text{deposition on rod}}$ increased by about fifty times when a field was applied, the apparatus was not suited to the study of changes in the mechanism of deposition, because of the considerable field-induced changes in flow pattern. Such flow-line distortion, forcing flame gases towards the walls of the tube and leading to vortex formation, tends to make results dependent on the particular geometry; experiments minimizing flame and gas flow distortion are necessary. This can be accomplished only in a system where lines of electrical force approach parallelity to the flow lines in the flame. The apparatus used in the next phase of this work is shown in Fig. 7.4. When a field is applied to the ethylene diffusion flame, the matrix and the collector plate acting as the electrodes, the shape of the lines of force must be approximately as shown

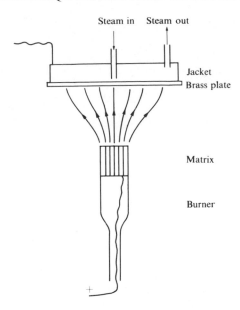

FIG. 7.4. Making flow lines approximately parallel to those of field intensity. From Payne and Weinberg.[18]

in Fig. 7.4, following reasonably closely the shape of the gas flow-lines. This is confirmed by photographs (Fig. 7.5(a), (b)) which show that, with the field applied in the sense causing positive ions to drift in the direction of the flow lines, very little change in flame shape occurs.

In the absence of a field, carbon is deposited on the collector plate in a uniform layer as shown by Fig. 7.5(d), in which no structure is visible (the lines in the deposit being scratches with a pin point to illustrate this). With the collector plate negative, a bulky deposit builds up rapidly, although the mass-rate of deposition is somewhat smaller than that in the absence of a field. The great increase in volume is due to decreased density and the micrographs for varying current densities, increasing in the sequence (e)–(g), show the reason. The effect of the field is to build loose, bulky agglomerates, separated by bare patches. Reversing the field causes the flame to distort (Fig. 7.5(c)) with parts of it lapping the burner mouth. Under this condition very little carbon is deposited on the top plate; instead it grows in the form of long filaments of up to 1 cm on the matrix and burner rim, a more continuous layer being formed on the outside of the burner top.

The first conclusion is that the carbon, or at least the great majority of the particles, behaves as a positive charge carrier in this system and can therefore be directed and confined to the negative electrode. The reason for positive charging will become obvious later; it is to do with both thermionic emission (which always charges the emitter positively) and attachment (which charges

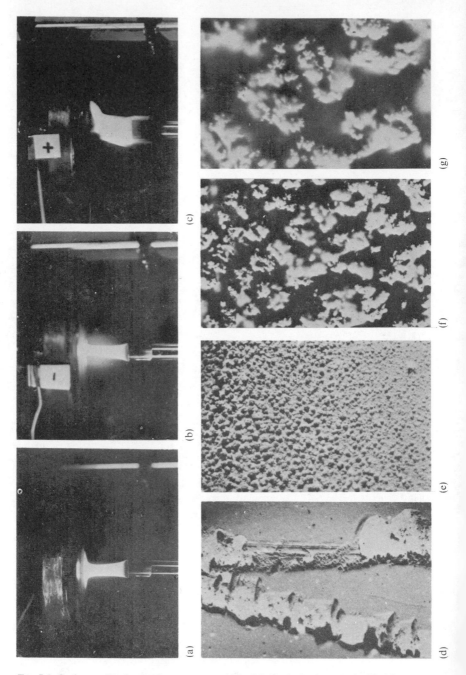

FIG. 7.5. Carbon collection in the apparatus of Fig. 7.4. Optical micrographs (d)–(g) at constant magnification of approx. 30 ×. From Payne and Weinberg.[18]

to the sign of the local ion flux) always tending to reinforce, in a system where the flame surface approaches the burner mouth and thereby shields the pyrolysis zone from the applied field. The form of deposition follows directly from the effect of curvature on charge density and field strength. Thus, the first specks deposited increase curvature and concentrate lines of force upon themselves, so that they collect particles arriving subsequently, which might otherwise have travelled to adjacent sites. The longer such a filament grows, the larger is the surrounding area which it protects from further deposition —hence the bare patches and reduced density.

In order to study in detail anything beyond the process of deposition, a different burner system is required yet again. It is desirable that it should be symmetrical, that the flame surface should not short-circuit either electrode space by approaching the burner rim closely, and that the incipient carbon particles to be collected for study should not travel to the electrode through hot oxidizing flame gases. The counter-flow diffusion flame[19] using the burner system developed to produce a plane flame[20] (see Chapter 3) seemed[1,2,21] particularly well-suited for this type of work. The matrices that streamline the approach flow can be used as electrodes parallel to the flat flame and cooled by transpiration of the reactants. Light gauzes can be placed over them when it is desired to weigh the deposit. Depending on the direction of the applied field, either positive or negative charge carriers from the flame zone pass through the pyrolysis zone, and bring down any carbon particles to which they attach on the fuel-side electrode and out of contact with oxygen. It is only if particles somehow acquire a charge opposite to that with which they are bombarded that they may cross the flame and deposit on the other electrode. In addition to providing a nearly unidimensional configuration, the system has the advantage of minimizing ionic wind effects because these occur in opposition to the gas flows.

Two burners of this type were employed, differing in the fuel and oxidant supplies, the electrode spacing, and the arrangement of the flanges at the burner mouths. The objective of the difference was that the second system (referred to as the 'oxygen burner') allowed much larger ion currents to be drawn and covered the whole range of flame currents, including saturation and secondary ionization. It thus gave much additional information, whose complexity, however, it would have been difficult to resolve had not the much smaller range of the original 'air burner' results been interpreted first. Measurements carried out included the dependences on applied potential of parameters such as mass rate of carbon deposition in various positions, particle size, light emission, and current density. Thus Fig. 7.6 shows rates of deposition against applied potential for four systems—the combinations of 'air' or 'oxygen burner' with the fuel-side electrode positive or negative. All but the smallest fields ensure that the pyrolysis zone is permeated by a flux of charge carriers of sign opposite to that of the fuel-side electrode.

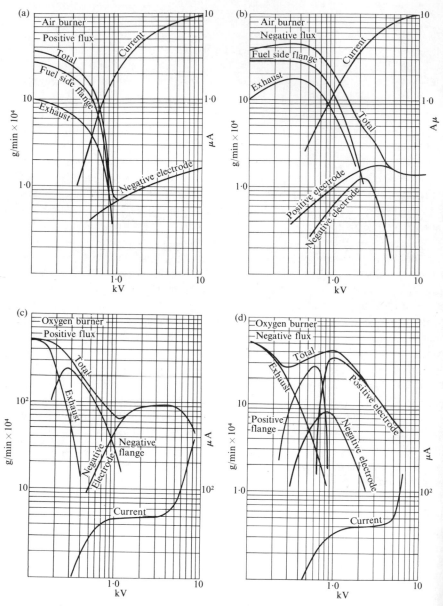

FIG. 7.6.

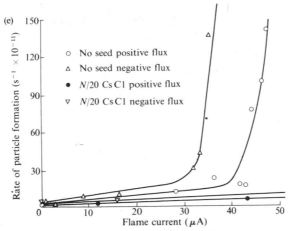

FIG. 7.6. Rate of soot collection plotted against applied voltage. (a) Air burner, positive flux; (b) air burner, negative flux; (c) oxygen burner, positive flux; (d) oxygen burner, negative flux. From Place and Weinberg.[21] (e) Number rate of particle formation against current. From the work of Mayo.

It will be convenient to summarize the conclusions of this work first, since it was found that electric fields affect all three phases of the process—nucleation, growth in the pyrolysis zone, and deposition—and these tend to affect the results in different ways. As regards the former, it has been suggested that the centres on which growth occurs may be charged (at least in systems where other nuclei are scarce) and can thus be manipulated electrically. Particle size and mass are governed by the period of residence in the pyrolysis zone and this can be varied within wide limits by applied fields, since all the particles are charged from the earliest stages.[1] Deposition occurs, usually on the negative electrode (although it will be shown that negatively-charged carbon can be produced under special circumstances) and the form of aggregate depends on the current density, as was discussed above.[18]

The contention that all the carbon particles are charged is borne out by the changes in deposition pattern at quite low fields. It will be apparent from Fig. 7.6 that gas-borne particles, which constitute the majority of the carbon in the absence of a field, disappear, and ultimately all the carbon that is formed collects on the electrodes in the burner mouths. Since particles that deposit there must have travelled against the flow stream, they could never be found on the electrodes at all, in the absence of a charge.

The processes by which a particle can acquire charge (see Chapter 2) can be divided into two categories: attachment of, or charge transfer from, another charge-carrier, and electron emission. The former includes 'diffusion' and 'bombardment' charging, as detailed earlier, but the important point here is that, while thermionic emission leaves the particle positively charged irrespective of field direction, all the other mechanisms charge the particle according to the polarity of the cloud of charge carriers that surrounds it.

Since the cloud of ions from the flame which is drawn through the pyrolysis zone tends to bring the carbon down on the fuel-side electrode irrespective of polarity, the relative importance of the two mechanisms can now be assessed by examination of Fig. 7.6. In (a) and (c) the pyrolysis zone is in a flux of positive ions and the two mechanisms reinforce. As would be expected, no negatively-charged particles are found. In (b) and (d), however, the two mechanisms compete. It is found that, at low fields and low negative fluxes, some particles become charged positively and cross the flame to deposit on the negative (oxygen) electrode. This implies that, as they ascend the temperature gradient in travelling towards the flame, their thermionic positive charge increases more rapidly than their negative charge acquired by attachment. As the field and flux of negative charge-carriers increases, however, attachment charging occurs more rapidly until, eventually, the positive charge is neutralized and all carbon particles are negatively charged. Thus, in the absence of attachment charging, it is anticipated that all the carbon particles would be charged positively due to the thermionic effect alone.

Having considered the charging mechanism, it will be simpler to discuss the next effect—variation of residence time—in a regime of positive flux, where both mechanisms reinforce to charge positively, for example Fig. 7.6(a). Electron micrographs show that very small fields suffice to reduce particle size drastically, by reducing the period of residence in the pyrolysis zone, as shown by Fig. 7.7(a) and (b), taken at a magnification of 40 000 at applied

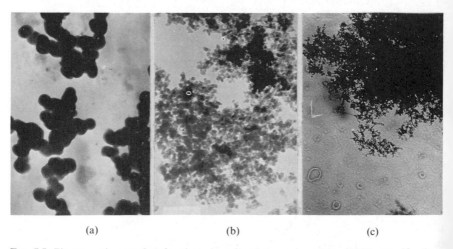

(a)   (b)   (c)

FIG. 7.7. Electron micrographs of carbon deposits at approximately ×40 000 magnification. (a) No applied field; (b) applied voltage 1 kV; (c) applied voltage 10 kV. From Place and Weinberg.[21]

potentials of 0 and 1 kV respectively. At the same time, a striking decrease occurs in the mass collected. At 1 kV this is only 2 per cent of the carbon formed in the absence of an applied field—see Fig. 7.6(a). It can be shown that, in this regime, all the decrease in mass can be accounted for by the reduction in size of the particles.

The decrease manifests itself even to the casual observer by a very great reduction in flame luminosity—see Plate I(i). The yellow luminosity due to carbon particles in the flame virtually disappears by the time the potential has reached 1 kV. The drop in carbon formation and luminosity is thus not caused simply by precipitating the fully-formed carbon particles from the flame but rather by rapid removal of the growth centres, thereby greatly reducing the total amount of carbon formed. This is possible because particles seem to acquire charge very early in their life. Just how early this occurs is indicated by spectroscopic measurements.[1] The emission from species such as C_2 and CH exhibits a similar decrease in intensity, as does the continuum radiation from carbon particles when a field is applied.

The authors next set out to achieve the opposite effect—i.e. to increase residence times by means of applied fields. Since the flow lines are opposed to those of the electric field, the motion of the charged particle is determined by the balance between the drag forces and the interaction of the field with the charge on the particle. Thus, it was found that in the positive flux case there is a critical applied voltage at which particles can be held almost stationary against the axial component of the gas flow. When this was reached, macroscopic particles and filaments would suddenly appear in the pyrolysis zone, and eventually deposit on the burner flanges to produce a network of carbon filaments extending right into the flame[1]. A similar and more controllable result could be obtained in the negative-flux case at a slightly higher voltage. In this case the net charge on the particle was smaller as a result of the competing positive and negative charging mechanisms and the electric field required to balance the drag force was accordingly larger.

In the above experiments, electrical effects were used mainly as a method of manipulating particles in a reaction zone. Other methods, though rather less versatile, can produce some similar effects. Thus, Homann and Wagner[22] were able to cause the growth of macroscopic carbon particles by local recirculation in the reaction zone of a premixed flame—an alternative method of increasing residence times.

As regards nucleation the matter is rather different. Here, hypotheses are based on much more circumstantial evidence since, unlike carbon particles, the precursors have not been collected and examined (except perhaps by mass-spectrometry—see refs. 25 and 26). Furthermore, the counterflow diffusion flame system is unusual in that it provides a minimum of external nuclei in the pyrolysis zone. This makes it particularly suitable for studying nucleation by ions, if this does occur, so long as it is not assumed that ions are the *only* possible nuclei—they may, in fact, be much less important where flame products have freer access to the pyrolysis zone.

The effects described earlier occur in the lower range of applied voltages, when the pyrolysis zone is permeated by only relatively small fluxes of ions from the flame. Although residence-time effects still occur at higher voltages and the size of the particles collected continues to decrease with larger field strengths (see Fig. 7.7(c)) the main interest in this regime lies in the number of particles produced. This does not vary greatly at low potentials and it is for this reason that it is possible to resolve the effects of fields on growth from those on the initiation of particles.

Considering first the case where a large flux of positive ions crosses the pyrolysis zone, it is found that the mass collected increases somewhat after passing through the minimum that is caused by the shortening residence time. Since the particle size continues to decrease, this rise in mass represents a considerable increase in the numbers of particles formed in the flame. An eightfold increase over the rate of particle generation with no applied field was obtained at the highest field strengths. Since the only factor that is changing under these conditions is the ion current, which is proportional to the number of positive ions penetrating the pyrolysis zone, it was thought likely[1,2] that the ions coming from the flame act as nuclei on which carbon formation can take place; a good correlation between the number of particles formed and the ion current was in fact found.

Such a nucleation effect could be purely physical—for example, analogous to condensation on charged particles in the supersaturated atmosphere of the cloud chamber. However, the positive ions postulated as produced by the chemi-ionization or subsequent charge transfer processes in hydrocarbon flames, which were discussed in Chapter 6—species such as CHO^+ and $C_3H_3^+$ (see, for example, Sugden and Green[23] and Kistiakowsky and Michael[24])—would appear to make suitable nuclei for carbon growth from a chemical point of view also. This is especially plausible in view of the results of Knewstubb and Sugden,[25] who identified ions of the type $C_nH_n^+$ in flames where n varied from 3 to 10, which suggests that such ions are capable of undergoing polymerization reactions. Again, all this does not imply that ions are the only possible nuclei.

The evidence from the spectroscopic work mentioned above is also compatible with this view. The emitting C_2 and CH are, of course, neutral, so that their decrease with increasing field strength must be due either to a charged parent being removed by the field or, equivalently, to a decrease in the number of uncharged parents. In either case, the emitting species is likely to arise only in the early stages of carbon formation as has been discussed.[1]

To gain more information regarding the role of positive ions in the nucleation of flame carbon, the effects of an additive ion—caesium—which was unlikely to take part chemically in the carbon formation, was examined.[2] Caesium chloride was evaporated into the oxidant supply to the burner and, in the regime where seeding was effective, a fivefold increase in current over that from the unseeded flame could be obtained before the increased wind effects made the flame too unstable. The effect of dragging the additional positive charges through the pyrolysis zone was to produce a twofold to threefold increase in the mass of carbon deposited. The peak in the mass collection coincided with that in the ion current. Above a certain voltage, the normal flame current was sufficient to precipitate the CsCl particles fed in, before they could reach the flame, and the carbon deposition fell with the current to the normal unseeded level as this critical voltage was attained. Preliminary measurements for the negative-flux case with seeding indicated that the increased negative charge in the pyrolysis zone did not tend to increase carbon formation. It was therefore suggested that positive ions act as nuclei even when they have no organic structure, while negative charges have no such effect. However, the results with Cs^+ may be due to charge exchange with hydrocarbon fragments—see Miller[26].

The role played by negative charges was thus thought to be different.[1,2] This is because even at the highest negative charge fluxes represented by Fig. 7.6(b) and (d) there was no tendency for the mass collected to increase. In the absence of a complete survey of particle size by electron microscopy however, no very firm test of this hypothesis could be made at the time, because under the conditions represented by Fig. 7.6(b) and (d), residence time effects are still important, due to the competing charging processes by thermionic emission and by bombardment with negative charge. Since then, in very recent work with Dr. P. J. Mayo, we have been able to obtain electron micrographs in much greater numbers than previously and this has made it possible to resolve the amount of carbon formed into the number-rate of particle generation and the particle volume, over the entire range of flames with oxygen. Figure 7.6(e) shows this number rate plotted against current through the pyrolysis zone. It will be seen from the slopes that the 'nucleating efficiencies' of positive and negative charge carriers in unseeded flames are much closer than had been supposed and both rise steeply as saturation is approached. That of Cs^+, on the other hand, is smaller, irrespective of the increased total amount of carbon formed. The similarity of the curves, particularly the steep rise at saturation, is more significant than their relative positions—it must be borne in mind that a given current represents different field intensities in the flame, depending on the mobilities of the charge carriers and other parameters—see Chapter 8.

At the time of writing, this part of the subject is in such a rapid state of flux that any conclusions must be regarded as highly tentative. Preliminary measurements on particle mobilities suggest that each particle carries only one, or a very few, positive electronic charges, so that most of the current is carried by flame ions. One possible explanation of the remarkable increase in slope is that, as the flame approaches saturation, larger charge carriers are drawn out of the top of the pyrolysis zone and it is these—probably species such as the $C_nH_m^+$ identified by Knewstubb and Sugden[25] and by Miller[26]— which act as chemically effective nuclei. Another hypothesis, which complements rather than contradicts the above, is that because all electrons are removed at saturation, the charged precursors can no longer aggregate as they do normally after one of them has been neutralized by an electron. If so, the small particles withdrawn under saturation conditions have attained their terminal size entirely by heterogeous reaction from the gas phase†.

The considerable interest of these fundamental and largely unresolved early steps of the process should not be allowed to obscure the fact that the major physical effects discussed before are probably of more immediate practical significance. As regards varying residence times by electric fields, the control of particle size and amounts of soot formed may be applicable to, for example, improving combustion efficiency, modifying radiant heat transfer, maximizing the size of carbon agglomerates, and so forth. Concerning deposition, keeping some surfaces clear of soot may be as useful as concentrating its collection at specified points. The change in the form of the deposit from compact solid films to loose bulky agglomerates of low density may also have its practical value. The nucleation work falls on the borderlines of chemical and physical effects. It may be noted that additives that influence carbon formation usually also have low ionization potentials (for example, the alkali metals) or low work-functions (for example, particles of barium oxide). These processes and their respective chemical and physical factors are being studied in many laboratories at the present time.

Emphasis has been given to the case of flame carbon, because this is the field in which most of the work to date has appeared in the open literature. Other reactions, including the formation of metal oxide particles during the combustion of volatile metal compounds, have obvious similarities with the above. A particular example here is the formation of pigmentary titanium dioxide in the gaseous oxidation of the tetrachloride.

'Direct' generation of electricity

In conventional methods of generating electricity from the heat of combustion of a fuel, steam is raised in a boiler at high pressure. The steam drives a turbine that is connected mechanically to an alternator from which the electric power is ultimately drawn. Modern techniques have greatly improved the boiler, turbine, and alternator efficiencies; the overall thermodynamic efficiency of the cycle is limited, however, in the first instance by

† *Footnote added in proof.* See also ref. (119). However, Mayo's recent calculations suggest that the increase in number rate may be explicable simply in terms of removal of growth surface.

the maximum steam temperature that can be tolerated by boilers at high pressure and, in the second, by the maximum temperature at which turbine blades can operate (c. 750°C).[27] The former is at present the more usual limitation; the latter becomes limiting only in the case of gas turbines directly fed with combustion products. For these reasons, methods of generating electrical energy from thermal, which do not require the boiler/turbine part of the cycle, are currently of considerable interest. A large number of such direct generation schemes have been proposed†. The best known generators fall under the general headings of magnetohydrodynamic (MHD), electrogas-dynamic (EGD), calorelectric, thermoelectric, and thermionic. The fuel cell that involves the direct conversion of chemical into electrical energy without the requirement of a flame is not relevant here. The other direct energy converters mentioned are heat engines and could in principle obtain their energy from combustion. However, only those devices that can use combustion products as a working fluid are relevant in the present context. These are MHD, EGD, and calorelectric generators.

MHD generation of electricity

The principle of the MHD generator is illustrated in Fig. 7.8. Ionized gas is

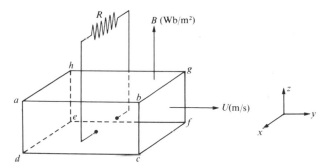

FIG. 7.8. D.C. MHD generator.

passed at a velocity of U m/s in the y-direction across a magnetic field of B Wb/m² in the z-direction. In accordance with current usage in the literature on MHD, the MKS system of units is employed in this section.

A charge, e, within the gas experiences a force $(eU \times B)$ owing to its motion through the magnetic field. For the given configuration, this is equivalent to an effective electric field of (UB) volts per metre in the x-direction. Thus if the faces *abcd* and *efgh* are made electrodes and connected through a load, R, a current will flow under the action of the induced field, i.e. power will be generated. The mechanical work done by the fluid is calculated as follows. If j is the local vector of current density then the force per unit volume of gas

†*Footnote added in proof.* See also ref. (120).

resulting from the flow of charge is the vector product, $j \times B$ N/m^3. The braking force in the y-direction is equal to $j_x B$, so that the fluid does mechanical work at the rate of $j_x BU$ W/m^3, part of which is converted into electrical energy as discussed below. When comparing the MHD generator to the conventional kind, it is clear that in the former the charge carriers (electrons) act simultaneously as the turbine blades and the conductors of the generator rotor. From one point of view the comparison is very much to the detriment of the gas, in view of its low electrical conductivity compared to that of a copper strip. In practice this means that it will be necessary to use gases at high temperature and seeded with alkali metals, or treated in some other manner, to raise the electron concentration to levels at which the conductivity is adequate.

On open circuit no current flows, i.e. the net force on the charges is zero at each point. This state is reached in the following manner. Charges of opposite sign migrate in opposite directions under the influence of the motional e.m.f. and collect on the electrodes until the surface charges are sufficient to neutralize the motional e.m.f. at each point. If E_{∞} is the electric field due to the surface charges on open circuit then, vectorially,

$$E_{\infty} + U \times B = 0, \tag{7.2a}$$

or, in components,

$$(E_{\infty})_x = -UB, \tag{7.2b}$$

$$(E_{\infty})_y = 0. \tag{7.2c}$$

Once current is allowed to flow, the field pattern is changed. Consider an electron moving with velocity components $U + v_y$ and v_x in the y and x directions, respectively. It will experience an effective net field, of components E'_x and E'_y, which is the result of both the motional field and the electrostatic field due to surface charges.

$$E'_x = E_x + (U + v_y)B, \tag{7.3a}$$

$$E'_y = E_y - v_x B. \tag{7.3b}$$

The motion of the electrons relative to the gas is determined by their mobility, K_e, and the net field, i.e.

$$v_x = -K_e E'_x = -K_e\{E_x + (U + v_y)B\}, \tag{7.4a}$$

$$v_y = -K_e E'_y = -K_e\{E_y - v_x B\}. \tag{7.4b}$$

Solving eqns (7.4a and b) for v_x and v_y gives

$$v_x = -K_e \frac{(E_x + UB + K_e BE_y)}{(1 + K_e^2 B^2)}, \tag{7.5a}$$

$$v_y = -K_e \frac{\{E_y - K_e B(UB + E_x)\}}{(1 + K_e^2 B^2)}. \tag{7.5b}$$

Electron mobilities are so much greater than those of ions that, to a good approximation, only the contribution of the electrons to the current need be considered. A flux of electrons in one direction implies, by the usual convention, a current in the other. Thus, in terms of the electron conductivity σ—(see eqn (4.55)),

$$j_x = -nev_x = \frac{\sigma}{1 + K_e^2 B^2} (E_x + UB + K_e BE_y), \tag{7.6a}$$

$$j_y = -nev_y = \frac{\sigma}{1 + K_e^2 B^2} \{E_y - K_e B(UB + E_x)\}. \tag{7.6b}$$

The term $K_e B$ is usually known as the Hall parameter, β. Now $K_e = e/M_e v_e$ (see eqn (4.51)). Therefore

$$\beta = K_e B = \frac{Be}{M_e v_e} = \omega_b \tau_e, \tag{7.7}$$

where ω_b is the cyclotron frequency (Chapter 4) and τ_e the mean free time between electron collisions. The value of β is very important in determining the type of generator to be used, and differs greatly for different gases; see, for example, Table 7.1.[27]

TABLE 7.1

The Hall parameter β

Pressure: 1–10 atm; B: 1–5 W/m^2; T: 1500–2500°K

	N$_2$, Combustion products	Helium	Argon
β	0·15–10	0·2–15	2–250

It will be noticed that there are currents in both the x and y directions. The y component arises because an electron with a velocity component v_x experiences a force $-v_x B$ in the y direction.

Before considering different types of MHD generators, it is useful to define a quantity δ that is equal to the ratio of the operating potential to that on open circuit. The output potential, equal to the electrostatic potential difference between the electrodes, is equal to $E_x L$ volts for the configuration of Fig. 7.8, where L is the duct width, i.e. by eqn (7.2b)

$$\delta = \frac{E_x L}{(E_{oc})_x L} = -\frac{E_x}{UB}. \tag{7.8}$$

The negative sign arises because E_x is itself negative.

Faraday generator. By convention, a Faraday generator is of the kind in which the current that delivers power to the load flows transverse to the gas flow in the generator, i.e. the x component of the current in Fig. 7.8. The Hall generator, which is discussed below, delivers power by the longitudinal current (Hall current). Consider Fig. 7.8; the continuous electrode running along the duct will tend to short-circuit the longitudinal electric field. To a first approximation this effect can be described by the condition $E_y = 0$. From the definition of δ, E_x is equal to $-\delta UB$; therefore from eqns (7.6a and b)

$$j_x = \frac{\sigma UB}{1+\beta^2}(1-\delta), \tag{7.9a}$$

$$j_y = \frac{\beta BU}{1+\beta^2}(1-\delta). \tag{7.9b}$$

The useful power output per unit volume is given by

$$W = -E_x j_x = \frac{\sigma U^2 B^2}{1+\beta^2}\,\delta(1-\delta) \ \ \text{W/m}^3. \tag{7.10}$$

The negative sign arises because, since power is generated rather than dissipated, the components E_x and j_x point in opposite directions and their product is therefore negative. The maximum power occurs when $\delta = \frac{1}{2}$. The total mechanical work, W_T, done by the fluid per unit volume is $j_x BU$. Hence the fractional conversion of mechanical to electrical power is equal to

$$\eta = \frac{W}{W_T} = \frac{-E_x}{UB} = \delta, \tag{7.11}$$

the remainder, $W_T(1-\delta)$ W/m^3 is dissipated in Joule heating of the gas.

It is clear from eqn (7.9a) that β should be made as small as possible for high power densities. The deleterious effect of large values of β can be offset to a considerable extent by preventing the flow of the Hall current. This can be achieved by dividing the electrodes *abcd* and *efgh* into a number of insulated segments, individual loads being placed across separate pairs. Ideally the condition $\beta = 0$ should be reached, there being no continuous outside path to complete the circuit for the internal Hall current. From eqns (7.6a) and (7.8)

$$j_x = UB\sigma(1-\delta), \tag{7.12}$$

i.e.

$$W = -E_x j_x = \sigma U^2 B^2 \delta(1-\delta) \tag{7.13}$$

and

$$\eta = \frac{W}{W_T} = \frac{-E_x}{UB} = \delta. \tag{7.14}$$

The maximum output occurs at $\delta = \frac{1}{2}$, as before, but now the performance of the generator is independent of the parameter β.

When β is very large, as, for example, in the case of argon (Table 7.1) segmentation of the electrodes may not be sufficient to suppress the axial currents. In such a case the Hall current itself could be used to deliver power to the output.

Hall generator. One suitable configuration for a Hall generator is shown in Fig. 7.9. The electrode arrangement is the same as for the segmented Faraday

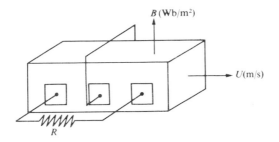

FIG. 7.9. Hall M.H.D. generator.

generator, from which it differs only in the external connections. In this case all the electrodes are short-circuited across to make E_x small—ideally zero—the end pairs acting as the output terminals.

On open circuit, j_y is zero: hence from eqn (7.6b)

$$(E_y)_{\infty} = \beta U B, \tag{7.15}$$

i.e.

$$\delta = \frac{\text{output potential}}{\text{open circuit potential}} = \frac{E_y}{\beta U B}. \tag{7.16}$$

Thus,

$$j_x = \sigma U B \frac{(1 + \delta \beta^2)}{(1 + \beta^2)}, \tag{7.17a}$$

$$j_y = -\sigma U B (1 - \delta) \frac{\beta}{1 + \beta^2}. \tag{7.17b}$$

The electrical output per unit volume is

$$W = -E_y j_y = \frac{\sigma \beta^2 B^2 U^2 \delta (1 - \delta)}{(1 + \beta^2)}. \tag{7.18}$$

As before, the maximum power output occurs at $\delta = \frac{1}{2}$. In this case, however, W increases with β, being equal to the value for the segmented Faraday generator for β equal to infinity. The total mechanical work done by the gas per unit volume is here

$$W_{\mathrm{T}} = Uj_xB = \sigma U^2 B^2(1+\delta\beta^2)/(1+\beta^2), \tag{7.19}$$

i.e.

$$\eta = \frac{W}{W_{\mathrm{T}}} = \frac{\delta(1-\delta)\beta^2}{(1+\delta\beta^2)}. \tag{7.20}$$

By differentiation of eqn (7.20) it is easily shown that the maximum efficiency occurs when

$$\delta = \delta_{\mathrm{m}} = \{\sqrt{(1+\beta^2)}-1\}/\beta^2 \to 1/\beta \tag{7.21}$$

and

$$\eta = \eta_{\mathrm{m}} = 1-2\delta_{\mathrm{m}} \to 1-2/\beta, \tag{7.22}$$

the arrows on the right-hand side indicating the trend as β tends to large values.

The characteristics of the three types of generator can be shown on a single graph, Fig. 7.10. In all cases

$$E = N\delta \quad \text{and} \quad j = M(1-\delta), \tag{7.23a and b}$$

i.e.

$$E = \frac{N}{M}(M-j) \quad \text{and} \quad W = \frac{N}{M}j(M-j), \tag{7.24a and b}$$

where N and M have the values shown in the figure.

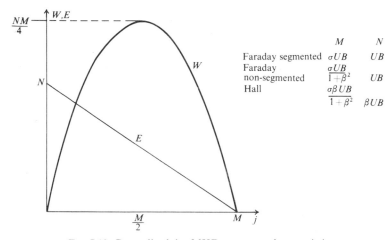

	M	N
Faraday segmented	σUB	UB
Faraday non-segmented	$\dfrac{\sigma UB}{1+\beta^2}$	UB
Hall	$\dfrac{\sigma\beta UB}{1+\beta^2}$	βUB

FIG. 7.10. Generalized d.c. MHD generator characteristic.

The theoretical models presented are necessarily simplified but are adequate to describe the overall current/potential characteristics of MHD generators, as can be seen from Figs. 7.11 and 7.12.[29] According to eqn (7.23b) the current should vary as $(1-\delta)$ for both Faraday and Hall generators. This relationship is well borne out by experiment.

The generators described are intended merely to illustrate the two kinds of operation, namely the Hall and the Faraday modes. Many different configurations for d.c. generators have been suggested; one example is the vortex generator[30] whose method of operation will be obvious from Fig. 7.13. The principle, however, is the same in all.

The generators discussed up to this point produce d.c. at relatively low voltage which would require conversion to a.c. prior to being transformed up to the potential necessary for distribution. It is possible, however, to

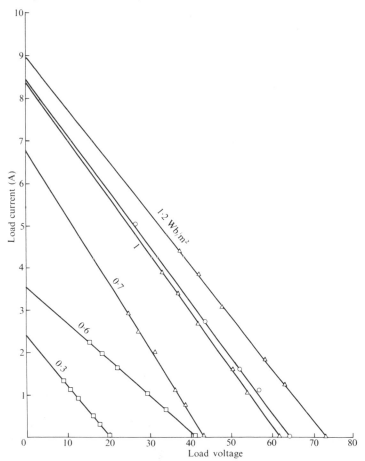

FIG. 7.11. Load characteristics of a Faraday-type generator. From Lindley.[27]

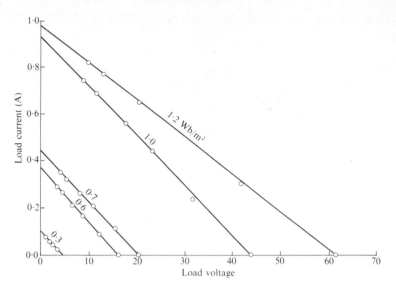

FIG. 7.12. Load characteristic of a Hall-type generator. From Lindley.[27]

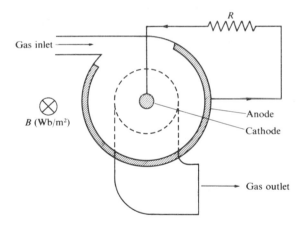

FIG. 7.13. Vortex d.c. MHD generator.

produce a.c. directly in the generators described, by either modulating the magnetic field or the gas conductivity, the latter being the easier alternative. Thring[31] has suggested the use of pulsed combustion for this purpose. For example, fuel could be intermittently injected into the oxidant stream in order to produce alternating layers of hot, ionized and cooler, non conducting gas. An alternative approach is to use inductively coupled devices,[32] which have the advantage of also being electrodeless. A series of coils are laid along the duct in such a way as to produce a travelling magnetic field—the principle is identical to that by which a rotating magnetic field is produced in an induction motor. The gas, flowing at a velocity faster than that of the field, is slowed down and power can be withdrawn inductively. In order to achieve an acceptable power output, the conductivity required is considerably above that available from ionized gas flows originating from combustion sources,[33] which is a serious impediment to the use of this kind of system.

The practical difficulties encountered in MHD are best illustrated by reference to the equations for volumetric power generation. In all cases this quantity has a maximum value, which is given by

$$W_{max} = \tfrac{1}{4}\sigma U^2 B^2. \tag{7.25}$$

Clearly it is desirable to make σ, U, and B as large as possible. Having regard to the maximum practical values of U and B ($c.$ 10^3 m/s and 10 Wb/m^2 using superconducting magnets), it is necessary for σ to be of the order 10–10^2 mho/m for economical operation. Lower values require the use of prohibitively large ducts. Neither combustion products nor inert gases, both of which are proposed working fluids, naturally achieve such high conductivities below temperatures of many thousands of °K. Seeding with materials of low ionization potential is required in order to obtain the desired conductivity in equilibrium at temperatures achievable by the combustion of fossil fuels ($c.$ 3000°K) or by nuclear reactors, the latter being much the lower. Frost[34] has calculated electrical conductivities at atmospheric pressure over a wide range of temperatures for seeded gas in equilibrium— Fig. 7.14. It is obvious from the figure that if MHD generators are to depend upon thermal ionization, especially in the presence of combustion products, the outlet temperature from the duct will need to be very high (>2000 to 2500°K). Such high temperatures make severe demands upon materials of construction, especially in the presence of highly corrosive seeded combustion products. Another consequence of the high outlet temperature is that these devices can only be used as 'toppers' to conventional generation cycles when used for large-scale power production.

Various suggestions based upon nonequilibrium processes have been put forward to overcome the temperature limitation. Kerrebrock[35] has pointed out that the presence of an electric field in the duct might appreciably increase the electron temperature and consequently the level of ionization. The details of the process are discussed in Chapters 4 and 6. McNab and Lindley[36] have proposed that high levels of nonequilibrium ionization could be frozen into the driver gas by sudden expansion through a nozzle. Originally the ionization might come from a discharge, a flame—perhaps augmented with a discharge—a chamber irradiated with ionizing radiation, etc. It is known, for example, that there is considerable nonequilibrium

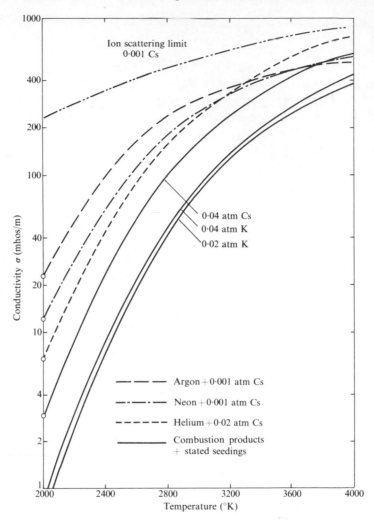

FIG. 7.14. Conductivity of seeded gases. From Coombe.[27]

ionization in the reaction zone of a flame (see Chapter 6). However, in view of the rapid rates of recombination, it seems unlikely that it would be possible to sustain the required levels of ionization for the traverse time in the duct, which is of the order of 10^{-2} s. From eqn (6.5), the half-life of ionization, τ, is equal to $\frac{1}{2}\alpha n$. If $\alpha = 10^{-9}$ cm^3/s—a rather low value as recombination coefficients go—then τ ranges between 0.5×10^{-3} and 0.5×10^{-5} s for ion concentrations in the desired range, viz. 10^{12}–10^{14} cm^{-3}. One possibility is to generate nonequilibrium ionization continuously within the duct. An example of this is provided by the work of Sugden and his collaborators, described in Chapter 6, in which the addition of chlorine to an alkali laden $H_2/O_2/N_2$ flame kept the electron concentration above the equilibrium level for times of the order of milliseconds, i.e. so long as there was an excess of free hydrogen atoms in the products.

MHD ducts powered by combustion products are necessarily open cycle, that is, the working fluid is ultimately discharged to the atmosphere. This leads to many problems apart from those associated with the corrosive and possibly abrasive nature of the seeded products of combustion. First there is the question of seed recovery. It is too expensive to discharge all the seed to the atmosphere. (It is interesting to note here that if the ionization of flame carbon could be used, the problems of seed recovery would disappear and that of corrosion would be greatly reduced.) Second, combustion products have relatively large electron collision cross-sections and have a lower conductivity than, for example, argon with the same electron concentration. This implies working temperatures up to 500° K higher than some inert gases would require, when account is also taken of the fact that an inert gas working in a closed cycle in which there is no loss of seed can use caesium (ionization potential 3·89 eV) whereas combustion products, necessarily working in an open cycle, must use potassium (ionization potential 4·32 eV) for economic reasons. Lastly, again for economic reasons, the operating pressure of the open duct must be considered in relationship to atmospheric pressure. This provides a further constraint upon the open-cycle system.

Numerous experiments have been carried out on combustion-fired MHD generators of the Faraday type. Brogan and others[37,38] have described experiments carried out using kerosene with oxygen or oxygen enriched air. With a total thermal throughput of 20 MW the initial output was reduced to 100 kW, by Hall-current losses. After suppression of the Hall currents, outputs of 600–700 kW were obtained. Way and Hunstad[39,40] have studied a rig running on a mixture of diesel oil (43 per cent), butyl cellosolve (12 per cent) and potassium-2-ethyl hexoate (45 per cent) burning in N_2/O_2 mixtures of mass ratio 1:2, the oxygen equivalence ratio varying from 0·85 to 0·95. In their measurements they found the actual power to be about 25 per cent below the theoretical value. In view of the uncertainty in equilibrium data used in the calculation of the conductivity, and in the temperature distribution within the duct, this agreement is considered to be good. The literature on the subject of combustion-driven MHD generators is extensive. Further useful information can be found in the proceedings of conferences quoted in refs. (29)–(40), in ref. (27), and in the proceedings of the International Symposium on Magnetohydrodynamic Electrical Power Generation, Paris, 1964, published by the European Nuclear Energy Agency.

For completeness it is worth mentioning that closed-cycle[27] generators, working on an inert gas seeded with caesium, pose fewer material problems because of both the lower working temperature and the less corrosive nature of the fluid. Also there is no loss of seed and much greater flexibility in the choice of gas composition and operating pressure. In addition, in the case of a nuclear fuel, the reactor could be used both as the source of heat and, possibly, of ionization.

One important problem not mentioned so far concerns the erosion of the electrodes. There is a strong tendency for electrode spots of high current density to form, which results in local vaporization and rapid deterioration of the electrode. The importance of this effect can perhaps be judged by the reluctance of some establishments engaged in MHD research to disclose their results on this aspect of the subject.

EGD generation of electricity

The principle of EGD generation is illustrated in Fig. (7.15). There is a source of ions, or charged particles, at one end of a duct through which gas

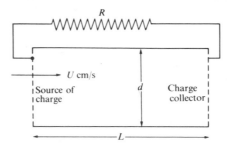

Fig. 7.15. EGD generator.

flows at U cm/s. The charges are carried along by the gas and deposited upon a collector electrode. A load between source and collector completes the circuit. The field set up by the space charge of ions or charged particles is in a direction such as to exercise a force on them acting towards the source. As in the case of the MHD generator, the charges here act as the counterparts of both the blades of a turbine and the moving conductors of the conventional system. The power is generated by the charges being driven up a potential gradient by the flowing gas. The actual source of charge can take many forms: a thermionic emitter, a corona discharge, an electrically sprayed mist, a cloud of droplets or particles charged by passage through a corona, a sooting flame to provide charged carbon particles, etc. In general these devices produce very high output potentials and require multistage operation.

If the charges are positive, the field will be in the direction shown. As in the case of MHD generators, several geometries have been proposed. However there are two basic types. One is the short wide generator, in which $L \ll d$ (plate generator), the other is the long and thin one, in which $L \gg d$ (tube generator).

Plate generator.[41-44] In this case, because the distance between the electrodes is small compared to the diameter of the duct, one can make the assumption that the field is parallel to the flow lines, i.e. $E = E_x$, the electrodes acting as the plates of an infinite plane capacitor. Thus

$$\frac{dE}{dx} = -4\pi ne \tag{7.26}$$

and

$$j = (U - KE)ne. \tag{7.27}$$

The negative sign arises in (7.26) because E and x point in opposite directions. On substituting eqn (7.27) into (7.26) and integrating from $x = 0$, $E = E_0$ it follows that

$$[UE - \tfrac{1}{2}KE^2]_{E_0}^{E} = -4\pi jx. \tag{7.28}$$

The field is a maximum at $x = 0$ and falls steadily. Given a value of U, then, for any current density, the power will be a maximum when the potential drop is a maximum. This implies that the generator should be as long as possible (i.e. $E = 0$ at $x = L$) and that the field should be as large as possible. If $E = 0$ at $x = L$ the power per unit area of electrode is[41,42]

$$W = \frac{UE_0^2}{8\pi}\left(1 - \frac{2}{3}\frac{KX_0}{U}\right) \text{ ergs/s/cm.} \tag{7.29}$$

To generate maximum power, the mobility of the charge carriers should be very small; for this reason it has been usual to employ charged aerosols in experimental work.[43,44] If $KE_0 \ll U$, the power is given by

$$W = \frac{UE_0^2}{8\pi} \text{ ergs/s/cm}^2. \tag{7.30}$$

The maximum value of E_0 is E_B, the breakdown field under the prevailing conditions. It was shown in Chapter 4 that, over a wide range of pressure, Paschen's law can be written approximately as

$$E_B = \rho\left(\frac{E_B}{\rho}\right)_0, \tag{7.31}$$

where ρ is the density and the subscript zero refers to standard conditions at 0°C and 760 mmHg. Therefore the power generated at breakdown for $KE_0 \ll U$ is given by

$$W_B = \frac{1}{8\pi}\rho^2 U\left(\frac{E_B}{\rho}\right)_0^2 \propto \rho^2 U. \tag{7.32}$$

It is also easily shown that the current density and output potential at break-down are given by

$$(jL)_B = \frac{UE_B}{4\pi} \text{ e.s.u./s/cm}^{-2}, \qquad (7.33)$$

$$V_B = \tfrac{1}{2}E_B L \text{ e.s.u. V} \qquad (7.34)$$

Let us imagine now that the gas driving the generator is allowed to expand in a nozzle from stagnation conditions (P_1, T_1). It will pass through a condition at which $\rho^2 U$ and hence W_B is maximized. If the nozzle is given a constant area at this point and electrodes are introduced, the gas will deliver power at the maximum possible rate for the given initial conditions. The analysis has been carried out[41,42] for a perfect gas:

$$W_{max} = \frac{1}{8\pi}\left(\frac{E_B}{\rho}\right)_0^2 \left[\frac{2k}{k+3}\left\{\frac{4}{(k+3)}\right\}^{4/(k-1)} \right]^{0.5} \frac{P_1^2}{(RT_1)^{1.5}} \text{ ergs/cm}^2/\text{s}, \quad (7.35)$$

where P_1 is in dyn/cm^2, T_1 in °K, E_B is e.s.u. V/cm, ρ in g/cm^3, and k is the polytropic expansion coefficient (for a reversible adiabatic process, $k = C_p/C_v$). The velocity at which W maximizes is in the subsonic region.

The total energy of the gas is given by its specific stagnation enthalpy h_1. If the gas is perfect,

$$h_1 = \frac{\gamma R}{\gamma - 1} T_1, \qquad (7.36)$$

where $\gamma = C_p/C_v$. The total energy flux, $\dot{H}$, through the generator is equal to the product of the local mass flux and h_1. Thus the fraction of the power turned into electrical energy at maximum output is given by

$$f_{max} = \frac{W_{max}}{\dot{H}} = \frac{1}{8\pi}\left(\frac{E_B}{\rho}\right)_0^2 \left(\frac{4}{k+3}\right)^{1/(k-1)} \frac{\gamma-1}{\gamma} \frac{P_1}{(RT_1)^2} . \qquad (7.37)$$

$R = 8.4 \times 10^7/M$ ergs/°K/mol, where M is the molecular weight.

It is immediately obvious that the fractional conversion of the total into electrical energy is favoured by high pressure and low temperature. Consider, for example, air at $T_1 = 500°K$ and 10^3 atm; then $f_{max} \approx 10^{-2}$, whereas at 1500°K and 10^2 atm $f_{max} \approx 10^{-4}$. The addition of electron-attaching substances could be used to increase the breakdown field. For example, an increase of a factor of 3 increases the power and f_{max} by an order of magnitude. For the first conditions quoted, the power per stage is 0·22 MW/cm^2 of electrode. In a duct of 1 cm length this would imply a power density of 0·22 MW/cm^3, at 25 mA/cm^2 and 9 MV. Such high potentials would be very difficult to handle. Thus, rather than use additives to maximize the field strength, it would be better to use them to reduce the pressure at which the required potential output could be achieved. A value of the order of hundreds of kilovolts would be the highest useful. This could, of course, be achieved by reducing the length of the duct; however lengths of less than about 1 cm are unlikely to be practical, especially in view of the deformation of the electrode that is likely to occur. For example, at breakdown in air at 10^3 atm and $500°K$, the pressure on the electrode, given by $E_B^2/8\pi$, would be of the order of 200 atm, according to eqn (7.31).

The above calculations, although approximate, serve to indicate both the advantages and disadvantages of this kind of generator. Analyses taking into account compressibility effects within the EGD duct[44] are not significantly different from the aforegoing incompressible treatment even in the supersonic region. Generally, in order to obtain a high fractional conversion per stage it is necessary to have high field-strengths in the vicinity of the entrance electrode. This implies very high potentials even for short ducts and therefore severe insulation problems. The alternative is a great number of lower potential stages (c. 100 kV) each with its own charge supply.

The limitations due to extremely high potentials arise in the case of the plate generator because all the lines of force originating from the space charge terminate on the source electrode. Thus the total body force on the gas is limited by the space charge that can be accommodated between the electrodes before breakdown occurs at the source. The tube generator, to be described next, has the advantage that the space charge it contains is no longer subject to the same limitation. As a consequence, the net body-force on the gas at each stage can be much larger without incurring such high fields.

Tube generator. The merits of this type of generator, in which $L \gg d$, were first pointed out by Gourdine and other workers[45–47] who also originated the following analysis. On the assumption that changes in the axial field strength are small compared to those in the radial direction, Gauss' Law becomes

$$\frac{1}{r}\frac{\partial}{\partial r}(rE_r) = 4\pi ne, \tag{7.38}$$

where r is the radius and E_r the radial component of the field. If it is assumed that n is uniform across the channel, for example, owing to turbulence, then integration of (7.38) gives that at the wall, i.e. at $r = d/2$,

$$E_r(x) = E_r^* = \pi ned. \tag{7.39}$$

Because of the continuity of current,

$$\frac{1}{r}\frac{\partial}{\partial r}(rj_r) + \frac{\partial(j_x)}{\partial x} = 0, \tag{7.40}$$

where

$$j_r = KE_r ne. \tag{7.41}$$

Upon integration of (7.40) between $r = 0$ and $r = d/2$

$$\frac{\pi d^2}{4} \frac{\partial j_x}{\partial x} = -\pi K E_r^* ned. \tag{7.42}$$

The expression on the left-hand side of (7.42) is equal to the rate of loss of charge to the duct walls per unit length of duct.

Now

$$j_x = ne(U - KE_x). \tag{7.43}$$

Assuming that $(U - KE_x)$ is approximately constant—this would be the case if either $KE_x \ll U$ or E_x is itself constant—substituting (7.39) and (7.43) into (7.42) and integrating from $x = 0$ to $x = L$, it is found that

$$j_x(L) = j_x(0)\left\{1 - \frac{j_x(0)4\pi KL}{(U - KE_x)^2}\right\}^{-1}. \tag{7.44}$$

Gourdine did not calculate E_x explicitly but assumed that, given a load resistance R, the current can be raised to any value i by increasing the current from the ionizer, and that E_x, assumed to be uniform, automatically attains the value necessary to make $E_x L$ equal to iR, i.e. that output and load potentials are balanced.

It is instructive to compare the slender with the wide channel generators. If it is assumed that particles of a mobility such that $U \gg KE_x$ are used, then at each plane j_x is approximately equal to Un and eqn (7.44) simplifies to

$$\frac{j_x(L)}{j_x(0)} = \left(1 - \frac{4\pi KLne}{U}\right)^{-1}. \tag{7.45}$$

Substituting for ne from (7.39)

$$\frac{j_x(L)}{j_x(0)} = \left(1 - \frac{4L}{d} \frac{KE_r^*}{U}\right)^{-1}. \tag{7.46}$$

For the loss of charge to the walls to be less than about 10 per cent

$$U \geq 40\frac{L}{d} KE_r^*. \tag{7.47}$$

Taking the breakdown value for air at n.t.p. (30 000 V/cm), the maximum velocity as sonic for air at n.t.p. (30 000 cm/s), and the mobility as 10^{-3} cm^2/s/V (e.g. a particle of 20-μm diameter carrying 3500 electronic charges) one finds from eqn (7.47) that

$$L/d \leq 25. \tag{7.48}$$

A value of 25 for L/d will be used in the calculations that follow. Let us now consider the power output of a single stage. If the inequality of (7.48) is satisfied, then $U \gg KE_x$ even up to breakdown and, since there is little loss of charge to the walls, n is constant throughout the duct. The force on the gas per unit area of duct is given by

$$F = \int_0^L E_x ne\,dx \simeq E_x neL \text{ dyn/cm}^2, \tag{7.49}$$

i.e. the power output is given by

$$W = FU = UE_x neL = \frac{UE_x E_r^*}{\pi} \frac{L}{d} \text{ ergs/s/cm}^2, \tag{7.50}$$

there being negligible ohmic losses if $U \gg KE_x$. If $E_x = E_r^* = E_B/\sqrt{2}$, i.e. the net field $\sqrt{(E_x^2 + E_r^2)}$ is equal to E_B, then

$$W_B = \left(\frac{UE_B^2}{8\pi}\right)\frac{4L}{d}. \tag{7.51}$$

It follows that the output per stage of the slender generator is up to $4L/d$ times that of the plate generator. Using the estimated value of L/d, this is an improvement by a factor of 100. Taking air at 10^2 atm and 1500°K flowing at Mach 1, the power per stage is 0.14 MW/cm^2 of electrode. Setting $d = 0.1$ cm, this corresponds to a power density of about 7 MW/cm^3 at a potential of 1.75 MV and a current of about 3.9 A/cm^2, the fractional conversion in this case being 1 per cent.

Several assumptions are involved; for example, that E_x is constant, that E_r^* and E_x can both be made to run at 70 per cent of the breakdown value at the same time, that surface charges are unimportant and that turbulence is not an important factor in transferring charges to the wall.[48] Nevertheless, it is clear that the work output per stage is much larger for the tube than for the plate generator. This is a very great advantage in that far fewer charge injection points are required. The differences between the two geometries are best appreciated by consideration of the force exerted in a single stage, per unit area of electrode; see eqn (7.49). Ideally E_x, L, and n should be large in order to obtain the largest possible pressure drop and consequently the greatest output per stage, equal to $U\Delta P$ ergs/cm^2/s. In the case of the plate generator, E_x and n are linked by Gauss's Law, eqn (7.26), and making L and n large leads to large longitudinal fields. Thus for a high output it is necessary to run at high pressure (because of breakdown) and to accept extremely high potentials over relatively short distances. In the case of the tube generator, however, large values of n and L do not necessarily lead to large values of E_x, because the lines of force originating in the space charge can 'escape' radially. Thus in this case long generators with a high charge density can be used which give a high output per stage without the necessity of working at extremely high potentials. Clearly, to a first approximation,

E_x, which determines the potential developed per unit length of generator, and n, which determines the current density, are independent. The result is a greater flexibility of operation.

EGD generators at the present stage of knowledge appear to offer a very flexible source of electrical power. There is, however, a need for more detailed theoretical and experimental studies before their true capabilities can be assessed. The conclusion that charged particles of low mobility are required is significant from the combustion point of view because these arise naturally in flames as charged soot and fly-ash.

Calorelectric effect

It has been shown by Klein[49,50] and confirmed by von Engel and Cozens[51] that, when two electrodes at different temperatures are placed in a flame, a potential difference develops between them which increases with increasing temperature difference, the cooler electrode being positive with respect to the other. The apparatus used by von Engel and Cozens is shown in Fig. 7.16. The electrodes were metal rings 2 cm thick, of 3·5-cm outside

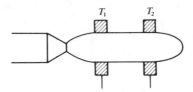

FIG. 7.16. Apparatus used in the study of the calorelectric effect.

and 1·2-cm inside diameters. They were placed around a coal gas flame, and separated longitudinally by about 2 cm. The flame, which could be aerated, burned at atmospheric pressure. When one electrode was maintained at 20°C and the other at 500°C, a potential difference of 2·2 V was developed on open circuit.

It was mentioned in Chapter 6 that von Engel and Cozens had suggested that electrons might reach temperatures greatly in excess of that of the neutral gas as a result of collisions with vibrationally and electronically excited species that arise from the flame reactions. These investigators have also advanced an explanation of the calorelectric effect which is based upon such an effect. It is worth recalling at this point the conclusions of the experimental work discussed in Chapter 6 which are that, although elevated electron temperatures have occasionally been reported in some flames, they are not found in all flames, and experimental evidence is still scanty. In the theory of Cozens and von Engel the electrons are taken to be at some elevated temperature, T_e, in the flame gases. As they diffuse with the positive ions towards the electrodes they pass through cool boundary layers within which they lose energy by collisions. The cooler electrode, having the thicker thermal boundary layer, will receive electrons that have lost more of their excess energy than have those reaching the hotter surface. In consequence, the floating potential V_f (see Chapter 5) assumed by the cooler surface will be less negative, the difference in floating potentials constituting the calorelectric force.

The expression used by von Engel and Cozens for the floating potential is eqn (5.9), which is strictly only applicable at very low pressures:

$$V_f = \frac{-kT_e}{2e} \ln\left[\frac{M_+ \cdot T_e}{M_e \cdot T_+}\right] \text{volts.} \tag{5.9}$$

Assuming that the gas and ion temperatures are equal to the temperature of the electrode, T_w, Fig. 7.17 shows floating potential as a function of mean electron energy, kT_e, for $T_w = 300$ and $800°K$, the value of M_+/M_e being set equal to 20×1840. Using the double-probe technique of Johnson and Malter discussed in Chapter 5, von Engel and Cozens measured the electron temperature in the vicinity of the probes at 20 and 500°C and found values of kT_e of 0·4 and 0·75 eV, respectively. The calculated value of the difference in wall potentials using these values is 2·1 V, which is very close to the measured calorelectric potential of 2·2 V. When loads were connected across the electrodes, currents of the order of microamperes could be drawn, showing that the calorelectric effect provides a means of direct generation of electricity. Unlike MHD and EGD, the electrical energy is not obtained

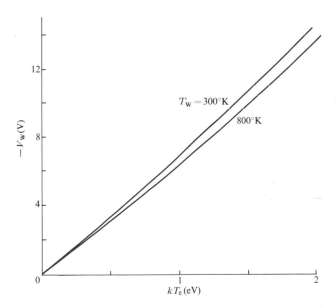

FIG. 7.17. Wall potential V_f as a function of mean electron energy kT_e for two wall temperatures T_w. From von Engel and Cozens.[51]

from the organized mean motion of the gas but from the random thermal motion of the molecular species which may or may not be in equilibrium. Thus the generation here is even more 'direct' than in the other two cases†. It is possible that the very low currents could be increased by addition of alkali salts to raise the electron concentration. Using eqn (5.9) one can estimate the maximum possible value of the calorelectric e.m.f. This occurs for the greatest difference in electrode temperatures ($T_{w1} = 300°K$, $T_{w2} = 2500°K$), the largest possible value of kT_e at the hot electrode (say 1 eV) and the lowest at the cold (say $T_e = T_w$). With these values $-V_f$ is 6 V. In fact its magnitude is very insensitive to all the assumed temperatures except T_e at the hot electrode, to which it is approximately proportional.

It should be noted that there are other explanations for the calorelectric effect. This is particularly significant in view of the fact[52] that elevated electron temperatures have not been found in all hydrocarbon flames. One possible explanation is simply that in the vicinity of the cold probe the electrons attach to form negative ions,[53] whereas at the hot electrode they remain free. This must always exist, at least as a contributory effect. In this case for a kT_e of 0·25 eV (2900°K), i.e. thermal electrons, the calorelectric e.m.f. equal to the floating potential of the hot electrode, would be 1·5 V. Considering the uncertainties in the use of eqn (5.9) at high pressure, the agreement with experiment on the basis of this hypothesis is also adequate.

Movement of charge in methods of detection and measurement

The measurement of ion currents as a method for the study of charge generation, or of other electrical properties of the flame itself, has been discussed in Chapter 5. In this section we are concerned with flame ionization being used as an auxiliary in studies that are not themselves concerned with flame ions. These methods are usually based simply on the greater electrical conductivity of the reaction zone and/or of the product gases, as compared with the cold reactants or with the surrounding atmosphere. This can be used, for example, to cause the passage of a current between two closely-spaced electrodes of an 'ionization gap' at quite small potential differences and thus to give a virtually instantaneous indication of the presence of the flame at a known location of small extent.

It is not always easy to draw a distinguishing line between the use of ionization probes to measure currents (and hence conductivities and electron concentrations) as against their use to signal the arrival of a shock, flame, or detonation wave. The reason is that several workers primarily concerned with the latter type of measurement have also used the amplitude of the current signals to estimate electron concentrations, generally for comparison with calculations of equilibrium ionization. The experimental

† *Footnote added in proof.* See also ref. (121).

requirements corresponding to the two aims, however, tend to conflict. Conductivity measurements demand current paths long enough to minimize the contributions of the cool boundary layers and quenching zones surrounding the electrodes, which have appreciable thermal inertia. Thus in the measurement of product electron concentration[28] in a detonation tube it is quite appropriate to measure currents between the tip of an axial probe and the walls of the surrounding tube. On the other hand, the signalling of times of arrival, in order to measure, for example, velocities of propagation of shock, flame, or detonation waves by ionization in a tube, benefits from small electrode spacings to minimize error in location, and from their mounting flush with the wall so as not to interfere with the propagation of the wave. Ideally, the separation between the electrodes should then be too small to measure ionization reliably.

This type of localized measurement has been used to determine velocities of shock waves, flames, and detonation waves, and to study pre-detonation conditions, effects of attenuating materials on detonation, etc.; see, for instance, Knight and Duff,[54] Hecht et al.,[55,56] Stern et al.,[57] Bollinger and Kissel,[58] Kistiakowsky and Zinmann,[59] and Evans et al.[60] In internal combustion and diesel engines it has given time-resolved information[61-64] concerning the propagation of normal flame fronts and the onset of knock. Several of these publications[54-56] discuss mechanical and electronic experimental details for accurate recording, approaching reliabilities down to 10^{-8} s. Pin-type probes with electrode separations of 0·1 inches or less have been used with potential differences of tens or hundreds of volts applied from capacitors or batteries, with resistances in circuit. In most of the cases referred to, it was desired to display successive signals from ionization gaps along a detonation tube on a common time base, using a single cathode-ray tube. Cathode followers were used in feeding signals to a cable and hence to the recording station. In order to compress the display of the time scale, a roster-type sweep, similar to that used for television representation, has been used. A typical record[54] is shown in Fig. 7.18.

In the work of Knight and Duff,[54] subsequent to applying the method to detonations and strong shocks (temperatures greater than 3000°K), amplification of the signals to detect waves of much lower conductivities was considered. An amplifier is described which can be used to fire a thyratron and thus to extend the range of applicability of the method to very weak detonations (for example, in a mixture of $15H_2:85 O_2$ which is close to the limit of detonability) and to shocks producing temperatures as low as 1000°K.

Thyratrons in channels of varying gain were also used in the work of Hecht, Laderman, Stern, and Oppenheim,[55-57] which was concerned primarily with determining the flame velocity during the development of detonation. After demonstrating that the principal ionization peak is caused by the flame, these workers carried out a statistical analysis of the fluctuations in

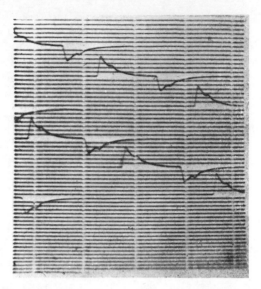

FIG. 7.18. A typical oscilloscope trace from an acetylene–oxygen detonation, using the pulse shaper. From Knight and Duff.[54]

flame velocity. The results were compatible with a model of the ionized wave fluctuating randomly about its mean position during the progress of the flame along the tube—a concept that is of some importance in the theory of pre-detonation conditions.

This provides a convenient transition point in the discussion to the study of 'steady' turbulent flames, i.e. flames whose mean position and shape is stationary with respect to the observer. Such flames occur on burners under conditions of turbulent flow and, as discussed in Chapter 3, can be thought of as rapidly fluctuating wrinkled fronts, the turbulent flame brush seen being the 'time exposure' recorded by the retina of the eye. They therefore also provide a suitable subject for study by ionization probes, whose rapid response and precise location permits statistical analysis of the fluctuations of the ionized zone. The 'electrical noise' induced by turbulence when electrodes are immersed in such a flame, can be correlated with the parameters of turbulence[65] Two more specific techniques have been described by Karlovitz et al.[66,67] The first is based on the higher level of ionization that exists in the flame (reaction) zone as compared with the hot product or cold reactant gases. Thus, a probe insulated except at its tip and approaching the premixed turbulent flame from its products-side can be made to signal contact with the flame front exclusively. In order to eliminate from the record all but such events, a signal-height discriminator was used, after amplification. The signals remaining after the discriminator were used to control an on–off tube, which was made conducting only when the flame-front was not in contact with the probe; the average d.c. current through the tube was thus proportional to the fraction of the time for which the flame front was not in a particular region. The number of contacts was also counted, using an electronic counter, connected to the on–off tube. The probe was made to traverse the flame brush.

The parameters deduced from these measurements are of only incidental interest here, but are of considerable significance to the theory of turbulent flames. In particular, the velocity of the fluctuating motion, the region of the most probable position of the flame and the r.m.s. deviation from this, were determined. In another part of the investigation, the incidence of holes in flames was observed, as an indication of the onset of flame instability. It may be of interest also for other applications to note that when the authors wished to eliminate small-amplitude fluctuations, they simply removed the high-frequency component by inserting a pulse-width discriminator circuit between the counter and the on–off tube. The performance of the system could be checked by subjecting the flame to sound waves of a known frequency.

A second method mentioned[66] was based entirely on the ionization in the hot products (assumed to be uniform in space and hence in time, as the flame fluctuates). Here the current from a probe at a low potential was used as an index of instantaneous gas velocity past the probe, much in the manner of the role of heat-flow in a hot-wire anemometer.

Before leaving the subject of turbulent flames, another type of electrical measurement will be considered which, strictly, does not employ ionization probes as here defined. This is the measurement of saturation current density, j_s, that was discussed in the context of measuring rates of ion generation in laminar flames and other ion sources in Chapter 5. There, a unidimensional front was visualized between, and parallel to, plane electrodes at such potentials that current no longer increased with increase in field intensity. The value of flow velocity is then immaterial and ion generation in the products is usually negligible in comparison with that in the reaction zone. The current is, however, proportional to the area of flame front contained between unit area of electrodes, and varies as the secant of the angle of inclination of a flat flame to the electrodes. The 'wrinkled flame' concept of the burning of mixtures in turbulent flow presupposes that no change in structure or chemistry of the instantaneous front takes place, as compared with its laminar equivalent. The ratio (turbulent j_s/laminar j_s) should therefore equal the ratio of the instantaneous flame areas, provided that the electrodes are locally parallel to the time-mean position of the flame front; if they are not, the above mentioned secant enters as a correction factor. On the simple theory of the wrinkled flame-front (which, however, cannot be absolutely correct[68]) this ratio is also equal to that of the turbulent to the laminar burning velocity. This is being studied experimentally.

Unfortunately, the method is limited by space charge considerations to flames that are relatively weak ion sources. The maximum j_s that can be drawn, theoretically, without breakdown is about 0.25 mA/cm^2 (Chapter 8) and this result applies at an electrode separation from the flame of 0.5 cm. This separation would be inadequate for all but the smallest scales of turbulence and the maximum j_s is inversely proportional to it. At the same time, saturation currents for stoichiometric hydrocarbon-air mixtures[69] are smaller by factors of only about 2.5 (C_3H_6), 3 (C_2H_4), and 10 (CH_4) than this theoretical maximum. Such measurements therefore have to be confined to specific fuels, or to mixtures well away from the stoichiometric.

Returning briefly to ionization probes, there is much qualitative and semi-quantitative work, which is not published in the scientific literature, on their use as flame detectors in warning and switching devices. They are also used as detectors in gas chromatography. In this application hydrogen, as the carrier gas in the analysis of hydrocarbons and other organic compounds, after traversing the column, enters a small burner provided with the ionization probe. As the various separated species are conveyed to the diffusion flame, they alter its ionization, which is quite low for pure hydrogen. If their volumetric flow rate is appreciable, as compared with that of the carrier hydrogen, they modify the shape and height of the flame also, and hence the position of the probe with respect to the flame zone. This change in location can be minimized by using a premixed Meker-type burner. These factors produce a change in signal that can be used to infer the arrival of a separated component of the mixture and can be calibrated, to some extent, in terms of mixture composition.

Movement of charge in oscillating fields

The use of attenuation of electromagnetic waves by a plasma has already been discussed as a method of measuring electron concentrations. Another practical aspect of this effect, relevant to combustion, is the interaction of flame gases, particularly those from rocket exhausts, with radio signals. It is often essential to maintain radio communication with rockets for the purposes of tracking and guidance. Under certain circumstances the highly-ionized exhaust plume emerging from the motor could be a serious handicap in this respect. In the case of transmission to the rocket the signal must traverse the plume, which is greatly expanded at low pressures. In consequence the signal may suffer so much attenuation, refraction, or reflection as to prevent its arrival in sufficient strength at the receiver. Transmission from the rocket can be disturbed in two ways. The antenna pattern itself may be distorted and, in addition, the emerging signal undergoes further interaction with the exhaust plume; the combined effect may be to cast a 'communication shadow' over the ground station. The complementary practical consideration arises when a highly-ionized rocket exhaust is deliberately used as a reflector. Thus 'holes' blown in the ionosphere by high altitude nuclear explosions could be filled in this way. A special solid propellant composition giving rise to very high electron concentration[70] has been discussed in Chapter 3.

The theory of reflection, refraction, and attenuation of electromagnetic waves was discussed in Chapter 5. There it was shown that for plasma frequencies greater than the angular frequency of the wave, the plasma acts as a good reflector and is termed 'overdense'. It is the overdense part of the plume that has the greatest effect upon the atenna pattern.[71] In practice, the effect can be tested on an 'antenna range' with a scale model of the rocket casing, using a metallic cone to simulate the overdense part of the plasma. Figure 7.19 shows schematically the effect on the antenna pattern. The

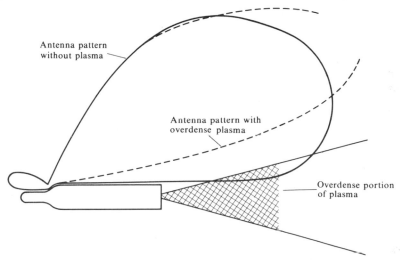

Fig. 7.19. Antenna pattern change due to overdense exhaust plume. From Balwanz.[71]

effect of the plume upon the propagation of the emitted waves can be analysed by ray-tracing techniques, if sufficient is known about the shape and electron concentration distribution of the exhaust. The plume is divided up by contours of constant electron density, i.e. constant refractive index, μ, and attenuation coefficient, χ. The path of a ray as it traverses the plume can be calculated from the equation for its local radius of curvature, R, in the refractive index field,[72]

$$\frac{\mu}{R} = |\text{grad } \mu| \sin \phi, \tag{7.52}$$

where the right-hand side is equal to the component of the refractive index gradient at right angles to the ray. The value of μ is determined from the

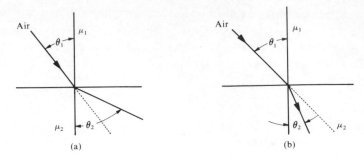

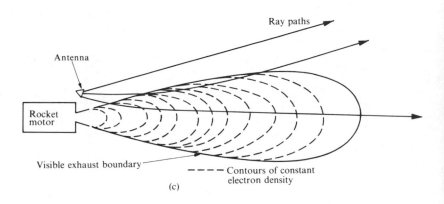

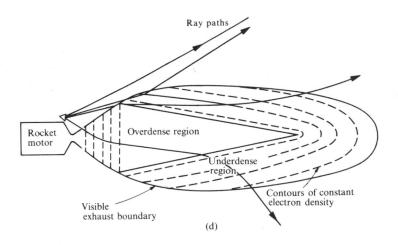

FIG. 7.20. Ray tracing through theoretical models of exhaust. (a) Underdense plasma; (b) over-dense plasma; (c) underdense rocket exhaust; (d) exhaust with both overdense and underdense regions. From Balwanz.[71]

theory of Chapter 5. It should be noted, however, that paths passing through or near the transition region between the over- and under-dense regions are difficult to establish. Once the paths are known, the attenuation can be calculated by integration along the ray. Thus the factor in e^{-xz} in eqn (5.67) is replaced by $\exp\{-\int_0^z x(z)\,dz\}$. One of the main difficulties lies in obtaining realistic theoretical models for the exhaust plume over the range of flight conditions encountered at various altitudes. Figure 7.20, taken from the work of Balwanz,[71] gives an example of ray tracing for different models.

The detection of flames, shocks, and detonations by d.c. probes has been discussed in the previous section. It is also possible to use a.c. fields for this purpose with the advantage that it is unnecessary to contact the plasma directly with a material electrode. For example, the measurement of the Doppler shift of microwaves reflected from a shock wave has been used to determine its velocity.[73] In Chapter 5 it has been shown that the resonant properties of electrical circuits are significantly affected by the proximity of sufficiently ionized gases. This effect can be used as a means of detection. For example, a resonating coil wound around a flame, shock, or detonation tube would detect the arrival of the pulse of plasma by the change in Q of the circuit. This should be of value in automatic warning and/or triggering devices for non-metallic ducts carrying inflammable gases.

MOVEMENT OF NEUTRAL GAS

Electric fields

This section is concerned with practical effects due to ionic winds, and the first step must be to attempt to distinguish these from other possible consequences of the movement of the ions themselves. In historical studies of the ionic wind[74,75] and in all subsequent work involving ionization not accompanied by chemical reaction, this problem never arises. In combustion research, however, the distinction has often not been very clearly drawn, particularly in the early work. One reason is that the wind is an inevitable consequence of the ion current, when a field is applied to a flame, and it must have been rather tempting to consider some observations in terms of specific chemical effects of certain ionic species rather than as due to an overall gas flow. Indeed the growth of flame carbon on ions, as discussed earlier, may well be a chemical effect and, in general, it is never possible to demonstrate conclusively that the ion flux and associated chemical reactions do not play *some* part in an observed effect. The most that can be done is to derive theoretically the maximum rate of ionic mass flow, demonstrating how small it is (unless the charge-carrier is an aggregate of many molecules), and compare practical observations with the corresponding calculated wind effects. While this leaves the possibility of specific chemical effects being induced by transposing, for example, very small amounts of ionized active species, it is then possible to conclude in most cases that the gas flow accounts for the observations, and it is not necessary to postulate 'chemical' effects of ions in order to account for the experimental results. This is the basis of our subdivision. In the case of the electrical control of flame carbon, for instance, the effects discussed were evidently caused directly by the movement of reacting charge carriers. The contents of this section can be accounted for in terms of wind effects alone.

The theory of ionic winds has been discussed in Chapter 4 and their maximum magnitudes are calculated in Chapter 8. The principal conclusions are that the force acting on unit volume of the neutral gas is the current density divided by the ionic mobility of the charge carrier. The maximum value is approximately $800 \, dyn/cm^3$, prior to breakdown. This force is about 800 times greater than the maximum body force (ρg) for natural convection in air at s.t.p. A flow pattern is set up under the influence of these forces, the gas flowing from the ion source to each electrode, in a symmetrical system. It entrains surrounding gas when free to do so and, at the electrode, attains a velocity of up to 550 cm/s, in the ideal case. When entrainment is confined to specific apertures, velocities up to 1100 cm/s can be produced at the flame. These quantities apply to single stage devices and can be increased by aggregating stages in series. They fall rapidly when secondary ionization sets in at any electrode.

In practice, maxima are somewhat reduced by the entrainment of hot flame gases, deviations from unidimensionality at the electrodes, and other factors discussed on p. 154. Since the forces in the two electrode spaces oppose each other, and since they are often unequal, a net flow at the ion source sometimes results. The inequality can arise from, or be deliberately introduced by making the flame-electrode separations asymmetric, or by utilizing a difference in mobility between the two charge-carriers; the combination of both is illustrated in Plate I(ii). However, only for a completely closed system does this result in an overall undirectional flow such that the driving pressure head is calculable simply by subtracting the pressure heads on the two sides. In open systems with impermeable electrodes, a pattern of interacting vortices (see Fig. 4.13) is set up and tends to produce a variety of flame distortions and deflections that have been observed in many investigations reviewed below. It will be shown that the pronounced unidirectional effects observed in much of the early work are caused by incidental differences in the mobilities in otherwise geometrically symmetrical systems.

The effects produced on flames by fields applied between electrodes at high potentials are quite spectacular and were observed early on. In 1814, Brande[76] recorded that 'when the flame of a candle is placed between two surfaces of opposite electrical states, the negative surface becomes most heated ...' and '... the smoke and flame of the candle were visibly attracted towards it'. Malinowski,[77] in 1924, found that he could extinguish flames by application of a suitable field. The flames propagated in rich mixtures of light petroleum and air, the field being applied between co-axial cylindrical electrodes. Bone, Fraser, and Wheeler[78] and Thornton[79] applied longitudinal fields to $CO + O_2$ mixtures. In the former study, the gases had been dried (over P_2O_5) for periods of the order of a year and variations in propagation rates along a tube were observed even though the flames must have been exceedingly poor ion sources. Flames in these dry mixtures would, in fact, propagate repeatedly, up to four times, at intervals of 24 h, leaving a final product mixture of CO_2 ; CO : O_2 in the ratio 9·3 : 60·4 : 29·3.

The early work of Malinowski was later extended by Malinowski and Lawow,[80] using the same apparatus but a wider range of hydrocarbon flames. They concluded that flames were arrested only when a carbon deposit was observed to form, while in other mixtures retardation of the flame propagation occurred, except in the case of hydrogen flames, which gave indefinite results. Haber[81] used plane electrodes and combined his work with spectroscopic observations. He concluded that the retarding effect of an electric field became appreciable only when the flame exhibited a well-defined Swan (C_2) spectrum.

It would be easy to construe this conclusion and those of Malinowski and Lawow as implying some chemical effect of the ions. However, we have seen already that flame-carbon or soot provides large charge carriers which produce characteristic increases and asymmetries in wind effects. An example was given in Fig. 7.3 showing the increase in field-induced heat transfer as the hydro-carbon content of a mixture increased and, in particular, as the flame became luminous. The work on the effect of fields on flame carbon[1,2] discussed earlier has shown, in addition, that field-induced changes in carbon yield are paralleled by changes in emission spectra of those species which appear to be involved, if not as intermediates then as by-products, in the formation of soot. This compares with the findings of Kinbara and Nakamura[82,83] that 'ions in a flame are produced at the carbon formation stage', though here again it is worth noting that the interpretation of probe measurements, as in the work of Nakamura,[84] involves the assumption of a mass ratio of the positive charge carrier to that of the electron. The former has been assumed to be 20–40 proton masses whereas, if it changes to that of a carbon particle at some height in the flame, a peak may be observed due to this change alone. The picture that emerges is that the 'interaction of chemistry and physics' in the above experiments is confined to the formation of the charge carriers, especially if they are particulate. Thus the field intensity and mixture composition can determine the number, size and charge of, for example, carbon particles and this, in turn, will influence wind effects.

Guenault and Wheeler[85] found that the growth of a flame initiated between two condenser plates was greatly influenced by an electric field. They suggested at the time (1931) that ionic body forces may be the cause and subsequently (1932) studied flame propagation in vertical tubes, using Malinowski's experimental system, in an attempt to resolve the matter. They concluded, from observations of changes in flame shape and velocity of propagation, that flame arrest or retardation by the electric field could be explained by an aerodynamic effect of the flow of flame ions. This was such as to cause flame movement towards the negative electrode. No hypothesis involving a 'chemical effect' was necessary.

It is worth observing with respect to the classical studies above, some of which have been reviewed by Lewis,[86] and indeed some of the subsequent researches, that the absence of an effect in the case of certain mixtures does not imply that no effect can be induced. As previously mentioned, the effects of positive and negative ions of similar mobility will tend almost to cancel in approximately symmetrical systems, in so far as any net flame displacement is concerned (see Chapter 4). Because of the small effect of ionic mass on mobility, for small ions, this is likely to occur whenever the positive ion is not a carbon or other particle and the path lengths in colder gas are long enough for electrons to become negative ions by attachment. Thus the appreciable effects observed in the classical studies were discovered almost accidentally when the system happened to be sufficiently asymmetrical either geometrically or, usually, because the positive charge carrier was a large carbon particle. Once the theory is understood, however, it becomes unnecessary to rely on such differential effects, which are confined to specific mixtures. So long as any charge carriers are present, maximum wind-effects can be generated deliberately, and examples will be given below of achieving this by contacting the flame with one electrode and/or by suitable ducting of the wind.

Pre-mixed burner flames in transverse electric fields were studied by Calcote,[87] with and without the addition of alkali metal salts. The shape of the flames and the trajectories of small particles, introduced centrally in a burner of rectangular cross-section, were recorded. The treatment of the results is somewhat marred by two assumptions—one practical and one theoretical. The former is that the particles follow the gas flow lines, which neglects the effect of the charge they are likely to acquire from the ions and electrons, in the presence of a field. The particles were of metallic aluminium —the theory of charge acquisition and subsequent mobility of particles under such conditions[88] is discussed in Chapter 2. Fig. 7.21[87] shows that, on application of a field, some particles in fact increase their velocity component in a direction opposite to that of the net induced gas movement. The second assumption is that there is some 'energy increase of the flame' due to the energy drawn from the field by the charge carriers in the flame zone alone. This regards the flame as something more tangible than a state of the gas and entirely neglects the main forces acting on the long ion columns between flame and electrode together with the gas flow they induce. The misconception is a very understandable one, in view of the larger ion concentration in the flame reaction zone. However, as has been shown (Chapter 4 and 8), the field intensity is least there and the co-existing charges of opposite polarity pull in opposite directions. Even when the influence of one charge carrier is negligible because it is an electron and has a high mobility, it is in fact the current density that determines the body force, and this is constant all the way to the electrode (in a unidimensional system) so that the contribution of the relatively thin flame zone is very small.

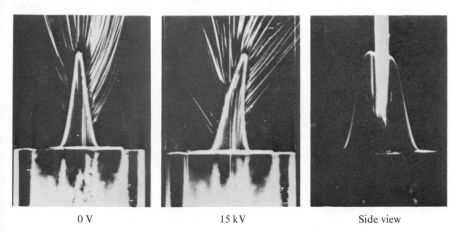

0 V   15 kV   Side view

FIG. 7.21. Particle tracks in 2·6 per cent butane–air flame. Transverse field. From Calcote.[87]

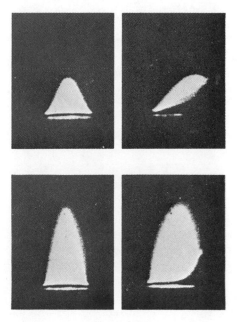

FIG. 7.22. Transverse electric field on inner cone of *n*-butane–air flames. From Calcote.[87]

Nonetheless, the paper made a useful and important contribution, in attempting to account for observed effects in terms of purely mechanical factors and in discounting the role of electron diffusion ahead of the flame which had been associated with the mechanism of flame propagation by some earlier workers. It also demonstrated that the flame base can be made to lift from the burner mouth under the action of a sufficiently high ionic wind velocity. This is shown in Fig. 7.22. We attribute it to the entrainment of cold air from under the burner mouth (Chapter 4), which is somewhat favoured over entrainment of gas from above by the burner stream momentum and by the lower viscosity of the cold gas.

In a subsequent publication, Calcote and Pease[89] studied the effects of longitudinal fields, i.e. fields parallel to the initial direction of flow lines, on inner bunsen cones. They investigated the influence of fields on stability to blow-off,† on dead space (between burner lip and flame base), and on flame pressure, in order to assess the practical potentialities of using fields as a means of stabilizing flames and also to help resolve the issue of chemical versus wind effects. Their apparatus is illustrated in Fig. 7.23 and, adopting their convention of naming fields according to the potential on the top ring, the principal findings were as follows.

Positive fields produced large and consistent increases in flame stability to blow-off. An example is shown in Fig. 7.24. This was accompanied by the corresponding effects of decrease in dead space, with considerable overhang for lean mixtures, and increase in flame back-pressure. Negative fields produced results that were considered to be much less definite and reproducible. Thus the dead space increased at first, until the flame began to 'jump on and off'. Against this, at even higher flow rates, application of a field was found to stabilize a lifted flame. The general pattern was a decrease in stability at low fields, followed by an increase at high ones.

The authors, while recognizing the effects of positive fields as due to ionic winds, tended to attribute the reversal at high negative fields to chemical effects of electrons, or at least suggested that electrons acquire enough energy to excite, or even ionize, the gas in a zone where increased reaction rate would enhance flame stability. They quote in support the increase in current under the higher field conditions as evidence of local breakdown (even though the power dissipation is still minute). While such an hypothesis can never be disproved—as discussed earlier—it seems unnecessary. The observations are comprehensively explicable on the basis of mechanical effects alone, using the theory of Chapter 4. We must recognize, in this case, that the events occur in a duct in which body forces from the ion source towards the two electrodes act in opposition to each other and the larger one will determine the direction of the net effect. If this is such as to cause ambient gas to flow on the outside of the burner mouth in the same direction as that within, it will result in decreased stability and back pressure and an

† *Footnote added in proof.* For more recent observations see ref. (122).

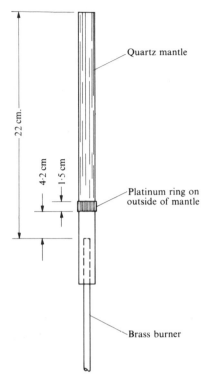

FIG. 7.23. Burner, mantle, and electrode arrangement. Quartz mantle 2·5 cm o.d., 0·15 cm wall thickness; brass burner 0·66 cm o.d., 0·12 cm wall thickness. Electric field is 'positive' when ring is positive with respect to burner and 'negative' when this polarity is reversed. From Calcote and Pease.[89]

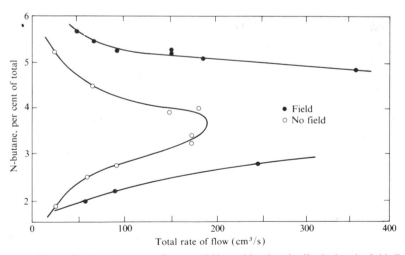

FIG. 7.24. Blow-off of n-butane–air flame, 15 kV, positive longitudinal electric field. From Calcote and Pease.[89]

increase in dead space; and vice versa. Now the relative magnitudes of the opposing forces are given by the quantity (electrode separation/mobility) for a constant current density—as previously shown—up to secondary ionization potentials. When breakdown occurs in either electrode space, the body force there falls rapidly.

In the case of positive fields, many of the negative charge-carriers remain electrons in the hot flame gases, the net effect being a downward ionic wind opposing blow-out, because of the current of positive ions of very much smaller mobility. For negative fields, the electron traverses the cold dead space without attachment only if its trajectory is short enough. This is not the case if the flame lifts sufficiently high, for then electrons can attach and a net downward force can result. It will be apparent that this also accounts for the 'jumping' flames for, as the flame base moves downward, the probability of electron attachment decreases, the wind reverses, and so an oscillation is set up. When the field is increased sufficiently for breakdown to occur, this would be expected to set in first at the top electrode for the negative-field case. Every factor favours this: the lower breakdown-field at higher temperatures, the lower mobility of the positive ion, and the larger ion source-electrode separation there (for detailed theory, see Chapter 8). As soon as secondary ionization conditions are approached, the body force towards the top electrode falls steeply and it can no longer compete against the wind due to the negative charge-carriers (even if this is small), which tends to stabilize the flame. All the incidental observations of this investigation are similarly explicable and it proves unnecessary, once more, to invoke chemical effects of the charge carriers. Nor is the above interpretation based on purely theoretical speculation. Comparison of the results of the experimental ionic wind study[90] discussed in Chapter 4 show unequivocally not only the disappearance of ionic wind at breakdown—Fig. 4.10— but also the behaviour of the negative charge carrier as an ion, after path lengths in cold gas of a few millimetres. It was indeed found in that work that the 'positive and negative winds' were quite equivalent, for equal, cold, electrode spaces, so long as the positive charge-carrier does not become a particle of carbon or of some other material. Until these measurements were obtained, the authors, in their first, essentially correct, quantitative analysis of these phenomena[18] tended to attribute any net gas flow in the absence of particles to the negative charge-carrier being an electron. This seems to apply only if the negative charge-flow travels in hot products all the way to the electrode.

Some results on flame distortion from this earlier work are of interest: Figures 7.25(b) and (c) show the effect of neutralizing each ion column in turn by a corona discharge from a point. Figure 7.25(a), which is of an ethylene diffusion flame in a strong field, shows not only the net wind in the direction of movement of the positively-changed carbon particles but also a small effect near the base of the flame in the opposite direction, due to the

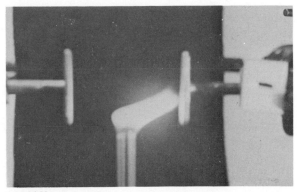

(a) 13 kV

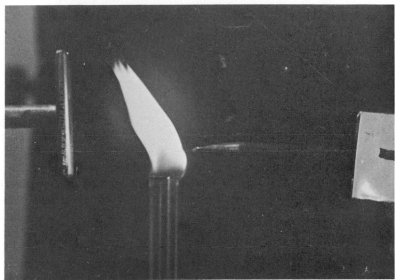

(b) 7 kV

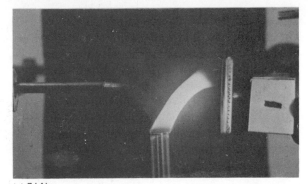

(c) 7 kV

FIG. 7.25. Ethylene diffusion flames subjected to fields. From Payne and Weinberg.[18]

particular flow pattern (illustrated in Fig. 4.13(c)) induced by the two opposing ion streams. Optical methods based on visualizing refractive index
changes can also be used to demonstrate this effect, by showing the distortion
of boundaries between hot and cold gas. This is not quite the same as displaying overall gas-flow, the appearance of the record being influenced by
the steepness[72] of the temperature gradients. Figure 7.26[91] shows a
schlieren record of an ethylene diffusion flame in a field between two co-axial
ring electrodes. Figure 7.27 is a series of phase contrast photographs[92]
(particularly high sensitivity to refractive index variations[72]) of a propane
flame stabilized on a Meker burner between electrodes 5 cm apart, at the
applied voltages shown. The essential symmetry, modified by differences in
mobility, is clearly visible.

The use of suitably directed wind effects to alter flow and mixing patterns
in a desired way is the obvious next step. One early manifestation which has
possible practical applications is the reduction in flame temperature,
discovered by Lewis and Kreutz,[93] when a transverse field was applied.
The effect may be understood in terms of the entrainment theory of Chapter 4.
The possibility of decreasing the volume required for combustion (i.e.
increasing combustion intensity) was demonstrated already along with flame
deflections; this is shown most easily in a system where burning is normally
very slow, such as in a laminar diffusion flame. Figure 7.28, from the work
previously referred to,[18] illustrates the effect on an ethylene diffusion flame
of 4kV applied between the ring-shaped electrodes. The top-electrode is
positive to minimize the upward directed body force by providing electron
paths in hot gas and thereby avoiding early attachment; the hour-glass-
shaped lines of force cause combustion to be completed in a small fraction
of the former height, see also Fig. 7.26. The centre of the 'button' flame
produced is blue in colour, resembling a pre-mixed rather than a diffusion
flame. Applications of this kind have advanced considerably with the
introduction of the principle of entrainment confined to specified regions, as
discussed later.

If, instead of a burner mouth, the source of fuel gas is a burning droplet,
analogous methods using fields can be applied to the combustion of droplet
suspensions, whether or not dispersed electrically. In order to increase
combustion intensity here, by increasing relative velocities between a cloud
of droplets and gas, it may be preferable to superimpose transverse oscillating
fields, as mentioned previously.

Increasing the relative velocity between droplets and surrounding oxidant
can, of course, have the opposite effect—to the point of extinguishing the
flame. Which of these will occur depends ultimately on whether the rate
of vaporization and heat release is increased more than the rate of heat
removal, and this is determined by parameters such as the relative velocity,
the oxygen content and temperature of the surrounding gas, and the

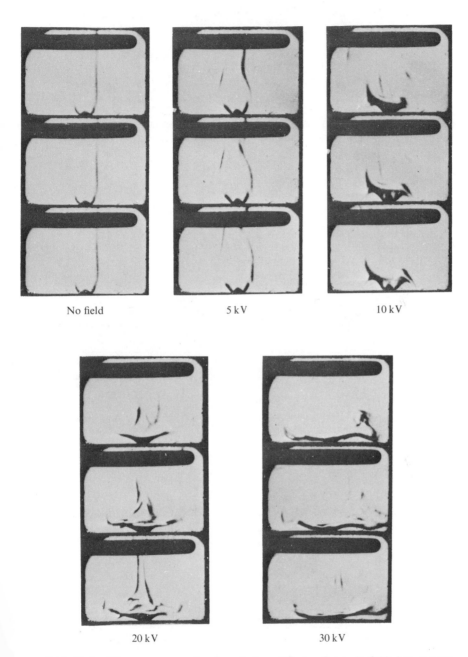

No field   5 kV   10 kV

20 kV   30 kV

FIG. 7.26. Ciné schlieren sequences showing ethylene diffusion flames in fields between two co-axial ring electrodes (lower ring just outside frame). From Watermeier.[91]

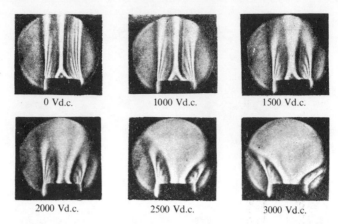

| 0 Vd.c. | 1000 Vd.c. | 1500 Vd.c. |

| 2000 Vd.c. | 2500 Vd.c. | 3000 Vd.c. |

FIG. 7.27. Phase contrast records of propane flames on Meker burners subjected to transverse fields. Electrodes, at potentials shown, separated by about 2 in; negative on right. From Saunders and Smith.[92]

(a)   (b)

FIG. 7.28. Effect on ethylene diffusion flames of field between co-axial ring electrodes. From Payne and Weinberg.[18]

volatility of the fuel. Thus d.c. fields have been used to extinguish flames surrounding fuel droplets.[84] Nakamura extinguished burning droplets of alcohol in a system fully described by Fig. 7.29. These droplets were, of course, larger by several orders of magnitude than those produced by electrostatic spraying.

Ionic winds have also been used to increase heat transfer from flames to solid surfaces.[18] This work must be distinguished from the use of fields to increase heat transfer in gases and liquids involving no ion sources.[94,95] The latter depends on dipole effects in non-uniform fields and applies to polarizable fluids only. In the former, the source of heat and ionization are the same (i.e. the flame) and the effects produced are very much greater. The system used in the work on flames is illustrated in Fig. 7.30. The heat exchanger B consisted of two co-axial brass tubes joined together at top and

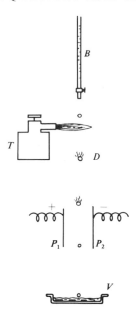

Fig. 7.29. Extinction of alcohol droplet flame by means of electric field. B, burette; T, torch; D, alcohol droplet; P_1, P_2, plane parallel electrodes; V, catchpot. From Nakamura.[84]

bottom by brass end-pieces, leaving an annular space of 0·3 cm between them. Inlet and outlet pipes for water had their axes tangential to the circular cross-section in order that the swirling motion imparted to the water might assist mixing, and thermocouples were sealed through the outer tube in a line at 4·5-cm intervals. This 'stirred' water jacket, in which mixing was further helped by packing with coarse lead shot in certain regions, was used as a large number of constant flow calorimeters in series, to determine heat transfer from the flame to the jacket as a function of height, in the presence and absence of an applied field. The jacket was earthed through a micro-ammeter, the positive (4–5kV) electrode being a spring-loaded nichrome wire along its axis, passing through the flame into the burner.

Flames of ethylene, coal gas, and hydrogen were burnt on a cylindrical glass burner, i.d. 6 mm, at flow rates up to 60 ml/s. The composition range in the case of premixed ethylene flames was 9·4 to 15·1 volume per cent fuel, when excessive soot deposition in the presence of a field made further increases in fuel concentration impracticable. Experiments were then extended to premixed and diffusion flames of coal gas (47 to 100 per cent fuel) as well as to pure diffusion flames of hydrogen, in order to ascertain to what extent the effect was dependent on carbon content.

The results of this study are summarized in Fig. 7.31. The horizontal axis represents height along the jacket. The vertical scales for the full lines are

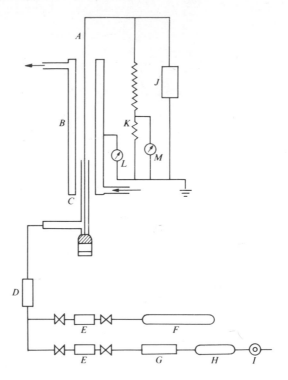

FIG. 7.30. *A*, Central wire electrode; *B*, calorimeter; *C*, burner; *D*, mixing tower; *E*, orifice flow meters; *F*, fuel cylinder; *G*, drying tower; *H*, buffer volume; *I*, compressor; *J*, h.t. unit; *K*, resistance chain; *L*, microammeter; *M*, electrostatic voltmeter. From Payne and Weinberg.[18]

the integrated heat transfer rates (cal/s) up to the height of each thermocouple. The local heat transfer rate per unit area of surface (cal cm^{-2} s^{-1}) is given by the slope, at each height, of the graph of the above quantity plotted against height, on dividing by the circumference of the inner cross-section of the calorimeter. The results are, in effect, curves of total heat transferred per second to a short heat exchanger, plotted against heat exchanger length, in the presence and absence of field (full lines). In each case the upper full curve represents the case with a field applied. The broken lines represent local percentage improvements, on the same distance scale. Note, however, that the percentage scales are not the same for all the graphs.

The following conclusions were drawn. The application of a field always results in an appreciable increase in heat transfer rates, for the geometry used. The increases themselves are always greater than the power dissipated in producing them by at least two orders of magnitude because the currents involved are so small. As regards the mechanism, any contribution due to (a) the increase in thermal conductivity occasioned by the increased transfer

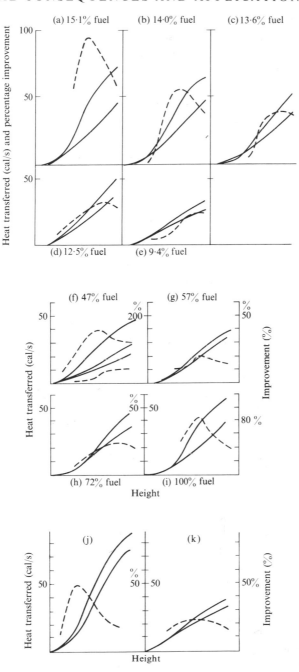

FIG. 7.31. Heat transferred (cal/s) versus height in calorimeter: (a)–(e) ethylene-air; (f)–(i) coal gas—air; (j)–(k) hydrogen diffusion flame. From Payne and Weinberg.[18]

capacity due to ion movement in the field direction and (b) the heat released on the electrode surface by recombination of ions after discharge, could be ruled out as negligible by order-of-magnitude calculations. With one exception—the higher curve in the leanest coal gas/air flame, which will be discussed separately—no obvious change in flame shape was observable in the symmetrical system. This left, as the principal effect, changes in the gas flow pattern induced by ionic winds. When electrons had to pass through cold gas, their path in it was always less than 3 mm, while that of the positive ions was, at the very least, three times greater. The gas, following the path of the heavier positive ions, moves radially outward, thus bringing fresh hot reaction products to the calorimeter walls. Gas so removed from the centre must be replaced by inward flow in other regions, thereby setting up a vortex system ideally suited to its task, since the distribution of the field-induced outward force follows the ion distribution. This is a maximum in the hot flame region, while the return flow occurs after discharge of the ions, i.e. after the hot gas has been cooled by contact with the walls.

In accordance with this picture, the effect increases with the proportion of large positive charge-carriers, being greatest for the richest flames and increasing with the carbon content of the fuel. The abrupt increase when ethylene flames become luminous has already been referred to (see Fig. 7.3). At the same time, the improvement does not fall much below 25 per cent even in the case of pure hydrogen diffusion flames.

In the case of the highest curve for the 47 per cent coal-gas flame, a change in flame shape appears to have been induced. This is fully discussed in the publication[18] but the point to note here is that in all the other flames, including the intermediate curve for the same mixture, the increase in heat transfer was achieved without feedback of the ionic wind to the reactant approach flow and that much greater effects become possible when fields are used to modify flame shape as well.

The above describes a somewhat preliminary investigation and it seems likely that the method can be refined further. It is interesting to note that parallel developments[96–98] have taken place in nuclear engineering where heat transfer from the ionized gases emerging from a nuclear reactor has been increased by applied fields. Since the major resistance to heat transfer is offered by a narrow boundary layer, attack by a general gas flow at right angles to the boundary layer is unlikely to produce the optimum effects. Thus if a small tube carrying a flow of water and acting as a calorimeter completely immersed in flame gases is made the electrode, no appreciable increase in heat transfer[18] to it results. Nor is there likely to be much change in the general flow-pattern around it, because gas cannot flow symmetrically to a point unless the point acts as a sink. By making only part of an otherwise nonconducting surface the electrode, however, the facility of controlling the wind direction may be usable to much better effect by causing boundary layer separation and reverse flow locally to increase heat transfer.

There is a group of combustion systems in which the rate of heat supplied to a stationary surface determines the rate of fuel supply to the flame. Examples are flames on wicks (providing the rate of transfer, by capillarity, of liquid to the tip of the wick is not rate controlling) the burning of liquid surfaces and, to some extent, that of solid compositions, including rocket propellants. In these one may expect that making the burning surface one electrode, or in other ways using wind effects to vary heat transfer to it, would induce profound effects on the rate of the combustion process. Some work on the effect of longitudinal fields upon diffusion flames on wicks was carried out by Kinbara and Nakamura and reported on[99] in a paper that is, unfortunately, not very generally available. A short summary appears in ref. (84). It was found possible to vary the rate of fuel consumption by this means and even cause the flame to extinguish. The decrease in heat transfer to the wick, and consequently in the rate of fuel supply, was attributed principally to the decrease in radiation, but convective effects are likely to have played a major part also.

A possible application of more practical interest is the control of solid propellant burning by means of applied fields. Once a mass of propellant has been ignited, its rate of burning, at constant temperature and pressure, cannot readily be changed from the course determined by its composition and geometry. The facility to do so is desirable both because it might permit the thrust of solid propellant rockets to be varied in flight and because it could be used in attempts to arrest destructive pressure oscillations encountered during combustion instability, by allowing the rate of burning to be oscillated so as to damp them.

In principle, there are two velocities that might be modified by the application of electric fields; the normal burning rate and the rate of flame-spread across surfaces. Preliminary experiments on small samples at atmospheric pressure[100] suggested that although both are affected, the range of variation is so much greater in the latter case that it is worth-while to think of methods of using this effect. As regards the mass burning rate, an increase of about 10 per cent was recorded, using aluminized propellants as the negative electrode, but this proved quite trivial by comparison with the effects that could be induced by controlling the burning area by means of ionic winds. Varying the rate of flame-spread over propellant surfaces has certain direct applications, the most obvious being the ignition of a propellant charge in a rapid and uniform manner. The setting-up of destructive oscillation is often traceable to events during the ignition process. However, in order to go further and use changes in this surface velocity to vary the overall rate of propellant consumption, it would be necessary to employ the principle of opening up some internal area at a variable rate. This would entail changing the conventional forms in which propellants are cast, but it was suggested that this could be worth-while because it was found very easy, for example, to increase the speed of flame-spread by a factor of 200 times, at atmospheric pressure. In order to make visible the boundaries between hot and cold gas and to observe their movement with time, cine-photography through a schlieren system was used; a typical set of records in the presence (a) and absence (b) of a field designed to accelerate flame spread is shown in Fig. 7.32. The igniting streamer of hot gas adjacent to the propellant surface is clearly visible in (a). Rates of flame-spread were deduced from records such as this, after a steady state had set in, by measuring the position of the schlieren record of the flame front as a function of time with respect to the initial ignition plane, plotting these points and measuring the slope of the best straight line. Contacting the flame with one electrode, for example a thin wire, and making the propellant surface the other, always produced an increase in the rate of flame-spread irrespective of geometry or polarity. This is because the flow of either charge carrier will entrain hot flame gases and force them into contact with the unignited propellant surface. The use of positive charge-carriers and suitable geometry will, however, further optimize this effect.

Inducing the opposite effect, i.e. decreasing flame-spread, would therefore have to make use of the circumstance that the system is always enclosed in practice, so that a net wind can be used to produce an appreciable velocity at the flame by outweighing the effect at one electrode space with that at the other. Thus, with an arrangement simulating channels in the propellant, a decrease of flame-spread velocity of approximately ten times could be induced. The reason why this is less than the effect in the opposite sense was thought to be that there is a certain component of the propagating mechanism that is due to non-convective effects, such as radiation. Thus the use of a field as an 'accelerator' is rather more effective than as a 'brake', and it was suggested that this could be compensated by casting the propellant so that its normal burning area decreases with consumption.

It was chiefly because of these large factors of about 2000 times overall and because of the promise, based on theoretical calculations, of even larger effects at the higher pressures relevant to rocket combustion chambers, that the authors speculated on what re-thinking of current concepts would be necessary to apply this principle to control of thrust. Thus, open cavities in casting are now avoided because they ignite 'spontaneously', and to ensure a continuous supply of fresh surface it might be necessary to cast the propellant so as to burn through to them as consumption proceeded. Again, if a source of colder gas were required, it might be necessary to allow an annulus between the propellant and the outer casing. It may well prove, in fact, that the most useful practical application of the principle will be to ignition, as mentioned previously, in which these problems do not arise.

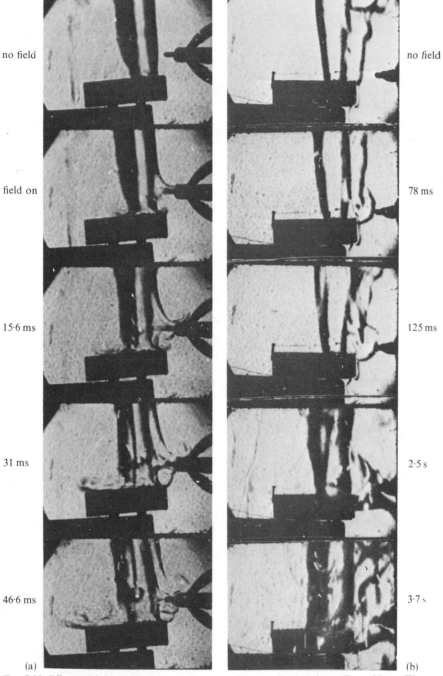

no field

field on

15·6 ms

31 ms

46·6 ms

(a)

no field

78 ms

125 ms

2·5 s

3·7 s

(b)

FIG. 7.32. Effects of fields on flame spread across solid propellant surfaces. From Mayo, Water-meier, and Weinberg.[100] (a) hot gas covers surface by 3×10^{-2} s; (b) half remains unignited after 3s.

With the exception of that part of this mainly empirical work concerned with retardation, no previous research appears to have concerned itself with gas entrainment confined to specific points of entry to the ion stream. Yet the use of free entrainment is quite unnecessarily unsophisticated, as mentioned in Chapter 4, the maximum velocity occurring at the electrode where it is least useful for combustion applications. No use has been made of the large entrainment velocity that can be realized at the flame when entrainment to a large ion stream is confined by a small aperture in the vicinity of the ion source. This omission is analogous to using buoyancy forces for improving combustion processes without employing the principle of a chimney or flue.

Following the development of the entrainment theory[101] discussed in Chapter 4, the aeration of diffusion flames and increasing combustion intensity by recirculating hot products were chosen as typical of the controllable effects that should be achievable by entrainment at the flame. The aeration of diffusion flames was considered an interesting example from a practical point of view also. Although diffusion flames have small volumetric rates of heat release and tend to soot, they are often used in practice for tasks that could be better performed by premixed flames, because of the simplicity of burner (a simple fuel jet) and their intrinsic safety; particularly their immunity from flash-back. If control of air entrainment could be achieved by applied fields alone, the danger of flash-back would still be avoided since no finite mixing volume would be involved and the need for a compressor and associated metering and mixing equipment would not arise.

The experimental design which was tested is illustrated in Fig. 7.33. In order to make mixing simultaneous with entrainment, the fuel flow was subdivided into several small streams. The momentum and separation of the jets, however, was quite insufficient to entrain any appreciable amount of air in the absence of a field. In that case the jets of fuel simply merged, producing a long diffusion flame as shown in Plate I(iii)a. The burner mouth was constructed of sawn-off hypodermic needles arranged in a symmetrical pattern, and a variable aperture placed around this assembly acted as the base for a quartz cylinder that supported a gauze electrode. On making this the cathode with respect to the burner, the body force integrated along the large column of positive ions opposed only by the very short column of negative charge-carriers (presumably largely electrons) could be used to entrain air into the flame. Plate I(iii)b demonstrates the effect of applying a potential difference of 30 kV between the electrodes for a fuel flow of 100 cm^3/s. It will be seen that in this case the flame height was reduced from approximately 11 to about 2 cm and all traces of sooting were removed. A somewhat simplified theoretical model set up for this system accounted for its behaviour fairly well.[101]

Various practical extrapolations of such a device were also investigated. Figure 7.34 shows a wick burner operating on the same principle, to

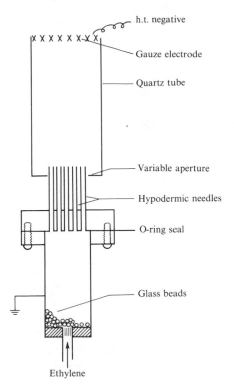

FIG. 7.33. Burner for the electrical aeration of a diffusion flame. From Lawton, Mayo, and Weinberg.[101]

demonstrate that a premixed flame could be obtained without either reactant being available in the form of a compressed gas. Figure 7.35 illustrates a modified gas burner, designed to preheat the incoming air by forcing it to pass over the radial copper fins of a heat exchanger. The purpose of this design was to increase combustion intensity and maximum flame temperature by recirculating some of the enthalpy. Again these aims could be met without incurring any of the penalties of the usual air supply, preheating- and mixing-systems.

As regards increasing combustion intensity, it was shown in Chapter 3 that the simplest method, without adding to the enthalpy from external sources, is by the recirculation of a proportion of hot products. The reason why this increases flame stability, within very wide limits, is that the reaction rate is greatly increased by the temperature rise (the dependence being of the form $e^{-E/RT}$)—the decrease brought about by the corresponding dilution of reactants with products being much smaller.

The use for this purpose of the torroidal vortices in the wake of solid bodies, as discussed in Chapter 3, has several disadvantages, notably the

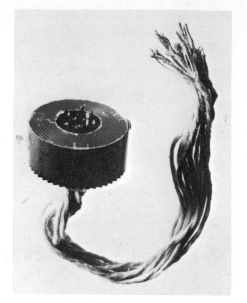

FIG. 7.34. Wick burner.

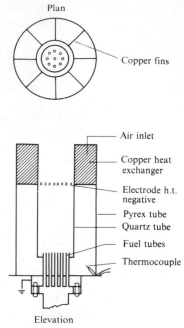

Plan

Copper fins

Air inlet

Copper heat exchanger

Electrode h.t. negative

Pyrex tube

Quartz tube

Fuel tubes

Thermocouple

Elevation

FIG. 7.35. Aerated diffusion flame burner with preheat. From Lawton, Mayo, and Weinberg.[101]

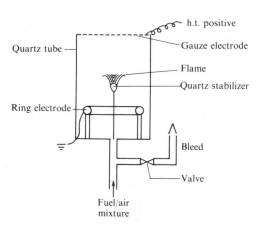

h.t. positive

Gauze electrode

Quartz tube

Flame

Quartz stabilizer

Ring electrode

Bleed

Valve

Fuel/air mixture

FIG. 7.36. High intensity burner. From Lawton, Mayo, and Weinberg.[101]

unprofitable part of the pressure loss which produces random turbulence, and the considerable heat transfer from flame gases to the obstacle, which constitutes a considerable enthalpy loss in the gas and causes deterioration of the stabilizer. The use of ionic winds for this purpose would be most attractive. Their flow velocity, though only of the order 10^2–10^3 cm/s for one stage, can be accurately directed, and electrodes constructed of, for example, high melting-point wire contacting the flame gases at suitable points, could be designed to cause little obstruction and be subject to negligible wear. Some effects on flame stability of applied fields in the wake of a bluff body have also been mentioned by Parker and Heinsohn.[102] Figure 7.36 shows a small experimental system that was designed in such a way as to direct the hot gasflow induced by the electric field downward into the reactants.[101] The flame was initially stabilized downstream of a small quartz sphere inside a quartz tube. The co-axial metal ring below the sphere acted as the cathode, while the positive electrode was a wire placed on top of the quartz tube. It will be observed that this arrangement contrived a flow field very nearly the opposite of that achieved by the apparatus shown in Fig. 7.33, the main drag in the gas column being due to the positive ions now travelling downwards towards the ring, the contribution of electrons being kept very small by ensuring that, at the high temperature, they do not attach to neutral molecules. Very considerable increases in flame holding were achieved in this manner; the phenomenon is shown in Plate I(iv). The idealized theory developed to account for these observations suggested, however, that for rates of heat-release peaking close to the final temperature, a system based on edge-stabilization of flames by recirculated hot products may well provide a more profitable use of the entrainment velocity, when this is limited to about ten times the burning velocity.

To conclude this section, it should be noted that all the experiments and devices mentioned make use of only a single stage. It is, however, quite feasible to sum the gas flow effects, and hence the theoretical maxima which apply to each stage, by aggregating stages in series—much as in EGD generation of electricity.

Crossed current and magnetic field vectors

Before we leave the subject of exercising forces on neutral gas by electrical methods, we should also consider what can be done by means of magnetic fields. The basic theory that underlies this principle has been discussed in Chapter 4, while that of the maximum effects obtainable is covered in Chapter 8. It follows from these theoretical considerations that, unless the orthogonal magnetic field is applied to a current considerably greater than that which can be obtained by applying electric fields to flame plasma without secondary ionization, the forces are negligible by comparison with those resulting from sub-breakdown electric fields alone. Except for the case of an ultra-high-velocity, highly-ionized plasma flow bathing the electrodes (exemplified by the idealized MHD duct), this implies the need for some form of discharge current at right angles to the magnetic field, and makes the subject rather marginal to the present monograph since the process is no longer dependent on flame ions. However, as the force that can be exercised on the gas in this manner is not confined to any absolute maximum (so long as the current is not limited), and since it can be used for combustion applications in much the same way as that discussed in the preceding pages, it is appropriate to summarize briefly some of its established and potential applications.

In the section on discharge-augmented flames (p. 88) we discussed a method of distributing the heat from the arc channel by causing it to rotate in a magnetic field. Quite a modest field parallel to the 'burner' axis and interacting with the radial component of the current caused[103] rates of rotation of the order of 10^5 rpm and served to distribute the electrical contribution to the final gas enthalpy throughout the burning stream. A somewhat inadequate record of this phenomenon is shown in Plate I(v). Thinking of the discharge as a state of the gas, i.e. regarding the arc channel as being perfectly 'porous' and offering no resistance to, or drag on, the gas is, however, somewhat naïve. It has been shown, on the contrary,[104] that most of the flow tends to bypass the arc channel, if only because of the very much reduced gas density within it. Considering also the great increase in viscosity of the hot gas, it becomes apparent that thinking of the discharge as a solid rotating spoke would be no more far-fetched than regarding it as a perfectly permeable phenomenon travelling through the gas. The truth lies somewhere in between these two extreme idealizations, the precise position depending on the current density and the relative speed of the arc with respect to the gas. Two points follow. First, the gas will be dragged along by the arc channel; for instance it will be caused to rotate in the case of the magnetically spun plasma jet. Second, the arc channel can generate turbulence on passing through the gas at high relative speed. The first of these consequences, unlike the second, is independent of the 'permeability' of the discharge channel: so long as the channel does not accelerate, the momentum is transferred to the neutral gas, much as was the case with ions drifting in an electric field.

If the plasma jet, or plasma jet-augmented flame, is surmounted by a co-axial heat transfer jacket, similar to that used in the apparatus illustrated in Fig. 7.30, these fluid-mechanical effects induced by the arc's rotation result in increased heat transfer to the jacket walls just as would be expected, for example, from stirring by a rotating spoke. In experiments carried out in these laboratories, the distributions of heat transfer have been plotted as functions of the current in the field coil, at constant power input to the burner. Some results due to J. Cox are shown in Fig. 7.37; it will be seen that large increases in rates of heat transfer result. The method is somewhat similar to the use for this purpose of electric field-induced ionic winds; it differs in that it is much more expensive in terms of power used but much less limited as regards absolute maxima. Flame deflections have been induced[105] by crossed current and magnetic field vectors at pressures of 20–100 mmHg, where appreciable currents can be induced to flow at much reduced breakdown fields. One objective of this study was the possibility of vectoring thrust from rocket nozzles.

It will be obvious that this method of generating forces on a gas is the converse of 'MHD generation of electricity'. In the latter, current is collected after a gas has been forced to do work on the crossed $j \times B$ vector; here feeding a current into the system allows work to be done on the gas. One would expect so general a principle to have found other aerodynamic applications. An example that does not take us too far into plasma dynamics is the electromagnetically-driven shock tube; see, for example Kolb.[106] The principle is

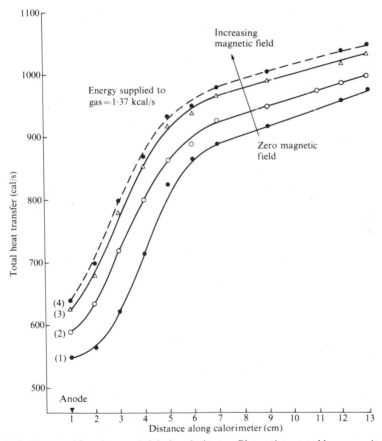

FIG. 7.37. Heat transferred versus height in calorimeter. Plasma jet rotated by magnetic field of increasing magnitude. (After J. B. Cox)

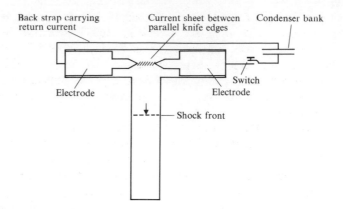

FIG. 7.38. T-tube for electromagnetic generation of shock waves. From Gaydon and Hurle.[107]

illustrated in Fig. 7.38.[107] The magnetic field at right angles to the sheet-discharge is provided by the 'back strap'—a stout lead carrying the return current from the discharge. By discharging a large condenser through the system, very strong shock waves (of at least Mach No. 100) can be produced.

The movement of arcs and the shape they assume under the influence of transverse magnetic fields—self-generating or separately applied—has been the subject of several aerodynamic studies; for example refs.[108]–[113]. Their use for heating the input air to wind tunnels has been considered (see, for example, Bunt, Olsen, and co-workers.[114–116] An arc between parallel rectilinear rails will propagate along them (because of the combined magnetic field due to the current flow along the conductors) and some of the configurations of looping with such electrodes and making them re-entrant[115] may prove relevant to augmenting flames.

It would appear to be a simple matter to design, on paper, various items of continually operating flow systems on this principle. Thus a radial arc in a flat cylindrical box, with a central inlet and a tangential outlet at its periphery, when placed in an axial magnetic field should act as a centrifugal compressor, the only moving part being a 'spoke' of discharge. The thermodynamic efficiency of such devices is invariably rather low because much of the energy is necessarily converted into heat. There may possibly be applications in which heat dissipated in this way need not be completely lost. Thus if some such electromagnetic compressor were to feed a jet engine, the additional enthalpy might provide enough ionization for driving a small MHD duct, to provide the current for the 'compressor'. Of course, this sort of qualitative speculation may well lose its attractiveness once economic considerations are taken into account. From the point of view of electrical control of combustion processes, however, it is interesting to show that, in principle, even the blades of compressors and turbine generators are replaceable by electromagnetic fields and currents and, although these replacements may have a high running cost, they have a low capital cost, zero inertia, require no lubrication, tolerate ultra-high temperatures and do not wear out. Whether or not they will become of practical importance will depend on how far the trend to ever higher working temperatures will continue.

REFERENCES

1. PLACE, E. R. and WEINBERG, F. J. *Proc. R. Soc.* A**289**, 192 (1965).
2. PLACE, E. R. and WEINBERG, F. J. *11th Int. Symp. Combust.*, p. 245. The Combustion Institute (1967).
3. HEINSOHN, R. J. and LAY, J. E. *A.S.M.E. Reprint*, 64-*ENER*-2 (1965).
4. REZY, B. J. and HEINSOHN, R. J. *Trans. Am. Soc. Mech. Engrs* **88**, 157 (1966).
5. LODGE, SIR OLIVER. *Electrical precipitation* (Institute of Physics Lecture). Oxford University Press (1925).
6. ROSEFF, E. and WOOD, A. J. *Introduction to electrostatic precipitation—theory and practice*. Constable, London (1956).
7. WHITE, H. J. *Industrial electrostatic precipitation*. Pergamon Press, London (1963).
8. ZELENY, J. *Proc. Camb. phil. Soc. math. phys. Sci.* **18**, 71 (1915).
9. VONNEGUT, B. and NEUBAUER, R. L. *J. Colloid Sci.* **7**, 616 (1952).
10. DROZIN, V. G. *J. Colloid Sci.* **10**, 158 (1955).
11. HOGAN, J. J. and HENDRICKS, C. D. *A.I.A.A. Jl* **3**, 298 (1965).
12. LORD RAYLEIGH. *Proc. Lond. math. Soc.* **10**, 4 (1878).
13. TAYLOR, SIR G. *Proc. R. Soc.* A**280**, 1383 (1964).
14. NAYYAR, N. K. and MURTY, E. S. *Proc. phys. Soc.* **75**, 369 (1960).
15. PESKIN, R. L. and LAWLER, J. P. Proceedings of the A.P.I. research conference on distillate fuel combustion. *A.P.I. Publication No.* 1702 (1963).
16. GRAF, P. E. Proceedings of the A.P.I. research conference on distillate fuel combustion. *A.P.I. Publication No.* 1701 (1962).
17. HENDRICKS, C. D. and SCHNEIDER, J. M. *Am J. Phys.* **31**, 450 (1963).
18. PAYNE, K. G. and WEINBERG, F. J. *Proc. R. Soc.* A**250**, 316 (1959).
19. POTTER, A. E. and BUTLER, J. N. *J. Am. Rocket Soc.* **29**, 54 (1959).
20. PANDYA, T. P. and WEINBERG, F. J. *Proc. R. Soc.* A**279**, 544 (1964).
21. PLACE, E. R. and WEINBERG, F. J. *Aeronautical Research Council* 25560, *CFCK*, 648 (1964).
22. HOMANN, K. H. and WAGNER, H. G. *Ber. Bunsenges. Phys. Chem.* **69**, 20 (1965).
23. SUGDEN, T. M. and GREEN, J. A. *9th Int. Symp. Combust.*, p. 607. Academic Press (1963).
24. KISTIAKOWSKY, G. B. and MICHAEL, J. V. *J. chem. Phys.* **40**, 1447 (1964).
25. KNEWSTUBB, P. F. and SUGDEN, T. M. *Nature, Lond.* **181**, 474 (1958).
26. MILLER, W. J. *11th Int. Symp. Combust.*, p. 252. The Combustion Institute (1967).
27. *Magnetohydrodynamic generation of electrical power* (edited by R. A. Coombe), Chapman and Hall (1964).
28. BASU, S. and FAY, J. A. *7th Int. Symp. Combust.*, p. 277. Butterworths, London (1959).
29. HARRIS, L. P. and MOORE, G. E. *3rd Symposium on the engineering aspects of magnetohydrodynamics*. Rochester, New York (1962).

30. DONALDSON, C. DUP. *2nd Symposium on the engineering aspects of magnetohydrodynamics* (edited by C. Mannal and N. W. Mather), p. 228. Columbia University Press, New York and London (1962).

31. THRING, M. W. *Properties and applications of low temperature plasma* (Plenary lectures), p. 329. Butterworths, London (1967).

32. BERNSTEIN, I. B. *et al. 2nd Symposium on the engineering aspects of magnetohydrodynamics* (edited by C. Mannal and N. W. Mather), p. 255. Columbia University Press, New York and London (1962).

33. JACKSON, W. D. and PEARSON, E. S. Symposium on magnetoplasmadynamic electrical power generation, Newcastle-upon-Tyne, 1962.

34. FROST, S. L. *J. appl. Phys.* **32**, 2029 (1961).

35. KERREBROCK, J. L. *2nd Symposium on the engineering aspects of magnetohydrodynamics* (edited by C. Mannal and N. W. Mather), p. 327. Columbia University Press, New York and London (1962).

36. MCNAB, I. R. and LINDLEY, B. C. *Advances in magnetohydrodynamics*, p. 27. Pergamon Press, London and New York (1963).

37. BROGAN, T. R. *et al. 3rd Symposium on the engineering aspects of magnetohydrodynamics*. Rochester, New York (1962).

38. BROGAN, T. R. *Conference on gas discharges and the electricity supply industry*. Butterworths, London and New York (1962).

39. WAY, S. and HUNSTAD, R. L. *8th Int. Symp. Combust.*, p. 241. Williams and Wilkins, Baltimore (1962).

40. WAY, S. *2nd Symposium on the engineering aspects of magnetohydrodynamics* (edited by C. Mannal and N. W. Mather), p. 166. Columbia University Press, New York and London (1962).

41. LAWTON, J. *Br. J. appl. Phys.* **15**, 935 (1964).

42. LAWTON, J. *Br. J. appl. Phys.* **16**, 753 (1965).

43. MARKS, A., BARRETO, E., and CHU, C. K. *A.I.A.A. Jl* **2** (1), 45 (1964).

44. BRANDMAIER, H. E. and KAHN, B. *I.E.E.E. Int. Conv. Rec.* Part 7, p. 28 (1966).

45. GOURDINE, M. C., BARRETO, E., and KAHN, M.P. *5th Symposium on the engineering aspects of magnetohydrodynamics*, p. 161. M.I.T. (1964).

46. DAMAN, E. K. and GOURDINE, M. C. International symposium magnetohydrodynamic electrical power generation, Salzburg, 1966.

47. GOURDINE, M. C. and MALCOLM, D. H. *19th Annual power sources conference*, p. 163. Atlantic City, New Jersey (1965).

48. DECAIRE, J. A. and LAWSON, M. O., 8th Symposium on the engineering aspects of magnetohydrodynamics, Stanford, March 1967.

49. KLEIN, S. *C.r. hebd. Séanc. Acad. Sci., Paris* **251**, 657, 2492 (1960).

50. KLEIN, S. *Proc. 5th Int. Conf. Ioniz. Phenom. Gases*, Vol. 1, p.806. North Holland, Amsterdam (1961).

51. VON ENGEL, A. and COZENS, J. R. *Proc. phys. Soc.* **82**, 85 (1963).

52. TRAVERS, B. E. L. and WILLIAMS, H. *9th Int. Symp. Combust.*, p. 657. Academic Press, London and New York (1963).

53. LAWTON, J. *AGARD conference proceeding No. 8*, p. 135 (1965).

54. KNIGHT, H. T. and DUFF, R. E. *Rev. scient. Instrum.* **26**, 257 (1955).

55. HECHT, G. J., LADERMAN, A. J., STERN, R. A., and OPPENHEIM, A. K. *Rev. scient. Instrum.* **13**, 1107 (1960).

56. HECHT, G. J., LADERMAN, A. J., STERN, R. A., and OPPENHEIM, A. K. *8th Int. Symp. Combust.*, p. 199. Williams and Wilkins, Baltimore (1962).
57. STERN, R. A., LADERMAN, A. J., and OPPENHEIM, A. K. *Physics Fluids* **3**, 113 (1960).
58. BOLINGER, L. E. and KISSEL, E. E. *J. Instrum. Soc. Am.* **4**, 170 (1957).
59. KISTIAKOWSKY, G. B. and ZINMANN, W. G. *J. chem. Phys.* **23**, 1889 (1955).
60. EVANS, G. W., GIVEN, F. G., and RICHESON, W. E. *J. appl. Phys.* **26**, 1111 (1955).
61. HASTINGS, C. E. *Tech. Notes natn. advis. Comm. Aeronaut., Wash.* No. 774 (1940).
62. VICHNIEVSKY, R. *C.r. hebd. Séanc. Acad. Sci., Paris* **214**, 216 (1942); **218**, 959 (1944).
63. VICHNIEVSKY, R. *Revue Inst. fr. Pétrole* **2**, 293 (1947).
64. KUMAGAI, S. and KUDO, Y. *9th Int. Symp. Combust.*, p. 1083. The Combustion Institute (1961).
65. MARSDEN, R. S. *4th Int. Symp. Combust.*, p. 683. Williams and Wilkins, Baltimore (1953).
66. KARLOVITZ, B., DENNISTON, D. W., KNAPSCHAEFER, D. H., and WELLS, W. E. *4th Int. Symp. Combust.*, p. 613. Williams and Wilkins, Baltimore (1953).
67. KARLOVITZ, B., DENNISTON, D. W., KNAPSCHAEFER, D. H., OXENDINE, J. R., and BURGESS, D. S. *J. appl. Phys.* **28**, 70 (1957).
68. FOX, M. D. and WEINBERG, F. J. *Proc. R. Soc.* A**268**, 222 (1962).
69. LAWTON, J. and WEINBERG, F. J. *Proc. R. Soc.* A**277**, 468 (1964).
70. FRIEDMAN, R. and MACEK, A. *10th Int. Symp. Combust.*, p. 731. The Combustion Institute, Pittsburg (1965).
71. BALWANZ, W. W. *AGARD conference proceeding No. 8*, p. 699 (1965).
72. WEINBERG, F. J. *Optics of flames*. Butterworth, London (1963).
73. HEY, J. S., PINSON, J. T., and SMITH, P. G. *Nature, Lond.* **179**, 1184 (1957).
74. ARRHENIUS, SV. *Wiedemanns Annln* **63**, 305 (1897).
75. CHATTOCK, A. P. *Phil. Mag.* **48**, 401 (1899).
76. BRANDE, W. T. *Phil. Trans. R. Soc.* **104**, 51 (1814).
77. MALINOWSKI, A. E. *J. Chim. Phys.* (*U.S.S.R.*) **21**, 469 (1924).
78. BONE, W. A., FRAZER, R. P., and WHEELER, W. II. *Phil. Trans. R. Soc.* A**228**, 197 (1929).
79. THORNTON, W. M. *Phil. Mag.* **9**, 260 (1930).
80. MALINOWSKI, A. E. and LAWOW, N. *Zh. Physik* **59**, 690 (1930).
81. HABER, F. *Sber. preuss. Akad. Wiss.* **11**, 1962 (1929).
82. KINBARA, T. and NAKAMURA, J. *5th Int. Symp. Combust.*, p. 289. Reinhold, New York (1955).
83. KINBARA, T. and NAKAMURA, J. *Combust. Flame* **3**, 227 (1959).
84. NAKAMURA, J. *Combust. Flame* **3**, 277 (1959).
85. GUENAULT, E. M. and WHEELER, R. V. *J. chem. Soc.* **1**, 195 (1931); **2**, 2788 (1932).
86. LEWIS, B. *J. Am. chem. Soc.* **53**, 1304 (1931).
87. CALCOTE, H. F. *3rd Int. Symp. Combust. Flames Explos. Phenomena*, p. 245. Williams and Wilkins, Baltimore (1953).
88. GUGAN, K., LAWTON, J., and WEINBERG, F. J. *10th Int. Symp. Combust.*, p. 709. The Combustion Institute (1965).
89. CALCOTE, H. F. and PEASE, R. N. *Ind. Engng Chem. ind. Edn.* **43**, 2726 (1951).
90. PAYNE, K. G. and WEINBERG, F. J. *8th Int. Symp. Combust.*, p. 207. Williams and Wilkins, Baltimore (1962).

91. WATERMEIER, L. A. D.I.C. Thesis, Imperial College, London (1965).
92. SAUNDERS, M. J. and SMITH, A. G. *J. appl. Phys.* **27**, 115 (1956).
93. LEWIS, B. and KREUTZ, C. D. *J. Am. chem. Soc.* **55**, 934 (1933).
94. SENFTLEBEN, H. and GLADISH, H. *Z. Phys.* **126**, 289 (1949).
95. SENFTLEBEN, H. and BULTMAN, E. *Z. Phys.* **136**, 389 (1953).
96. STACH, V. *Int. J. Heat Mass Transfer* **5**, 445 (1962).
97. BERGER, F. and STACH, V. *Proc. IInd Int. Conf. peaceful uses atom. Energy*, p. 7. Geneva (1958).
98. BERGER, F. and DERIAN, L. *Jaderna Energ.* **11**, 1 (1965).
99. KINBARA, T. and NAKAMURA, J. *Scient. Pap. Coll. gen. Educ. Tokyo* **4**, 21 (1954).
100. MAYO, P. J., WATERMEIER, L. A., and WEINBERG, F. J. *Proc. R. Soc.* A**284**, 488 (1965).
101. LAWTON, J., MAYO, P. J., and WEINBERG, F. J. *Proc. R. Soc.* A**303**, 275 (1968).
102. PARKER, T. A. and HEINSOHN, R. J. *Proceedings of the 3rd Conference on performance of high temperature systems.* Gordon and Breach, New York (1964).
103. CHEN, D. C. C., LAWTON, J., and WEINBERG, F. J. *10th Int. Symp. Combust.*, p. 743. The Combustion Institute (1965).
104. LAWTON, J. *Br. J. appl. Phys.* **18**, 1095 (1967).
105. DIMMOCK, T. H. and KINYEKO, W. R. *Combust. Flame* **7**, 283 (1963).
106. KOLB, A. C. *Phys. Rev.* **107**, 345 (1957).
107. GAYDON, A. G. and HURLE, I. R. *The shock tube in high-temperature chemical physics.* Reinhold, New York (1963).
108. SECKER, T. E. *Proc. Instn elect. Engrs* **106**A, 28, 311 (1959).
109. POWERS, W. E. and PATRICK, R. M. *Physics Fluids* **5**, 1196 (1962).
110. BLIX, E. D. and GUILE, A. E. *Br. J. appl. Phys.* **16**, 857 (1965).
111. ADAMS, V. W. The influence of gas streams and magnetic fields on electric discharges. *Royal Aircraft Establishment Reports, T.N. No. Aero* 2896 (1963); *T.N. No. Aero* 2915 (1964); *T.R. No.* 65273 (1965).
112. JEDLICKA, J. R. The shape of a magnetically rotated electric arc column in an annular gap. *Tech. Notes natn. advis. Comm. Aeronaut., Wash.* D-2155 (1964).
113. LORD, W. T. Some magneto-fluid-dynamic problems involving electric arcs. *Royal Aircraft Establishment Report, T.N. No. Aero* 2909 (1963).
114. BUNT, E. A., CUSICK, R. T., BENNETT, L. W., and OLSEN, H. L., Design and operation of the battery power supply of a hypersonic propulsion facility. *Appl. Phys. Lab.—The Johns Hopkins University, Report T.G.* 660 (1965).
115. BUNT, E. A. and OLSEN, H. L. *Proc. Instn. mech. Engrs* **180**, Part 3J (1966) (Convention on thermodynamics and fluid mechanics, Liverpool University, April 1966).
116. BUNT, E. A. and OLSEN, H. F. *Advanced energy conversion.* Pergamon Press, Oxford. In press.
117. HEINSOHN, R. J., WULFHORST, D. E., and BECKER, P. M. *Combust. Flame* **11**, 288 (1967).
118. WEINBERG, F. J. *Proc. R. Soc.* A**307**, 195 (1968).
119. HOWARD, J. B. *12th Int. Symp. Combust.*, Paper No. 89, Poitiers (1968).
120. SOO, S. L. *Direct energy conversion*, Prentice-Hall, New York, (1968).
121. SOO, S. L. and COLVER, G. M. *12th Int. Symp. Combust.*, Paper No. 1, Poitiers (1968).
122. HARKER, J. H. and PORTER, J. E. *J. Inst. Fuel*, **41**, 264 (1968).

8. Distribution of the Electrical Parameters and Limitations to Practical Effects

WITH a few exceptions, we are treating the onset of secondary ionization and breakdown as an upper limit to the phenomena considered in this monograph. The exceptions include 'augmented flames' (Chapter 3), where it does not matter if the directional momentum of the ions is degraded in various ways, since, at the discharge stage, electrical energy is used purely as a method of heating the gas. Another occurs in the application of the force resulting from crossed magnetic field and current vectors, where only the total flux of charge matters, irrespective of the source of the charge carriers. In all other cases, however, breakdown at an electrode constitutes a limit because it provides a source of alternative ions, unrelated to the combustion processes, which counterflow and neutralize those from the flame and destroy wind effects. Thus, as soon as secondary ionization by collision becomes a frequent event, the 'electrical' contribution of flames becomes trivial.

For a given charge-carrier travelling through a particular gas under sub-breakdown conditions, all practical magnitudes can be expressed in terms of current density alone; this covers the transport of mass and charge, the body force on the neutral gas and hence the ionic wind velocities (eqns (4.132) and (4.133)). The maxima to practical effects can therefore be expressed in terms of maximum attainable current densities. In principle, there are two possible factors limiting current density—the finite rate of charge generation by the combustion process and the onset of secondary ionization as the field strength is increased.

The former, which leads to the saturation current density, j_s, has been discussed in some detail for a plane laminar flame, as a method of studying ion generation in reaction zones (Chapter 5). However, unless the system is confined to that geometry and type of flame, the rate of generation of ions need not act as a current limitation. Thus the amount of flame area per unit area of electrodes can be increased by inclination of the flame front, by its corrugation under turbulence (see p. 52), or by increasing the number of individual flame fronts, using a suitably designed burner. It is also possible to augment the temperature and hence the rate of reaction, and, if variation of the ion species is permissible, to use various additives (Chapter 6). Finally, the current density can be increased purely geometrically by focusing lines of current in a convergent field (for example using a large flame and a point electrode).

The second kind of limitation, that due to local breakdown of the gas, is thus the ultimately limiting one. The relevant events now occur in the electrode regions. In the presence of space charge, the field strength, E, is not uniform, as has already been mentioned; this is simply a consequence of Gauss's Law. It will be shown below that the solution of this equation with that of charge conservation yields a field distribution of the general form shown by the heavy lines in Fig. 8.2a, which is such that the field strength rises continuously in going from the flame to either electrode and reaches its maximum value at the electrodes. It follows that breakdown occurs first at the electrode, or, in the symmetrical case, at both electrodes. Since the distribution of field strength depends chiefly on the current density, and since the relevant field strength at breakdown is solely a property of the gas in contact with the electrode(s), this criterion is largely independent of the properties of the ion source. The limiting current density at breakdown, j_B, is simply that j for which the distribution of space charge causes secondary ionization in the gas at one or both electrodes. j_B obviously depends on electrode spacing, since space charge continues to cause the field intensity to increase over the extent of these regions, unlike j_s, which is a property of the flame alone.

For flames that are such strong ion sources that $j_B < j_s$ for all reasonable electrode configurations, j_B is obviously limiting. For flames that are weak ion sources, $j_s < j_B$, and, with electrodes close to the flame, the saturation plateau in the current–potential curve may be quite extensive, i.e. the difference between the applied potential at which saturation first sets in and that value at which secondary ionization occurs, may be very large. Under these conditions j_s is limiting, but the limitation can always be overcome by increasing the ion supply per unit area of electrodes, using one of the methods mentioned above.

This a rather fortunate state of affairs. j_s, in common with other consequences of flame kinetics, is not predictable, but this does not arrest the development of the subject since j_s is not limiting. Its measurement becomes a valuable tool in the study of flame ionization. On the other hand, j_B, which is limiting in practice, is predictable because it is determined by the breakdown strength of air (or whatever gas surrounds the electrodes) and by the relationships between current, field strength, and distance from the flame, which are independent of the flame properties.

The development of the theory of field- and space-charge distribution around thin ion sources in electric fields is therefore necessary to predict practical maxima. As mentioned in Chapter 5, it is also necessary to the measurement of saturation currents, for the following reason. The mean field between the electrodes is always greatly in excess of that in the flame, because of the effects of space charge discussed above. If breakdown occurs before j_s is attained, i.e. if $j_s > j_B$, no saturation current is, of course, measureable. Since j_B and the fraction (field at the flame)/(field at the electrodes) depends on the separations of the two electrodes from the flame, the theory is necessary to calculate how breakdown may be delayed long enough for saturation to be attained first. Without special precautions, saturation currents are attainable with only the weakest ion sources. Analogously, the results of the theory tell us for what conditions and configurations the largest practical effects are attainable.

For a system in one dimension, x, consisting of two plane infinite electrodes with a plane, infinite ion source of finite thickness in between, all the planes being perpendicular to x, the following general equations apply (see Chapter 4). Gauss's Law gives

$$\frac{dE}{dx} = 4\pi(n_+ - n_-)e. \tag{8.1}$$

The relationship between field and potential V is

$$V = - \int E \, dx. \tag{8.2}$$

The growth of current density for each charge, due to the net number rate of charge generation per unit volume, r_n, is

$$\frac{1}{e}\frac{dj_+}{dx} = r_n = -\frac{1}{e}\frac{dj_-}{dx}. \tag{8.3}$$

On writing the recombination rate $(\alpha n_+ n_-)$ and r_c as the absolute volumetric number rate of charge generation, eqn (8.3) becomes

$$\frac{1}{e}\frac{dj_+}{dx} = -\frac{1}{e}\frac{dj_-}{dx} = r_c - \alpha n_+ n_-, \tag{8.4}$$

Since charges of opposite polarity arise in equal amounts, $dj_+/dx = -dj_-/dx$, as follows from (8.3), so that

$$j = j_+ + j_- = (K_+ n_+ + K_- n_-)Ee. \tag{8.5}$$

The theory in this chapter is based largely on parts of our previously published work.[1-3]

THE DISTRIBUTIONS OF FIELD AND CHARGE

Within the reaction zone

Although the limiting breakdown conditions occur in the electrode spaces, we require the theory within the flame to provide boundary conditions for the external distributions. It is also instructive to consider the effects of fields within flames from the point of view of 'flame conductivities', which have often been treated somewhat naively on an ohmic basis.

The simplest basis on which a representative theory may be developed is the following model. The flame, consisting of a single, plane, and infinite reaction zone is held between two parallel, plane, and infinite electrodes. It is of a constant thickness within which the rate of ion generation and the coefficient of ion recombination are constant. The 'reaction zone thickness' is thus defined as the thickness of that slab which combines the above properties with the net rate of ion generation of the real flame. This is a convenient basis for discussion rather than a necessary assumption. A suitable alternative model would be, for example, an infinitesimally thin zone of ion generation, followed by recombination over a more extended region into which the ions are convected. The limitations that these assumptions involve will be examined subsequently. In the spaces between the flame and the electrodes, no charge is generated and no charge is destroyed (because, when a large field is applied, only one kind of charge exists in each space). As has been mentioned, these electrode spaces are usually vastly greater than the reaction zone thickness and they can never be made smaller than the quenching distance.

(a) Slow ion removal

In the absence of a field, only recombination is deemed to rival generation, the effects of convection and diffusion being relatively small; the concentrations of negative and positive charges are equal. Thus as $E \to 0$, the system tends to become symmetrical as regards the two polarities, i.e., profiles are the same for positive and negative charge-carriers. In all other cases the difference in mobilities precludes this. Hence

$$r_c = \alpha n_+ n_- = \alpha n^2 \tag{8.6}$$

or

$$n = \sqrt{(r/\alpha)}. \tag{8.7}$$

Consider now the application of a field so small that this condition is virtually unaltered. By eqn (8.1)

$$dE/dx = 4\pi(n_+ - n_-)e = 0.$$

Thus the field is constant, which implies a potential drop proportional to distance, i.e.

$$V = V_0 + Ex. \tag{8.8}$$

The current is given by

$$
\begin{aligned}
j &= enE(K_+ + K_-) = eE(r_c/\alpha)^{0.5}(K_+ + K_-) \\
&= \{e(V - V_0)/x\}(r_c/\alpha)^{0.5}(K_+ + K_-) \\
&= \text{constant} \times (V - V_0).
\end{aligned}
\tag{8.9}
$$

Thus, under conditions approaching zero field and zero current, the flame behaves as an ohmic conductor of resistivity $(\alpha/r_c)^{0.5}/e(K_+ + K_-)$. Since, in the flame zone, the negative ion appears to be a free electron, K_- is 10^2–10^3 times K_+, so that K_+ becomes negligible in the above expressions. Drawing a current therefore immediately destroys the positive–negative symmetry referred to. If a value of K_- is assumed, the ion concentration under equilibrium conditions (or rather negligible departure from them) can be deduced from resistivity measurements, but this applies strictly only for infinitesimal potential differences.

(b) Saturation

The opposite extreme occurs when the applied potential has just reached a value beyond which the current no longer increases with increasing field. Ion removal by the field has now become so large as to make recombination negligible by comparison. Thus for the positive ions, for instance,

$$\frac{1}{e}\frac{dj_+}{dx} = r_c - \alpha n_+ n_- = r_c, \tag{8.10}$$

so that j_+ increases linearly across the width, X, of the reaction zone and j_- behaves symmetrically—Fig. 8.1(a)—because eqn (8.10) must apply equally to the flow of negative charge. Thus,

$$j_+ = r_c ex, \quad j_s = r_c eX \quad \text{and} \quad j_- = r_c e(X-x). \tag{8.11}$$

However,

$$j_+ = K_+ n_+ Ee,$$

and

$$j_- = K_- n_- Ee, \tag{8.12}$$

so that, where j_+ is symmetrical with j_-, the two concentration profiles

$$n_+ = \frac{r_c x}{EK_+} \quad \text{and} \quad n_- = \frac{r_c(X-x)}{EK_-} \tag{8.13}$$

must be entirely different because of the disparity in mobilities. This becomes important in calculating field distribution under the influence of a space charge that is now almost entirely due to positive ions. Thus, substitution into eqn (8.1) gives

$$\frac{dE}{dx} = \frac{4\pi r_c e}{E}\left(\frac{x}{K_+} - \frac{X-x}{K_-}\right), \tag{8.14}$$

in which the second term in the brackets is entirely negligible, except where x tends to zero. The contribution of these regions to the integral $\int E \cdot dE$ is obviously insignificant, so that, to a good approximation,

$$E^2 = E_0^2 + (4\pi r_c e/K_+)x^2. \tag{8.15}$$

It will be shown in the next section that, so long as the field does not exceed saturation conditions,

$$E_0 \simeq 0,$$

so that

$$E = 2(\pi r_c e / K_+)^{\frac{1}{2}} x. \tag{8.15a}$$

Substitution into eqn (8.13) gives the positive ion concentration as

$$n_+ = \tfrac{1}{2}(r_c / \pi K_+ e)^{\frac{1}{2}} \tag{8.16}$$

and an electron concentration less than 3 per cent of the above, which has been deemed negligible. The distribution of potential follows from (8.15a):

$$V = \int E \, dx = \left(\frac{\pi r_c e}{K_+} \right)^{\frac{1}{2}} x^2, \tag{8.17}$$

the potential drop across the flame being $(\pi r_c e / K_+)^{\frac{1}{2}} X^2$. These results are sketched in Fig. (8.1a).

(c) Sub-saturation

The case of charge withdrawal at a rate that is neither negligible in comparison with recombination, nor so large as to render the latter negligible, has been relegated to this stage in the argument because it proves to be composed of the two cases mentioned above. The equilibrium between generation and recombination (case (a)) is, in fact, meaningful only so long as an infinitesimal field is applied. The conclusions of eqns (8.8) and (8.9) are valid if one mobility is not so much greater than the other that a field large enough to cause significant movement of the larger ion will remove the smaller so fast as to reduce it to negligible concentration. This, however, is precisely what does happen if the negative charge-carriers in the flame are free electrons. The consequence is that, in the presence of any appreciable field, the concentration of electrons becomes negligible. Because their velocity is 10^2–10^3 times greater than that of positive ions at the same field strength, they can carry the same current while being present in concentrations 10^{-2}–10^{-3} times smaller (this is made clear by Fig. 8.1(a)). If their concentration is negligible, so must be their rate of recombination with positive ions.

On extending the idealization to an infinite negative mobility, each part of the reaction zone can be in only one of two states; either the field strength is finite, in which case there are no electrons and no recombination so that the region is saturated; or the field is zero and the region is in perfect generation–recombination equilibrium. Figure 8.1(b) represents a reaction zone from which current $j < j_s$ is drawn. Since positive ions flow across the plane at X, E is finite for x just less than X. In these regions, therefore, $n_- = 0$ and

$$\frac{1}{e} \frac{dj_+}{dx} = r_c - \alpha n_+ n_- = r_c,$$

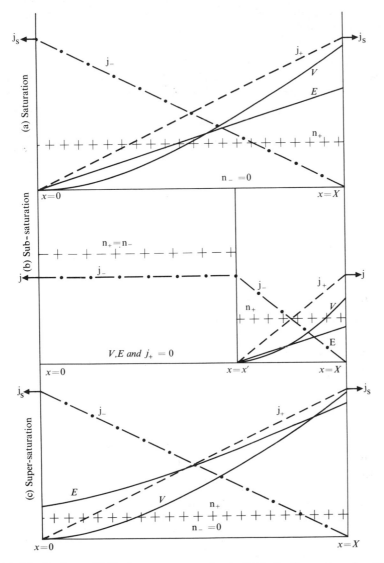

FIG. 8.1. Distributions of field, charge current, and potential in the flame. From Lawton and Weinberg.[2]

which is identical with eqn (8.10) and integrates to a line of the same slope. Hence j_+ must fall to zero at

$$x' = X(1 - j/j_s) \tag{8.18}$$

and all the current is accounted for.

In the region $0 < x < x'$ no net current may now originate. Therefore, since the field does not reverse direction anywhere, $E = 0$, no charge is removed and $n_+ = n_-$ there.

This is, of course, an idealization and the concept of infinite mobility raises some apparent anomalies. Thus for $0 < x < x'$, j_- is not zero even though E is; i.e. the negative current from $x' < x < X$ would 'seep across' a region of no field. (There is no mathematical inconsistency, because the infinite mobility would permit electrons a finite velocity.) Again, at x' there is an infinite gradient in n_- and the effect of diffusion, which is then not negligible, causes a spreading of the boundary. These consequences of the idealization have been examined theoretically[4] and can be shown to modify the structure only in detail and not as regards the main conclusions.

The distribution of charge, field, and potential (see Fig. 8.1(b)) follows from the above argument and the cases previously discussed. For $0 < x < x'$

$$n_+ = n_- = \sqrt{(r_c/\alpha)}, \qquad E = 0 \quad \text{and} \quad V = V_0.$$

In the region $x' < x < X$

$$n_- \simeq 0$$

$$E = 2(\pi r_c e/K_+)^{\frac{1}{2}}(x - x') = 2(\pi r_c e/K_+)^{\frac{1}{2}}\{x - X(1 - j/j_s)\} \qquad (8.15b)$$

$$= 2(\pi r_c e/K_+)^{\frac{1}{2}}(x - X + j/re),$$

by eqn (8.15a).

$$n_+ = \tfrac{1}{2}(r_c/\pi K_+ e)^{\frac{1}{2}}$$

as before (see eqn (8.16)), and

$$V = (\pi r_c e/K_+)^{0.5}(x - x')^2 = (\pi r_c e/K_+)^{0.5}(x - X + j/r_c e)^2, \qquad (8.17a)$$

which gives a potential drop of $(\pi)^{0.5} j^2 / K_+^{0.5}(r_c e)^{1.5}$ across the entire flame.

(d) Super-saturation

Increases in potential beyond the saturation condition cannot, of course, increase current, but they do alter the structure in terms of ion concentration and potential distribution. Thus eqns (8.10)–(8.15) remain intact, but E_0 in the latter is no longer zero. Hence

$$E = \sqrt{(E_0^2 + 4\pi r_c e x^2 / K_+)}, \qquad (8.15)$$

$$n_+ = \frac{rx}{(K_+^2 E_0^2 + 4\pi K_+ r_c e x^2)^{\frac{1}{2}}}, \qquad (8.19)$$

$$V = \left(\frac{4\pi r_c e}{K_+}\right)^{0.5} \left[\frac{x}{2}\left(\frac{K_+ E_0^2}{4\pi r_c e} + x^2\right)^{0.5} + \frac{K_+ E_0^2}{8\pi r_c e} \sinh^{-1}\left\{x\left(\frac{4\pi r_c e}{K_+ E_0^2}\right)^{0.5}\right\}\right]. \qquad (8.20)$$

These are illustrated in Fig. 8.1(c).

Effect of approximations

The three principal approximations involved are that of infinite electron mobility (in (b), (c), and (d)), together with the neglect of diffusion and of convection. None of these approximations are inevitable, but the complexity of the mathematics involved in avoiding them is not justified by the aims, nor indeed by the model chosen.

Estimates of the errors introduced by the assumption of infinite K_- can be obtained by successive approximations, using eqn (8.13), to determine the deviation of n_- from zero and hence correcting eqn (8.10) as well as the application of eqn (8.1) to the derivation of (8.14).

The neglected effects of diffusion and convection become significant only in very narrowly confined regions; the former where the approximate theory predicts infinite concentration gradients, the latter where the field is so small that (KE) becomes comparable with flow velocities. The effect is to round the contours of concentration as well as of current and field distributions and to alter the analysis somewhat, where zero fields are predicted. It is worth noting here that even 10^2 V/cm, i.e. about 1 per cent of the mean fields usually employed, will induce a positive ion velocity greater than burning velocities of any hydrocarbon-air mixtures. A more quantitative examination of these factors[4] confirms that the aforegoing theory yields an essentially correct account of the electrical structure.

Outside the flame

In the electrode spaces, the same general equations apply, but there is no generation or recombination, i.e.

$$r_c = 0, \quad \text{either } n_+ \text{ or } n_- = 0, \quad \text{and} \quad j_- = j_+ = j = \text{constant}$$

in the one-dimensional system. The other major difference is that, provided the electrode spaces are at room temperature, the negative charge-carrier is no longer an electron. Because of the association of electrons with neutral molecules during quite short path-lengths in air, as discussed in an earlier chapter, K_- may be expected to be of the same order as K_+, in view of the small dependence of K on mass number.

One exception, which can be important in particular cases, occurs when the electrode space is filled with hot product gases in which electron attachment is more gradual. This case will be discussed below.

For the general case,

$$j = j_\pm = K_\pm n_\pm eE. \tag{8.21}$$

(The plus or minus sign will henceforth be omitted on the understanding that the argument applies to either electrode space.) Substitution into eqn (8.1) gives

$$dE/dx = 4\pi ne = 4\pi j/KE$$

or

$$E^2 = E_0^2 + 8\pi jx/K. \tag{8.22}$$

The potential,

$$V = \int_0^x E \, dx = \pm \frac{K}{12\pi j} \left\{ \left(E_0^2 + \frac{8\pi jx}{K} \right)^{\frac{3}{2}} - E_0^3 \right\}. \tag{8.23}$$

These, together with the equations of the preceding sections, complete the general solution of the problem. For all sub-saturation conditions, E_0 may be taken as zero at the flame surface facing the anode (Fig. 8.1(b)) and so may the value, given by (8.15), at the opposite surface. The difference of the latter from zero is always negligible by comparison with the electrode fields because of both the small distance and the increased mobilities at flame temperatures. Equations (8.22) and (8.23) can then be integrated towards each electrode. If the field is not known anywhere between the electrodes, as when a 'super-saturation' potential is applied, E_0 can be solved by summing the contributions to potential (eqn (8.23)) and equating to the voltage applied. Once j becomes constant at j_s, eqns (8.22) and (8.23) can be written

$$E^2 - E_0^2 = \text{const.} \, x \tag{8.22a}$$

and

$$V = \text{const.} \, (E^3 - E_0^3). \tag{8.23a}$$

The field in the flame now rises rapidly with applied potential, but its variation across the flame zone remains negligible, for the reasons stated above.

Having established the form and method of solution, it is instructive to set down the effect of various parameters graphically.

Figure 8.2(a) illustrates a single flame surface that is rather a weak ion source ($j_s = 2 \times 10^{-5}$ A/cm^2, which is equivalent to 12.5×10^{13} ion pairs generated per unit area per unit time and corresponds, for example, to a mixture of 8.6 per cent CH$_4$ in air) placed symmetrically between electrodes 6 cm apart. The total potential applied is varied and the corresponding distributions of field (full lines) and potential (dashed lines) are shown. One of the field lines corresponds to the onset of saturation and another to the onset of secondary ionization (assumed here to occur when the maximum field exceeds the breakdown value of 3×10^4 V/cm for air at s.t.p.). The current densities drawn in each case are shown above the lines.

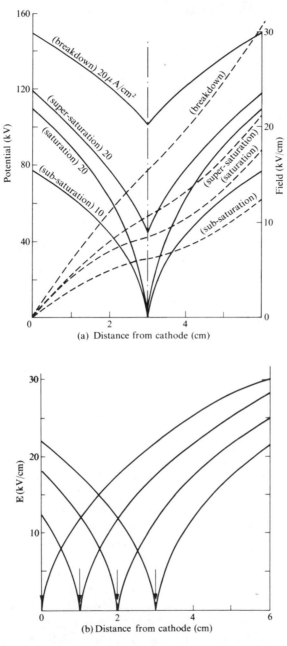

FIG. 8.2. (a) Field and potential distributions for a flame of saturation current 20 μA/cm^2 (e.g. 8·6 per cent CH$_4$ in air) placed midway between electrodes 6 cm apart in air at s.t.p. (b) Distribution of field strength for various flame positions. $j = 20\ \mu$A/cm^2: ↓ flame position. From Lawton and Weinberg.[2]

Figure 8.2(b) shows the effect on field distribution of varying the position of the same flame between electrodes at the same total separation with conditions maintained at the saturation point.

We may first draw conclusions concerning the optimum positioning of the flame with respect to the electrodes. The arguments that apply here are much the same as those considered in the case of saturation current measurements (Chapter 5), i.e. how to attain the maximum field in the flame without approaching breakdown fields at the electrodes. As the graphs show, the flame should be placed symmetrically between the electrodes and the separation between them should be kept as small as considerations of quenching will permit. The condition of symmetry arises because the mobility on either side of the flame has been assumed to be the same, for the reasons stated. When this is not so, for example, because one space is filled with hot products, the condition for the two electrode fields to be equal is, by eqn (8.22),

$$\frac{a_-}{a_+} = \frac{K_-}{K_+}, \tag{8.24}$$

where the a again denotes the distance to the electrode on the side indicated by the subscript.

Next, we must consider the question of generally maximizing the current and hence all the practical effects of applying fields. It will be convenient to illustrate this for a range of ion sources of different strengths. To ensure that all other conditions—for example, the identity of ions drawn—remain unchanged, we may think of this in terms of a particular premixed flame, which is corrugated by turbulence to various extents, to increase the current density available. According to the simple theory of the 'wrinkled flame front' (as discussed in Chapter 4), the ratio of the turbulent to the laminar burning velocity is equal to the ratio of the real fluctuating corrugated area, A_r, to the apparent time-mean area, A_a, and, if the latter is parallel to the electrodes, the maximum available current density is $(A_r/A_a)j_s$, where j_s is the saturation value for the laminar flame in that reactant mixture. It will be appreciated that this is merely a convenient illustration of an ion source of constant properties but continuously variable strength—the principle is quite unrelated to the wrinkled flame-front concept. As turbulence, and hence A_r is increased, the potential required for incipient saturation rises, but so long as each element of area remains just on the point of saturation, $E_0 \simeq 0$ in the flame and the field at the electrodes depends only on j (eqn (8.22)). Figure 8.3 illustrates the current–potential relationship for various ion sources—for example, variously corrugated flames in one mixture of reactants which gives $2 \cdot 5 \times 10^{13}$ ion pairs/s/unit laminar flame area (for example, $6 \cdot 7$ per cent CH_4 in air). The electrode separation is $3 \cdot 5$ cm and the flame is placed centrally. Line 1 is the j–V locus before saturation. It is thus

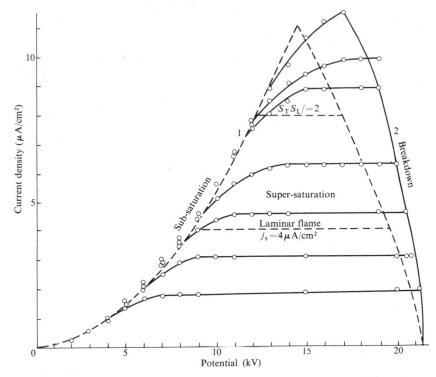

FIG. 8.3. Current-potential characteristic. –o– experimental, – – – – theoretical. From Lawton and Weinberg.[2]

eqn (8.23), integrated across the electrode space, with $E_0 = 0$. Line 2 is the boundary at which the electrode field attains its breakdown value, E_B. It is obtained by putting $x = a$ and $E = E_B$ in eqn (8.22) and substituting for E_0 in eqn (8.23). The intersection of these two lines gives the absolute maximum current of interest here. Up to that point j can be increased, for instance by further corrugating the flame, as shown. Beyond that point, secondary ionization would set in before saturation. The maximum useful current density is therefore given by inserting into eqn (8.22) both the conditions that $E_0 = 0$ and that $E = E_B$ at $x = a$, i.e.

$$j = E_B^2 K/8\pi a, \tag{8.25}$$

where E_B is the breakdown field in the electrode gas. The circles on Fig. 8.3 show experimental results obtained using the porous disc burner described for measuring saturation currents. The various ion source strengths were brought about by varying heat loss to the burner and hence final flame temperature. Considering the approximate nature of the numerical values assumed (for example, of E_B and K) in the theory, the agreement is good.

The value at the apex does, of course, depend on electrode separation from the flame, a. The criteria of optimizing this condition are those of minimizing the field at the electrodes and have already been discussed; the 'controlling' electrode space should be as small as is possible without quenching. If the breakdown field is to be attained simultaneously at both electrodes, eqn (8.24) must again be satisfied. Thus the absolute maximum useful current for an unperturbed flame is

$$j_{max} = E_B^2 K / 8\pi a_q, \tag{8.25a}$$

where a_q is the quenching distance, i.e. the least separation at which a cold surface will not interact with the flame process. It will be noted that the method of measuring saturation currents based on the porous plug burner does partly quench the flame, while using electron conduction in the other electrode space, and thus can exceed the j_{max} of eqn (8.25a). This equation is a most important relationship, which will be used to deduce absolute maxima to all practical effects numerically. It involves no 'flame variables' other than the quenching distance, but an adequate safety factor in the latter must in any case be allowed for.

Calculations similar to those above have been carried out for cylindrical and spherical flame electrode systems. Since the principle is identical and the algebra more complex, there seems little point in reproducing the theory here. The results, which are given below, are important for two reasons. First, some practical flame systems approximate to these geometries. Thus the heat transfer studies discussed in the preceding chapter, with the wire along the axis as one electrode and the circumferential jacket as the other, can reasonably be represented by cylindrical coordinates. Again, flames expanding from a central ignitor approximate to the spherical geometry. Another reason, however, is concerned with increasing the attainable currents and hence the maxima of all the practical effects that follow. This topic will be discussed fully later; only the effect of geometry is appropriate here.

The fundamental reason for the limitation imposed by eqn (8.22) is the growth of the field occasioned by the space charge that has been deduced for constant current density. Were the lines of current allowed to diverge, i.e. were the system not confined to unidimensionality, the increase in field strength would obviously be less steep. Divergence on one side of the flame often implies convergence on the other and, in the absence of some other change, a net loss in maximum current then results, because the convergent side becomes controlling. Such a change in circumstances is provided by the increase in mobility that results if the electrode space on the convergent side is filled with hot product gas into which electrons from the flame enter. The very much greater mobility of electrons decreases only gradually due to attachment in gases that are not much below flame temperature. If the attachment 'kinetics' is known, the foregoing theory can readily be modified to take into account the consequent variation of K with x for each component of the current density j. Thus if electrons attach according to a first-order law

$$\frac{1}{e}\frac{dj_e}{dx} = -\gamma n_e ,\tag{8.26}$$

it can readily be shown that the electron current j_e is given by

$$j_e = j \exp(-\gamma t),\tag{8.26}$$

where t is the time taken for the electron to travel a distance x from the flame. Equation (8.26) is independent of any convective effects, being a consequence of a rate of electron disappearance proportional to concentration only. From this equation, together with that of conservation of charge $(j_- = j_e + j_{\text{neg.ions}}$ in this instance) and Gauss's Law, the electrical structure of the electrode space can be deduced.[4] It is not worth presenting the detailed equations here, however, because numerical values of the attachment coefficient in mixtures of varying composition and temperature are not available. Measurements confirm that the variation of effective mobility, from that of an electron to that of an ion, occurs over sufficiently extended regions to cause a worthwhile reduction in field strength, when negative charge passes through hot products. Indeed, the effective mobility is increased to such an extent that the onset of breakdown in the positive ion region tends to become entirely controlling under these circumstances. For this reason, divergent fields on one side of the ion source are worth considering for the purpose of increasing practical effects, because the limitations are not symmetrical about the flame and any convergence can be tolerated more readily on the side where electron conduction can be utilized.

The field distribution on the divergent side of the two systems is given by: in the cylindrical case,

$$E^2 = \frac{2j'}{K_+}\left\{1 - \left(\frac{r_0}{r}\right)^2\right\}\tag{8.27}$$

and in the spherical,

$$E^2 = \frac{2}{3}\frac{j''}{K_+ r}\left\{1 - \left(\frac{r_0}{r}\right)^3\right\}.\tag{8.28}$$

Both cases apply for $E_0 = 0$ at the flame and j' is per unit length of electrode whilst j'' is the total current. The maximum currents are again deducible by putting $(r - r_0)$ and E equal to the quenching distance and the breakdown field respectively. Although these equations apply irrespective of flow orientation, the increase in maximum current densities and other effects arises only if the flame is propagating outward, as in the case of flame spread following point ignition. In the spherical case an abnormality occurs in that eqn (8.28) has a maximum when

$$r = (4)^{\frac{1}{3}}r_0 \tag{8.29}$$

unless the outer electrode is of that radius, or smaller, so that breakdown does not occur first at the electrode. The system is optimized by solving condition (8.29) with the quenching distance requirement, whereupon all its dimensions become fixed. The extent to which the various maxima can be increased by variations in geometry will be discussed below, together with the effects of other parameters on these limitations.

Maximum effects attainable by fields

Since conditions for absolute maxima are seen not to depend on flame properties, it is possible to estimate their numerical magnitudes, which then depend entirely on the choice of the extraneous parameters. Taking the case of the electrode regions in air at normal pressures and temperatures as an idealized example,[2] K was calculated for the H_3O^+ ion, for reasons of its abundance (see Chapter 6) and E_B was taken as 30 kV/cm. The latter is the breakdown field in air at s.t.p. for plane electrodes and presupposes that there are no points of large curvature on the electrodes due to, for example, unrounded edges or specks of carbon deposit, no entrainment of hot gas from the flame, and so forth. This results in a set of idealized maxima to which practical cases can only approximate. The various factors that detract from this ideal will be examined subsequently.

The flame-electrode separation, a, was chosen as 0·5 cm. This is in excess of normal quenching distances by a factor of the order of 10 times but is probably close to the least distance that will avoid serious practical difficulties for most burners. Some of the maxima however prove to be independent of this quantity. With these data, the maximum current density is

$$j_B = KE_B^2/8\pi a = 0·25 \text{ mA/cm}^2. \tag{8.30}$$

Thus 4000 cm^2 of electrode area is required for each ampere to be collected. This limitation must apply also to direct generation of electricity from hot gases unless (a) breakdown is allowed, locally at least, (b) electron conduction can be utilized down to distances of less than 0·5 cm from the electrodes, or (c) the operating pressure, or rather density, is much above atmospheric (see below).

Any applications based on the transport or collection of ions are similarly limited. For every 100 cm^2 of flame area, it is possible to collect approximately $M_i' \, \text{mg/h}$, where M_i' is the gram-molecular weight of the (singly) charged ion electrolyzed. This is quite appreciable if the 'ion' is a particle such as an agglomerate of carbon, but not otherwise. In the former case, a ratio of, say, 10^4 atomic masses of carbon per unit charge (Chapter 7) would lead to a collection rate of 2 g/min from a 100-cm^2 flame, provided the flame yielded that much soot, which indicates why such electrical phenomena in sooting flames were first discovered experimentally.

The maximum collection rate can be expressed as a fraction of the throughput, since, for premixed flames, the volume flow rate per unit area of flame is equal to the burning velocity, S_u. If the effective molecular weight of the reactants is M_R' and if they emerge from the burner at s.t.p., the mass flow rate per unit area of flame front is $M_R' S_u / 2 \cdot 24 \times 10^4$. Thus the fraction

$$\frac{\text{maximum mass of ions collected}}{\text{mass throughput}} = 5 \cdot 15 \times 10^{-5} \frac{M_i'}{S_u M_R'}. \tag{8.31}$$

The expression once more neglects the relatively small effect of M_i' on K. The smallest usable burning velocity is about 4 cm/s. Thus, if the ionic mass is of the same order as the molecular weight of the reactants, i.e. if the ion is not a particle of very many molecular masses (formed either during the flame process or deliberately introduced), the electrolysis of flames for desirable intermediates is not a process of appreciable yields. This is the basis of the discussion in Chapter 7 on the transfer of charged masses by the action of fields.

The maximum static pressure due to the ion current, the maximum wind velocity at the electrodes, and the maximum force/unit volume respectively are given by

$$P_B = E_B^2 / 8\pi = 400 \, \text{dyn/cm}^2, \tag{8.32}$$

$$v_B = E_B / 2\sqrt{(2\pi\rho)} = 550 \, \text{cm/s}, \tag{8.33}$$

$$F_B = E_B^2 / 8\pi a = 800 \, \text{dyn/cm}^3. \tag{8.34}$$

This body force is about 800 times greater than the maximum body force (ρg) in natural convection in air at s.t.p. The velocity in eqn (8.33) is the maximum mainstream velocity, i.e. the maximum velocity at the electrode for zero velocity at the flame. To obtain greater values it is necessary to use a multistage system, but even for a single stage, it is larger by an order of magnitude than burning velocities of stoichiometric hydrocarbon-air flames. We note, once more, that the theoretical prediction of an appreciable effect has been preceded by experimental discoveries of, for instance, flame deflections induced and heat transfer augmented by ionic winds. Maximum rates of convective heat flow that can be induced by applied fields, under the conditions of eqn (8.33) lie in the region of 10 to 40 cal cm^{-2}s^{-1}, depending on the kind of flame.

Just as the maximum mainstream velocity can be deduced by substitution of the maximum current (eqn (8.30)) into eqn (4.133d) so, for maximizing flow velocities in confined entrainment, j_B is substituted into expressions for the flow parameters in equations such as (4.141) and values of the optimum area ratio for entrainment are deduced. Referring to the particular case discussed on p. 152, the maxima occur as the ratio of radius to electrode distance tends to zero, but the variation over the practically useful range of this ratio is, in fact, less than 15 per cent. In this manner it is found that the maximum entrainment velocity

$$v_{e,B} = E_B/\sqrt{(2\pi\rho)} = 1100 \text{ cm/s}, \tag{8.35}$$

which is twice the maximum main stream velocity and occurs at $A_e/A = \frac{1}{2}$. Maximizing for entrained momentum flow gives 900 dyn/cm² at $A_e/A = 0\cdot55$, whereas the maxima of mass flow and main stream velocity occur at $A_e/A = 1$. Which of these is the more relevant depends on the application; thus maximum momentum is required for generating turbulence or inducing mixing, while maximum mass flows are required when a large proportion of a second gas is to be entrained and mixing is not controlling. The maxima for unit area of flame are:

$$\text{momentum flux} = 900 \text{ dyn/cm}^2, \tag{8.36}$$

$$\text{mass flux} = 1 \text{ g/s/cm}^2. \tag{8.37}$$

The latter corresponds to a velocity of 778 cm/s, i.e. $\sqrt{2}$ times that in the unconfined case. This is the basis of the discussion in Chapter 7 on the limitations of the practical use of ionic wind effects. For results applicable to axially symmetrical geometries the reader is referred to the original publications.[2-5]

It must be borne in mind that the only limitations considered here are purely 'electrical' ones. It is worth noting, however, that practical difficulties are considerably eased, in the case of wind effects, by the absence of a from eqns (8.32) and (8.33). Thus the magnitudes of these effects do not depend on close proximity of electrodes to flames. For these cases also the maxima can be greatly exceeded by aggregating stages in series.

The power dissipated at maximum current and wind effects is

$$W = KE_B^3/6\pi = 0\cdot092W \text{ cm}^{-2} = 2\cdot2 \times 10^{-2} \text{ cal cm}^{-2}\text{s}^{-1}. \tag{8.38}$$

The cost, in terms of power loss, of any of the single-stage applications is therefore negligible. It is, for instance, of the order 10^{-3} times the maximum increase in heat transfer for convection at the electrode.

In practice, effects induced by fields fall short of these idealized maxima. Measurements have been carried out[3] using wind velocity, as revealed by the interrupted trajectories of particles suspended in the gas stream, as a direct measure of the mechanical effects. It was found that while the theory was obeyed with very fair accuracy, the onset of breakdown at the electrode occurred at a considerably lower field strength than the 30 kV/cm given in tables for plane electrodes in air at s.t.p. There are several reasons for this, some of which are not independent of the flame itself, and these will now be examined in turn.

Perhaps the most fundamental inconsistency of the idealized theory is that it is impossible to have a perfectly permeable electrode, as regards pressure losses, which is at the same time unidimensional. In order to make an electrode permeable, it has to be provided with apertures; indeed, in all practical work in which vortices due to the return of gas discharged at the electrode had to be avoided, gauzes were used for electrodes. For any such geometry, lines of field intensity must converge upon the edges of the apertures and this convergence must, in turn, promote breakdown at a lower average current density than in the plane case. The process of breakdown at gauzes of different wire-spacing and radii has been studied theoretically and empirically.[5] This resulted in a relationship describing the attainable velocity of gas flowing through the electrode as a fraction of the above idealized maximum:

$$v/v_B = \tfrac{23}{30}\{f(2-f)\}^{0.11} r_g^{-0.054} \left[1 + \frac{1}{2C_D^2}\{(1-f)^{-4} - 1\} \right]^{-0.5} \qquad (8.39)$$

Here r_g is the radius of the gauze strands and $f/2$ is the ratio of this radius to the distance between strand centres. C_D is the discharge coefficient of the orifices, defined as the ratio of the actual volume flow to that for a lossless orifice and is a function of Reynolds number. The discrepancy from the idealized maximum need not be a very large one; for each wire radius there is an optimum spacing and the best results are obtained where this is combined with very fine strands. For example, fine wire of radius 10^{-3} cm woven at $f = 0.03$ would be expected to allow more than 80 per cent of the maximum velocity to be attained.

A somewhat similar reduction of the breakdown field occurs because of convergence of lines of force on to proturberances that will gradually form even on initially polished plane electrodes. It is one of the features of the process studied that any droplets or particles present as impurities, reactants, or products become highly charged[6] by the flame ions and thus will be deposited on the electrodes. In hydrocarbon flames, particles of carbon black are predominantly involved; see Chapter 7. Experiments carried out to ascertain the magnitude of this effect, by depositing soot from candle flames and from flames burning benzene on the gauze electrodes, showed reductions of 8–16 per cent in E_B. Since the former values were approached by deposits from candle flames while the latter were produced by those of benzene, it was concluded that the form of the aggregate on the electrode greatly influences the extent of the departure from the ideal. Although carbon deposition would be expected to occur only in certain special cases, the deposition of small specks of particulate matter in general is difficult to avoid. Moreover, the effect of deposits, e.g., of dust, formed under the action of a field is likely to be greater, because of their spiky form, than that of simply a dusty surface. This problem could only be overcome by using an electrode that is self-cleaning, protected by transpiration, or continuously renewed, for example by using a moving band.

Yet another important effect is the entrainment of hot flame gases. It has been shown[5] that it is difficult to maintain the gas in the electrode spaces entirely cold, even when the ion stream is 'filtered' (see p. 147) through a jet of air. (The limitation here is determined by the source of high potential available; entrainment can be avoided more effectively over large distances, but these require the application of large potential differences.) By applying Blanc's Law to the problem of mixing hot and cold gases it has been shown[5] that, on the basis of certain assumptions, the electrical forces before and after mixing are identical, for a given current. These forces can therefore be treated simply in terms of the changed temperature and composition at the electrode, irrespective of the extent of mixing. Again it can be prevented altogether only by some method of shielding the electrode spaces from flame gases, such as by transpiration of cool inert gas through the electrode itself.

Taking these various effects together, it seems unlikely, if no steps are taken to alleviate any of them, that the effects attained would ever exceed half the idealized maxima.

It now becomes desirable to invert the argument and consider whether any variations of the system analyzed could be used to give rise to larger maxima. Aside from using more than one stage, where applicable, the possibilities include changes in geometry, such as have already been mentioned, and environmental changes that would alter E_B—such as changes in temperature, pressure and composition, and the superimposition of a magnetic field. Regarding geometry, for reasons discussed, divergence of lines of force must always result in some improvement in the maximum conditions, and the improvements for the cylindrical and spherical geometries can be expressed in terms of the above theory. The results are set out as ratios in Table 8.1. They depend, of course, on the radius of curvature, reverting to the plane case as the radius tends to infinity. The values given are maxima, i.e. the limits to which the improvements tend as the inner radius tends to zero (except in cases, such as the static pressure in the spherical system, where the result is independent of radius). In the spherical system the outer electrode is, in addition, assumed to be situated at the radius of maximum field (eqn (8.29)). Maximum current densities are based on unit area of electrode, not of flame. The latter are considerably greater but less relevant

TABLE 8.1

Ratio of maximum effect to that in one dimension

	Plane	Cylindrical	Spherical
current density	1	2	1·48
static pressure	1	∞	1·72
wind velocity	1	$\sqrt{2}$	$\sqrt{3}$

to rates of collection. Mass rates of ion collection expressed as a fraction of reactant throughput (see eqn (8.31)) of course tend to infinity because the flame area, unlike the electrode area, tends to zero with the radius of curvature. A practical approach to this idealization, however, would be difficult experimentally.

Concerning changes in pressure and temperature at constant composition, the breakdown field may be treated as proportional to, and mobility as inversely proportional to density, for reasons discussed in previous chapters. Thus, keeping the electrode separation constant, i.e. neglecting possible improvements arising from changes in quenching distance, the maximum current density is proportional to the gas density, the maximum static pressure to its square and the maximum ionic wind velocity to its square root (eqns (8.30), (8.32), and (8.33)). At 10 atm, therefore, a flame of $100 \, \mathrm{cm}^2$ would yield a gramme of the predominant small ions in a few hours. As a fraction of the throughput this does, of course, remain an insignificant proportion. As a hypothetical method of, for instance, collecting free radicals, where these are ions, on to a low-temperature surface, this might just be worth considering, particularly in view of the considerable further increase in breakdown field brought about by the cold electrode space. As regards the wind velocity, although this is proportional to only the square root of pressure, the induced mass and momentum flows that are the relevant quantities for practical applications vary as $p^{\frac{3}{2}}$ and p^2 respectively.

Bathing the electrode spaces in cold and/or insulating gas is less attractive than might appear at first sight, when we bear in mind that the electrode separation has already been set close to the least practicable value. It is relevant here to mention again that gases which readily attach electrons and therefore have a large breakdown strength tend to be good inhibitors of combustion† (Chapter 3). Unless the reactants or products themselves contain gases of high breakdown strength (as would be the case with inhibited flames) the question therefore reduces to how much the quenching distance is altered by replacing the electrode as the quenching agency by a layer of cold and/or foreign gas. Similar reasoning applies to the case of neutralizing, completely or partially, the space charge on the controlling side. This has

† *Footnote added in proof.* See also ref. (9).

been demonstrated (see Fig. 7.25) but not applied in combustion; however it is one of the basic concepts in, for example, ion propulsion devices. These consist essentially of a plasma and a system of electrodes, first accelerating and then somewhat retarding the stream of positive ions. The last electrode is then usually surrounded by a ring-filament emitting electrons, which are entrained into the ion stream and thus prevent, or diminish, the build-up of space charge that would otherwise tend to cause the ions to return. The principle would thus be applicable here only if the positive ion column were not required to retain its net charge. In that case the calculated space-charge limitations would apply only to the region between the ion source and the electron-emitting filament or grid. We presuppose that electron conduction in the hot products is utilized on the other side. Unless the heat of the flame itself is used to maintain the filament at an adequate temperature, the question is governed simply by the effect on quenching distance of the temperature of a surface.

The effect on maxima of a superimposed magnetic field has also been considered.[3, 5] This is quite distinct from using a magnetic field directly as a method of exercising a force on the gas; here its purpose is a purely auxiliary one in that it can delay breakdown. This arises because, when a magnetic field is applied perpendicular to the electric one, it causes charged particles to move in curved trajectories, which effectively reduces the mobility in the transverse plane and hence has an effect similar to that of increasing the gas pressure, which in turn allows larger fields to be applied. The effect of this on the above maxima has been considered quantitatively; thus the maximum ionic wind velocity, for example, has been given for nitrogen, as

$$v_{\mathrm{B}} = 9 \cdot 8 \times 10^3 T^{-\frac{1}{2}} P^{\frac{1}{2}} \{1 + 0 \cdot 23 \times 10^{-10} (H/P)^2\}, \qquad (8.40)$$

where P is in atmospheres and H is in gauss. It will be seen that this effect is very small at atmospheric pressure, but increases with decreasing pressure and might just prove of value if electrical effects had to be applied at low pressures, where the maxima would otherwise be small. When crossed magnetic and electric fields act on a plasma the resultant forces on the gas can, of course be used beyond the effect of merely delaying breakdown. This is considered next.

THE USE OF MAGNETIC FIELDS

Several practical applications discussed, ranging from MHD generation of power to inducing turbulence and increasing heat transfer by spinning a plasma-jet discharge at high rates of rotation, involve the direct use of magnetic fields. It is therefore appropriate to consider how these effects are limited theoretically. Three possible theoretical models are involved, depending on the permitted current. There is, first, a purely 'magnetostatic' effect involving the susceptibility of the material but no continuous flow of current. Second, we must consider currents limited similarly as in the remainder of this chapter, i.e. flows of charge derived entirely from the flame—a regime that ends with the onset of secondary ionization and breakdown. Third, there are methods of exercising forces on flame gases by crossed current and magnetic field vectors in which the current may be derived from a discharge and is not limited to using flame ions.

The first of these possibilities was observed already by Faraday[7] who applied a magnetic field to a flame on a wax taper and observed its tendency to form an equatorial disk. Recently the mechanism has been considered theoretically by Cozens and von Engel[8] in an attempt to account for the deflection of flames in magnetic fields. The effect proposed as being the most significant is rather different from the body forces acting throughout the gas, considered hitherto: it is caused by the force acting on an interface separating gases of different diamagnetic properties. The pressure difference on such an interface is given by

$$\Delta P = (\chi_1 \rho_1 - \chi_2 \rho_2) H^2 / 2, \qquad (8.41)$$

where χ_1 and χ_2 are the mass susceptibilities of the two gases. This magnitude is usually extremely small; thus using the estimate of Cozens and von Engel, based on numerous approximations, such as that the flame gases consist entirely of CO_2, and that their average density is six to seven times smaller than that of the surrounding air, $\Delta P = 1 \cdot 5 \, \text{dyn/cm}^2$ for a magnetic field of 10 000 G. With a magnetic field of this order of magnitude it is just barely possible to produce a visible flame deflection of a small hydrocarbon diffusion flame, providing the momentum of the fuel flow is very small[5] —Fig. 8.4. It is not easy to make a significant comparison between a force acting on an interface and one acting per unit volume; however, the static pressure head driving the gases when a flame is placed in an electric field extending over a few centimetres is of the order 10^3 times greater than the above ΔP.

The principle changes entirely when a current is allowed to flow; the theory has been discussed in Chapter 4. Equation (4.10) gives the force acting per unit length of current $i_\perp$ (which is the component perpendicular to the direction of H) as

$$\frac{F}{\delta L} = i_\perp H.$$

In terms of the orthogonal current density, $j_\perp$,

$$\frac{F}{\delta V} = j_\perp H, \qquad (8.42)$$

where δV is a volume element within the current channel. This may now be compared directly with the 'electrical' force per unit volume. The ratio of

FIG. 8.4. Flame deflection in a magnetic field (applied in (b)). From Mayo.[5]

electric to magnetic force is

$$\frac{F_e}{F_m} = \frac{j/K}{j_\perp H/c} = \frac{c}{HK},$$ (8.43)

the velocity of light, c, being introduced in converting one system of units to another and the j's being cancelled on the understanding that one current is *along* the electric field while the other is *perpendicular* to the magnetic one. If $H = 10^4$ G and $K = 600$ cm^2/s/e.s.V., the numerical ratio is approximately 5×10^3. It follows that a direct comparison is most unfavourable to the magnetic case; if currents are to be limited by breakdown, then even 10 000 G would produce an effect only 1/5000 of that of the electric one. This state of affairs might improve somewhat if superconducting electromagnets, producing much larger values of H, became readily available.

This is the correct and obvious comparison for all applications concerned with flame ions themselves. Thus, if the aim were to collect flame ions by means of a magnetic field alone (the current being due to the convective flow of the charge carriers) the process would continue until the build-up of space charge would produce an electric field distribution similar to that discussed in preceding parts of this chapter. Ultimately, breakdown would result and the flame would cease to be the only, or indeed the major, source of ions. These limitations, of course, apply equally if an electric field is deliberately applied at right angles to the magnetic one. So long as the flame is to be the sole ion source, the previous restrictions apply and the effect of a magnetic field compares most unfavourably.

However, for certain applications, there are reasons for not considering the flame current as limiting in the presence of a magnetic field—unlike in the purely electrical case. Thus for exercising forces on neutral gas to generate wind effects, turbulence, etc., breakdown is limiting in the electrical case, because the force acting on unit volume of gas falls to zero when secondary ionization sets in. The basic reason for this is that the forces due to charges of opposite polarity pull in opposite directions. When breakdown occurs, free charges generated outside the flame region counter-flow those from the flame and the *net* charge per unit volume tends to zero. The current, and hence j and F in eqn (8.42), however, continues to rise, since positive and negative charges travelling in opposite directions constitute a unidirectional current that interacts with the magnetic field. The above-mentioned limitations therefore do not apply to this case. In order to exercise a body force on the gas comparable to the maximum due to electric fields alone ($\simeq 800$ dyn/cm^3) a current density of approximately 10 A/cm^2 is required at 1000 G. Although this is a far cry from the 0·25 mA and 0·09 W/cm^2 considered hitherto, the current density is not large when arcs are considered. In fact, this subject is very much on the border lines of our monograph since, in addition to the magnetic field, a discharge is required, flame currents being insufficient. Nevertheless there are two good reasons for mentioning it both here and in Chapter 7. The first lies in practical applications to combustion for the control of gas flows, as in the work concerned with increasing heat transfer from augmented flames (p. 309). The other is that, since none of the previously mentioned limitations apply to this mechanism, there is no limit to the force that can be exercised on the gas, provided the current can be made large enough. Since the discharge itself acts somewhat like a solid spoke, for reasons discussed in Chapter 7, very large effects can be induced in the gas (for example, in the above-mentioned application of rotating a plasma-jet, swirl rates of 10^5 r.p.m. can be readily obtained with fields of about 10^3 G) and we believe that this principle will prove to have important applications in future.

REFERENCES

1. PAYNE, K. G. and WEINBERG, F. J. *Proc. R. Soc.* A**250**, 316 (1959).
2. LAWTON, J. and WEINBERG, F. J. *Proc. R. Soc.* A**277**, 468 (1964).
3. LAWTON, J., MAYO, P. J., and WEINBERG, F. J. *Proc. R. Soc.* A**303**, 275 (1968).
4. LAWTON, J., Ph.D. Thesis, London University (1963).
5. MAYO, P. J., Ph.D. Thesis, London University (1967).
6. GUGAN, K., LAWTON, J., and WEINBERG, F. J. *10th Int. Symp. Combust.,* p. 709. Combustion Institute (1965).
7. FARADAY, M. *Phil. Mag.,* 53, **31**, No. 210 (1847).
8. COZENS, J. R. and VON ENGEL, A. *Adv. Electronics Electron Phys.* **20**, 99 (1964).
9. MILLS, R. M. *Combust. Flame* **12**, 513 (1968).

Author Index

Subject Index

A.C.
circuits, 182–3
electric field, 178–93, 282–5
Acetylene, 214, 218–20, 226–9, 238–9
Activation energy, 5, 14, 219–20, 232
ionization, 219–20, 232
Affinity, electron, 11, 236
Alkali
metal, 27, 121–2, 158, 218, 229–38, 267–70
seeding, 8, 20–3, 27, 121–2, 158, 218, 229–38, 267–70
Alkaline earth metals, 229, 233–4
Ambipolar diffusion, 118, 222–4
Analysis of flame structure, 49, 59
Application of crossed magnetic and current vectors, 308 ff
Applications
involving magnetic field, 340
limitations of, 315 ff
of fields, 242 ff
of ionic winds, 285 ff
Ar, 225, 261, 269–70
Arc discharge, 235
rotation in magnetic field, 309
Arcs, movement in magnetic fields, 310
Argon, 225, 261, 269–70
Atomic
chlorine 11, 236
flame, 227–8
hydrogen, 233
oxygen, 226–8
Atomization
electrical, 100, 244 ff
Attachment
dissociative, 11–12
electron, 3, 10–12, 22, 217, 236, 278
radiative, 10–11
three-body, 11
Attenuation
of detonations, 279
of waves in plasma, 187–8, 190–1, 282–5
Augmented flames, 55, 79, 308
and combustion intensity, 92
applications of, 90
based on rotating discharge, 89
heat transfer from, 90, 92
plasma jets, use in, 82
products and temperatures, 91

Augmented flames,—*continued*
quenching of, 92
temperature of, 45
uniform heating in, 83
Augmenting by recirculating energy, 42

Barium, 229, 233–4
Blow-off velocity gradient, 63, 65
Blow-out
diffusion flames, 59
in high-intensity combustion, 73
Bombardment charging, 29–30, 239
Boundary layers in mass spectrometry, 204
Branching chains, 45
Breakdown
at electrode, 326
causes for early onset, 333
limitations due to, 316 ff
methods for delaying, 344 ff
Burners, 61
for augmented flames, 79 ff
diffusion flames, 92
flat diffusion flames, 93 ff, 251
high-intensity combustion, 71
pre-mixed flat flames, 66
'porous cavity', 85
spherical flames, 70
stabilized detonations, 75
tubular, 61
using nozzles, 70
well stirred, 74
wetted porous sphere, 98
Burning
mechanism, 45
time of droplets, 99
velocity, 39, 47, 50–1
velocity, measurement of, 50
velocity of turbulent flames, 51

C_2, 214, 216, 224
Caesium, 218, 229–38, 267–70
Calcium, 229, 233–4
Calorelectric effect, 276–8
Carbon
dioxide, 226–8, 230, 236
effect on breakdown, 333
formation, 60